WIRELESS NETWORKING IN

THE DEVELOPING WORLD

Third Edition

Wireless Networking in the Developing World

For more information about this project please visit http://wndw.net

First Edition, January 2006
Second Edition, December 2007
Third Edition, February 2013

As we have discovered the developing world of wireless networking is all around us, the authors of this book have included projects in North America, Europe, and in Asia, South America, India and Africa. So we have come to the conclusion that most places have the potential to find affordable indoor and outdoor wireless networks useful. We hope you enjoy reading this book and use it as the starting point of a wireless project in your community.

ISBN-13: 978-1484039359

ABOUT THIS BOOK

This third version of this book was started as a BookSprint in September 2011 in the beautiful city of Copenhagen hosted by Sebastian Buettrich, who is one of the authors.

A core team of eight people then finished this version over the following months leading to publication in March 2013.

Throughout the project, the core group has actively solicited contributions and feedback from the wireless networking community worldwide. You can provide your own feedback or post technical questions to the authors at our Facebook page:
https://www.facebook.com/groups/wirelessu

This book is available as an eBook for your mobile device, or it is downloadable from the website http://wndw.net/ for free (high and low resolution available), or it can be ordered as a printed book from http://www.lulu.com/

We do give out a copy to every student who attends a wireless training course given by all of the Institutions we work with such as the International Centre for Theoretical Physics (ICTP), the Network Startup Resource Center (NSRC), the Asian Institute of Technology (AIT), The Internet Society (ISOC) and AirJaldi, to name just a few.
And we would all strongly encourage you to sign up for a local course.

For information about upcoming courses or if you would like to arrange a course in your region, please contact the editor, Jane Butler janesbutler@networktheworld.org

If you are planning a wireless project and you need a copy of this book and can neither download it as you have limited bandwidth nor afford to order it online, please send an email to Jane or send a message on Facebook and we'll mail a printed copy to you.

Core Contributors

Jane Butler, who is lead editor of this version of the book. Jane is currently President of the private Foundation called networktheworld.org which promotes and supports the growth of Internet connectivity around the world mainly by supporting wireless projects and training http://wirelessu.org. Jane is also head of industrial collaboration and outreach at University College London. Jane holds an Honours Degree in Engineering, is a Chartered Engineer and Fellow of the Institution of Electronics and Technology.
Jane can be reached at janesbutler@networktheworld.org

The editor would like to acknowledge and thank the core group of contributors who are listed below -

Ermanno Pietrosemoli. Ermanno is currently a researcher at the Telecommunications/ICT for Development Lab of the International Centre for Theoretical Physics in Trieste, Italy, and President of Fundación Escuela Latinoamericana de Redes "EsLaRed", a non- profit organization that promotes ICT in Latin America through training and development projects. EsLaRed was awarded the 2008 Jonathan B. Postel Service Award by the Internet Society. Ermanno has been deploying wireless data communication networks focusing on low cost technology, and has participated in the planning and building of wireless data networks in Argentina, Colombia, Ecuador, Italy, Lesotho, Malawi, Mexico, Morocco, Nicaragua, Peru, Trinidad, U.S.A. and Venezuela. He has presented in many conferences and published several papers related to wireless data communication and is coauthor and technical reviewer of the book "Wireless Networking for the Developing World" freely available from http://wndw.net. Ermanno holds a Master's Degree from Stanford University and was Professor of Telecommunications at Universidad de los Andes in Venezuela from 1970 to 2000.
Ermanno can be reached at ermanno@ictp.it

Marco Zennaro. Marco received his M.Sc. Degree in Electronic Engineering from University of Trieste in Italy. He defended his PhD thesis on *"Wireless Sensor Networks for Development: Potentials and Open Issues"* at KTH-Royal Institute of Technology, Stockholm, Sweden.
His research interest is in ICT4D, the use of ICT for Development.

In particular, he is interested in Wireless Networks and in Wireless Sensor Networks in developing countries. He has been giving lectures on Wireless technologies in more than 20 different countries. When not travelling, he is the editor of wsnblog.com. Marco can be reached at mzennaro@ictp.it

Carlo Fonda is a member of the Radio Communications Unit at the Abdus Salam International Center for Theoretical Physics in Trieste, Italy. Carlo can be reached at cfonda@ictp.it

Stephen Okay. Steve is a geek-of-all-trades with over 20 years of experience in systems/network programming and administration with a particular passion for free/open networks and software. He has deployed wireless networks in Laos, Malawi, Italy, and the United States. He is an Inveneo co-founder and has taught workshops on VoIP and wireless networking at Institutions around the world. He lives and hacks in San Francisco, California. Steve can be reached at steve@inveneo.org

Corinna "Elektra" Aichele. Elektra has been busy working on mesh networking protocols for the Freifunk community in Germany. Before inventing the B.A.T.M.A.N. routing protocol for wireless mesh networks in 2006, she was working on improving the OLSR routing protocol. She is one of the people behind the Mesh-Potato device, a rugged outdoor open-source and open-hardware WiFi router with an FXS port. She is part of the Villagetelco community, that strives to deploy mesh networks for VOIP and data. She lives in a solar-powered home in Berlin, Germany. The philosophy behind her ideas about ubiquitous communication for everyone is: "*The fact that you talk in your head doesn't mean that you think - but only that you speak with yourself*". Elektra can be reached at elektra@villagetelco.org
http://villagetelco.org
http://open-mesh.net/

Sebastian Buettrich. Sebastian is Research Lab Manager at the IT University of Copenhagen, http://pit.itu.dk
He works with embedded/pervasive systems, wireless technology, open source / free software and solar energy to build networks, systems, skills and capacity as a manager, developer, architect, consultant and teacher.
This work focused on '(but not limited to) developing countries and communities, especially in Asia and Africa. One current focus is to help

develop campus networks for research and education, with emphasis on global integration and sustainability. His current side affiliations are: http://www.nsrc.org - the Network Startup Resource Center http://wire.less.dk - NGO and company co-founded with Tomas Krag http://wirelessU.org - a group of dedicated professionals working towards a world-wide, people-centered, inclusive Information Society http://wndw.net/ - Co-author of the Wireless Networking in the Developing World book. Sebastian holds a Ph.D. in Quantum Physics from the Technical University of Berlin in Germany, with a focus on optics, radio spectroscopy, photovoltaic systems and scientific programming. He loves and plays music, is fascinated and engaged with text, language and poetry in many forms. Sebastian can be reached at sebastian@less.dk

Jim Forster. Jim is passionate about extending the Internet. He started at Cisco in 1988 when it was quite small and spent 20 years there, mostly in IOS Software Development and System Architecture, and becoming a Distinguished Engineer. While at Cisco he started working on projects and policies to improve Internet access in developing countries. Now he is engaged in both for-profit and non-profit efforts to extend communications in Africa and India. He founded networktheworld.org, a foundation dedicated to improving communications and Internet, especially in Africa and India. He is on several Board of Directors, including Range Networks / OpenBTS and Inveneo in the US, Esoko Networks in Ghana, and AirJaldi in India. Jim can be reached at jforster@networktheworld.org

Klaas Wierenga. Klaas works in the Research and Advanced Development group at Cisco Systems where he focuses on Identity, Security and Mobility topics, often in collaboration with the Research and Education Community. He is co-author of the Cisco Press book "Building the Mobile Internet". Prior to joining Cisco he worked at SURFnet, the Dutch Research and Education Network, where he created the global WiFi roaming service in academia called eduroam. He is also the Chair of the Mobility Task Force of TERENA, the European association of R&E Networks. Klaas participates in a number of IETF working groups in the fields of identity, security and mobility and chairs the abfab working group that deals with federated identity for non-web applications. He can be reached at klaas@wierenga.net

Eric Vyncke. Since 1997, Eric has worked as a Distinguished Engineer at Cisco in the field of security by assisting customers to deploy secure networks. Since 2005, Eric has also been active in the IPv6 area, he is notably the co-chair of the Belgian IPv6 Council and has a well-known site for monitoring IPv6 deployments: http://www.vyncke.org/ipv6status/ He is also Associate Professor at the University of Liège in Belgium. He participates in several IETF working groups related to security or to IPv6. Eric can be reached at eric@vyncke.org

Bruce Baikie. Bruce is a member of the Broadband for Good team at Inveneo as Senior Director Broadband Initiatives. He is leveraging his extensive experience in the energy and telecom industries, and 16 years at Sun Microsystems as telecom industry expert to advise on implementing solar powered ICT4D projects. His areas of expertise include: wireless networking, eco-data centers, DC telecom power systems, and solar power. Bruce has published numerous white papers and articles on green data center operations and solar power in ICT4D. His educational background includes a B.S. in Mechanical Engineering from Michigan Technological University and advanced studies in International Business from the University of Wisconsin. Bruce is also a guest lecturer on solar powered ICT4D at the Abdus Salam International Centre for Theoretical Physics in Trieste, Italy. During past two years, Bruce has been mentoring engineering students from Illinois Institute of Technology, University of Colorado-Boulder, San Francisco State University, and San Jose State University in ICT4D design and projects in Haiti, West Africa, and Micronesia. Bruce can be reached at bruce@green-wifi.org

Laura Hosman. Laura is Assistant Professor of Political Science at Illinois Institute of Technology. Prior to IIT, Professor Hosman held postdoctoral research fellow positions at the University of California, Berkeley and the University of Southern California (USC). She graduated with a PhD in Political Economy and Public Policy from USC. Her current research focuses on the role of information and communications technology (ICT) in developing countries, particularly in terms of its potential effects on socio-cultural factors, human development, and economic growth.
Her work focuses on two main areas: Public-Private Partnerships and ICT-in-education, both with a focus on the developing world.
Her blog, giving insights on her fieldwork experiences, is at http://ict4dviewsfromthefield.wordpress.com

Michael Ginguld. Founder, Director - Strategy and Operations, Rural Broad Band Pvt. Ltd.Co-Founder, CEO, AirJaldi Research and Innovation. Michael was born and raised in Kibbutz Kissufim in Israel. He has more than 20 years of experience working in ICT, community and rural development projects in India, Indonesia, Cambodia, Nepal, and Israel.

Michael worked in the non-profit and for- profit sectors with start-up grassroots organizations, advocacy groups, large international NGOs and commercial enterprises working in developing countries.

Michael worked and lived in Dharamsala between 1998 and 2002 and returned to India in the beginning of 2007 to join a rural connectivity initiative that eventually led to the creation of AirJaldi Research and Innovation, a non-profit organization dedicated to R&D and capacity building work in the field of Wireless networks in 2007, and of RBB, a for-profit working on the design, deployment and management of rural broad band networks in rural areas, in 2009. Michael holds a B.Sc. in Ag. Economics from the Hebrew University., Jerusalem, Israel, an MA in Development Studies from the Institute of Social Studies, the Hague, the Netherlands, and an MA in Public Administration from the Kennedy School of Government, Harvard University, Cambridge, USA. Michael is based in Dharamsala, Himachal Pradesh, India.

He can be reached at Michael@airjaldi.net

Emmanuel Togo. Emmanuel is from Ghana, and earned his first degree in Computer Science and Physics from University of Ghana in 1999. He currently works as the Head of the Networking Unit of the University of Ghana's Computing Systems (UGCS).

He is also a founding member of the Ghanaian Academic and Research Network's (GARNET) technical team working to build the national research and education network in Ghana. Emmanuel's current focus is designing and deploying an affordable, large-scale campus-wide WiFi network in Ghana. Emmanuel can be reached at ematogo@ug.edu.gh

The Open Technology Institute, (who provided a case study), strengthens individuals and communities through policy research, applied learning, and technological innovation.

Support

The editorial team would especially like to recognise the support of our technical illustrator, Paolo Atzori, who has over several months worked tirelessly to ensure the book has some wonderful, as well as accurate, easy to read illustrations. He has also ensured that we've been able to publish successfully several versions of the book in high and low resolution format.

Paolo Atzori. Paolo studied Architecture in Venice and Rome and Media Arts in Cologne. After working as an architect in Vienna, Paolo collaborated with the Cologne Academy of Media Arts (KHM); At NABA, Milan he was named the Director of the Master Digital Environment Design and Advisor of the PhD program of the Planetary Collegium, M- Node. He has created many theatrical and artistic projects, introducing new representations of space characterised by the dynamics of pervasiveness and interaction.
Paolo has also curated exhibitions dedicated to digital arts, directed educational programs, and published articles and essays on digital culture.
He has lived and worked in Venice, Rome, New York, Vienna, Cologne, Brussels, Tel Aviv. Since 2005 he has lived with his partner Nicole and their children Alma and Zeno in Trieste, Italy. In 2011 he founded with Nicole Leghissa the Agency "Hyphae".
http://hyphae.org
http://vimeo.com/groups/xtendedlab/videos
http://www.xtendedlab.com/
http://www.khm.de/~Paolo

Authors and editors of earlier versions of the book

Rob Flickenger. Rob has written and edited several books about wireless networking and Linux, including Wireless Hacks (O'Reilly) and How To Accelerate Your Internet (http://bwmo.net/). He is proud to be a hacker, amateur mad scientist, and proponent of free networks everywhere.

Laura M. Drewett is a Co-Founder of Adapted Consulting Inc., a social enterprise that specialises in adapting technology and business solutions for the developing world. Since Laura first lived in Mali in the 1990s and wrote her thesis on girls' education programs, she has strived to find sustainable solutions for development.

Laura holds a Bachelors of Arts with Distinction in Foreign Affairs and French from the University of Virginia and a Master's Certificate in Project Management from the George Washington University School of Business.

Alberto Escudero-Pascual and **Louise Berthilson** are the founders of IT +46, a Swedish consultancy company with focus on information technology in developing regions. More information can be found at http://www.it46.se/

Ian Howard. After flying around the world for seven years as a paratrooper in the Canadian military, Ian Howard decided to trade his gun for a computer. After finishing a degree in environmental sciences at the University of Waterloo he wrote in a proposal, "Wireless technology has the opportunity to bridge the digital divide. Poor nations, who do not have the infrastructure for interconnectivity as we do, will now be able to create a wireless infrastructure." As a reward, Geekcorps sent him to Mali as the Geekcorps Mali Program Manager, where he led a team equipping radio stations with wireless interconnections and designed content sharing systems.

Kyle Johnston, http://www.schoolnet.na/

Tomas Krag spends his days working with wire.less.dk, a registered non-profit, based in Copenhagen, which he founded with his friend and colleague Sebastian Büttrich in early 2002. wire.less.dk specialises in community wireless networking solutions, and has a special focus on low-cost wireless networks for the developing world. Tomas is also an associate of the Tactical Technology Collective http://www.tacticaltech.org, an Amsterdam-based non-profit "to strengthen social technology movements and networks in developing and transition countries, as well as promote civil society's effective, conscious and creative use of new technologies."
Currently most of his energy goes into the Wireless Roadshow (http://www.thewirelessroadshow.org), a project that supports civil society partners in the developing world in planning, building and sustaining connectivity solutions based on license-exempt spectrum, open technology and open knowledge.

Gina Kupfermann is graduate engineer in energy management and holds a degree in engineering and business. Besides her profession as financial controller she has worked for various self-organised community projects and non-profit organisations. Since 2005 she is member of the executive board of the development association for free networks, the legal entity of freifunk.net

Adam Messer. Originally trained as an insect scientist, Adam Messer metamorphosed into a telecommunications professional after a chance conversation in 1995 led him to start one of Africa's first ISPs. Pioneering wireless data services in Tanzania, Messer worked for 11 years in eastern and southern Africa in voice and data communications for startups and multinational cellular carriers. He now resides in Amman, Jordan.

Juergen Neumann (http://www.ergomedia.de) started working with information technology in 1984 and since then has been looking for ways to deploy ICT in useful ways for organizations and society. As a consultant for ICT strategy and implementation, he worked for major German and international companies and many non-profit projects. In 2002 he co-founded www.freifunk.net, for spreading knowledge and social networking about free and open networks. Freifunk is globally regarded as one of the most successful community-projects in this field.

Frédéric Renet is a co-founder of Technical Solutions at Adapted Consulting, Inc. Frédéric has been involved in ICT for more than 10 years and has worked with computers since his childhood. He began his ICT career in the early 1990s with a bulletin board system (BBS) on an analog modem and has since continued to create systems that enhance communication. Most recently, Frédéric spent more than a year at IESC/Geekcorps Mali as a consultant. In this capacity, he designed many innovative solutions for FM radio broadcasting, school computer labs and lighting systems for rural communities.

Contents

GLOSSARY

APPENDICES

CASE STUDIES

INTRODUCTION

This book aims to empower people to build DIY networks using wireless technologies. It has been compiled by a bunch of networking geeks that have been busy designing, deploying and operating wireless networks for quite some time, all of them actively participating in expanding the reach of the Internet all over the world.

We believe that people can have a significant stake in building their own communications infrastructure and also influence the wider community around them to make sure networks become affordable and available.

We hope to not only convince you that this is possible, but also show how we have done it, and to give you the information and tools you need to start a network project in your local community.

By providing people in your local community with cheaper and easier access to information, they will directly benefit from what the Internet has to offer.

The time and effort saved by having access to the global network of information translates into value on a local scale. Likewise, the network becomes all the more valuable as more people are connected to it. Communities connected to the Internet at high speed have a voice in a global marketplace, where transactions happen around the world at the speed of light.

People all over the world are finding that Internet access gives them a voice to discuss their problems, politics, and whatever else is important in their lives, in a way that the telephone and television simply cannot compete. What has until recently sounded like science fiction is now becoming a reality, and that reality is being built on wireless networks.

The country of Aipotu

Now lets for a moment look at a fictional country called 'Aipotu', in the developing world. Aipotu has been connected to the Internet merely by expensive VSAT links for a long time.

A brand new optical submarine telecommunications connection has finally arrived at the shore of Aipotu.

The challenge for Aipotu is now to roll out a complete communication infrastructure for a whole country from scratch.

The method of choice today is likely a three tier strategy. First and foremost Aipotu should try to roll out optical fibre lines wherever possible. Fibre lines offer the capability to transport a "sea of bandwidth".

The cost of optical fibre is very low, considering the capacity. By upgrading the optical transceivers the capacity of a optical fibre line can be upgraded without laying new cable. If Aipotu can afford to establish a fibre connection to every household there is no reason not to go for it.

This would make our three tier model obsolete and we could stop here. However, there are probably areas in Aipotu that cannot afford fibre lines.

The second tier that the people of Aipotu can use in order to connect remote villages or small cities are high speed point-to-point links between high points. It is possible to establish high speed links (40 Mbps) of 30 km or more between towers of 30 metre height on flat terrain.

If mountain tops, high buildings or hill tops are available, even longer links are possible. The network technology experts of Aipotu don't have to worry too much about the wireless technology that they are mounting on top of their towers - the cost lies mostly in building the towers, proper lightning protection, power supplies, power back-up and theft protection, rather than in the actual wireless equipment and antennas.

Like the technology of optical transceivers, wireless transceivers also keep advancing, but a wireless link will always be orders of magnitude slower than the capacity of optical fibre.

The third challenge for Aipotu is to solve the problem of the last mile(s): Distributing access to all the individual households, offices, production facilities and so on. Not too long ago the method of choice was to run copper wires but now there is a better choice. This third tier of our network model is clearly the domain of wireless networking technology.

Purpose of this book

The overall goal of this book is to help you build affordable communication technology in your local community by making best use of whatever resources are available.

Using inexpensive off-the-shelf equipment, you can build high speed data networks that connect remote areas together, provide broadband network access in areas where even dialup does not exist, and ultimately connect you and your neighbours to the global Internet.

By using local sources for materials and fabricating parts yourself, you can build reliable network links with very little budget.

And by working with your local community, you can build a telecommunications infrastructure that benefits everyone who participates in it.

This book is not a guide to configuring wireless for your laptop or choosing consumer grade gear for your home network. The emphasis is on building infrastructure links intended to be used as the backbone for wide area wireless networks as well as solving the last mile problem.

With those goals in mind, information is presented from many points of view, including technical, social, and financial factors.

The extensive collection of case studies included present various groups' attempts at building these networks, the resources that were committed to them, and the ultimate results of these attempts.

It is also important to note that all of the resources, techniques and design methodologies described in this book are valid in any part of the world. There are many rural parts all over the globe that remain unconnected to the Internet for reasons of cost, geography, politics and so on.

Deploying wireless networking can often lead to these problems being solved thereby extending connectivity to those who as yet have not achieved it. There are many community based networking projects emerging everywhere. So whether you live in the United Kingdom, Kenya, Chile or India or anywhere else, this book can be a useful practical guide.

Since the first spark gap experiments at the turn of the XIX century, wireless has been a rapidly evolving area of communications technology. While we provide specific examples of how to build working high speed data links, the techniques described in this book are not intended to replace existing wired infrastructure (such as telephone systems or fibre optic backbone).

Rather, these techniques are intended to augment existing systems, and provide connectivity in areas where running fibre or other physical cable would be impractical.

We hope you find this book useful for solving your communication challenges.

Fitting wireless into your existing network

If you are a network administrator, you may wonder how wireless might fit into your existing network infrastructure.

Wireless can serve in many capacities, from a simple extension (like a several kilometre Ethernet cable) to a distribution point (like a large hub).

Here just a few examples of how your network can benefit from wireless technology.

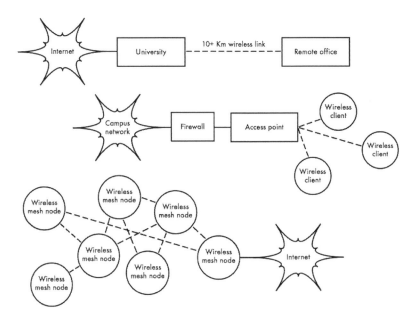

Figure I 1: Some wireless networking examples.

How this book is organized

This book has 4 main sections called -

Physics
Networking
Planning and Deployment
Maintenance, Monitoring and Sustainability

At the end you will find a **Glossary** as well as **Appendices** and **Case Studies**.

Within the 4 main sections there are chapters written by key experts with theoretical and practical hands on experience of the topics.

There are a broad range of topics in the chapters which have been selected as being key to enabling you to start and grow a real wireless deployment in your own community. Another resource you may find useful is here - http://wtkit.org/groups/wtkit/wiki/820cb/download_page.html

It is the set of presentation materials used by these same key experts to deliver wireless networking training classes around the world.

In addition all of the key experts who have written this book regularly check our Facebook page. So as you plan your deployment please do ask questions on our page - we do answer quickly.

https://www.facebook.com/groups/wirelessu

PHYSICS

1. RADIO PHYSICS

Wireless communications make use of electromagnetic waves to send signals across long distances. From a user's perspective, wireless connections are not particularly different from any other network connection: your web browser, email, and other applications all work as you would expect. But radio waves have some unexpected properties compared to Ethernet cable. For example, it's very easy to see the path that an Ethernet cable takes: locate the plug sticking out of your computer, follow the cable to the other end, and you've found it! You can also be confident that running many Ethernet cables alongside each other won't cause problems, since the cables effectively keep their signals contained within the wire itself.

But how do you know where the waves emanating from your wireless device are going? What happens when these waves bounce off objects in the room or other buildings in an outdoor link? How can several wireless cards be used in the same area without interfering with each other?

In order to build stable high-speed wireless links, it is important to understand how radio waves behave in the real world.

What is a wave?

We are all familiar with vibrations or oscillations in various forms: a pendulum, a tree swaying in the wind, the string of a guitar - these are all examples of oscillations.

What they have in common is that something, some medium or object, is swinging in a periodic manner, with a certain number of cycles per unit of time. This kind of wave is sometimes called a *mechanical* wave, since it is defined by the motion of an object or its propagating medium.

When such oscillations travel (that is, when the swinging does not stay bound to one place) then we speak of waves propagating in space. For example, a singer singing creates periodic oscillations in his or her vocal cords. These oscillations periodically compress and decompress the air, and this periodic change of air pressure then leaves the singers mouth and travels, at the speed of sound.

A stone plunging into a lake causes a disturbance, which then travels across the lake as a *wave*.

A wave has a certain *speed*, *frequency*, and **wavelength**. These are connected by a simple relation:

*Speed = Frequency * Wavelength*

The wavelength (sometimes referred to as **lambda, λ**) is the distance measured from a point on one wave to the equivalent part of the next (or, in a more technical way, to the next point that is in the same phase), for example from the top of one peak to the next.

The frequency is the number of whole waves that pass a fixed point in a period of time. Speed is measured in metres/second, frequency is measured in cycles per second (or Hertz, represented by the symbol **Hz**), and wavelength is measured in metres. For example, if a wave on water travels at one metre per second, and it oscillates five times per second, then each wave will be twenty centimetres long:

*1 metre/second = 5 cycles/second *W*
W = 1 / 5 metres
W = 0.2 metres = 20 cm

Waves also have a property called **amplitude**. This is the distance from the centre of the wave to the extreme of one of its peaks, and can be thought of as the "height" of a water wave. Frequency, wavelength, and amplitude are shown in Figure RP 1.

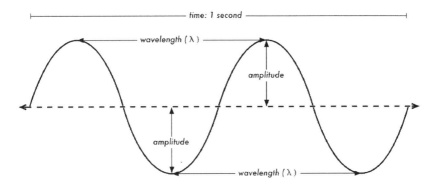

Figure RP 1: Wavelength, amplitude, and frequency. For this wave, the frequency is 2 cycles per second, or 2 Hz, while the speed is 1 m/s.

Waves in water are easy to visualize.

Simply drop a stone into the lake and you can see the waves as they move across the water over time. In the case of electromagnetic waves, the part that might be hardest to understand is: "What is it that is oscillating?"

In order to understand that, you need to understand electromagnetic forces.

Electromagnetic forces

Electromagnetic forces are the forces between electrical charges and currents. Our most direct access to those is when our hand touches a door handle after walking on synthetic carpet, or brushing up against an electrical fence.

A more powerful example of electromagnetic forces is the lightning we see during thunderstorms.

The *electrical force* is the force between electrical charges.

The *magnetic force* is the force between electrical currents.

Electrons are particles that carry a negative electrical charge. There are other charged particles too, but it is the electrons that are responsible for most of what we need to know about how radio behaves.

Let us look at what is happening in a piece of straight vertical wire, in which we push the electrons from one end to the other and back, periodically. At one moment, the top of the wire is negatively charged - all the negative electrons are gathered there. This creates an electric field from the positively charged end to the negatively charged one along the wire.

The next moment, the electrons have all been driven to the other side, and the electric field points the other way. As this happens again and again, the electric field vectors (represented by arrows from plus to minus) are leaving the wire, so to speak, and are radiated out into the space around the wire.

What we have just described is known as a dipole (because of the two differently charged poles, plus and minus, that are created in the straight vertical wire), or more commonly a *dipole antenna*.

This is the simplest form of an omnidirectional antenna. The moving electric field is commonly referred to as an electromagnetic wave because there is also an associated magnetic field. A moving electric field, such as a wave, always comes together with a magnetic field - you will not find one with out the other. Why is this the case?

An electric field is caused by electrically charged objects.

A moving electric field is produced by moving electrically charged objects, such as we have just described above in a dipole antenna.

Wherever electrical charges are moving, they induce a magnetic field. Mathematically, this is formulated in Maxwell's equations: https://en.wikipedia.org/wiki/Electromagnetic_field#Mathematical_description

Since the electrical and magnetic components are tied together in this way, we speak of an electromagnetic field.
In practical wireless networking, we focus in the electrical component but there be always a magnetic component as well.

Let us come back to the relation:

$$Speed = Frequency * Wavelength$$

In the case of electromagnetic waves, the speed is c, the speed of light.

$$c = 300,000 \ km/s = 300,000,000 \ m/s = 3*10^8 \ m/s$$

$$c = f * \lambda$$

Electromagnetic waves differ from mechanical waves in that they require no medium in which to propagate. Electromagnetic waves will even propagate through perfect vacuum.
The light from the stars is a good example: it reaches us through the vacuum of space.

Symbols of the international system of units

In physics, maths, and engineering, we often express numbers by powers of ten.
We will meet these terms again, and the symbols used to represent them, e.g. gigahertz (GHz), centimetres (cm), microseconds (µs), and so on.

These symbols are part of the international system of measurement **SI** (http://www.bipm.org/utils/common/pdf/si_brochure_8_en.pdf), they are not abbreviations and should not be changed.
The case is significant and should not be altered.

SI symbols

atto	10^{-18}	1/1000000000000000000	a
femto	10^{-15}	1/1000000000000000	f
pico	10^{-12}	1/1000000000000	p
nano	10^{-9}	1/1000000000	n
micro	10^{-6}	1/1000000	μ
milli	10^{-3}	1/1000	m
centi	10^{-2}	1/100	c
kilo	10^{3}	1000	k
mega	10^{6}	1000000	M
giga	10^{9}	1000000000	G
tera	10^{12}	1000000000000	T
peta	10^{15}	1000000000000000	P
exa	10^{18}	1000000000000000000	E

Knowing the speed of light, we can calculate the wavelength for a given frequency. Let us take the example of the frequency of 802.11b wireless networking, which is:

$$f = 2.4 \ GHz = 2,400,000,000 \ cycles \ / \ second$$

$$wavelength \ (\lambda) = c \ / f = 3*10^{8} \ / \ 2.4*10^{9} = 1.25*10^{-1} \ m = 12.5 \ cm$$

Frequency and therefore wavelength determine most of an electromagnetic wave's behaviour. It governs the dimensions of the antennas that we build as well as the effect of the interactions with objects that are in the propagation path, including the biological effects in living beings.

Wireless standards of course are distinguished by more than just the frequency they are working at - for example, 802.11b, 802.11g, 802.11n and 802.16 can all work at 2.4 GHz -, yet they are very different from one another.

The chapter called ***Telecommunications Basics*** will discuss modulation techniques, media access techniques, and other relevant features of wireless communications standards. However, the basic capabilities of electromagnetic waves to penetrate objects, to go long distances, and so forth - these are determined by physics alone. The electromagnetic wave "does not know or care" what modulation or standard or technique you put on top of it. So, while different standards may implement advanced techniques to deal with NLOS (Non Line of Sight), multipath and so forth - they still cannot make a wave go through a wall, if that wall is absorbing the respective frequency. Therefore, an understanding of the basic ideas of frequency and wavelength helps a lot in practical wireless work.

Phase

Later in this chapter, we will talk about concepts like interference, multipath and Fresnel zones. In order to understand these, we will need to know about the ***phase*** of a wave, or rather, ***phase differences*** between waves. Look at the sine wave shown in Fig RP 1 - now imagine we have two such waves moving. These can be in exactly the same position: Where the one has its peak, the other one also has a peak. Then, we would say, they are in phase, or, their phase difference is zero. But one wave could also be displaced from the other, for example it could have its peak where the other wave is at zero. In this case, we have a phase difference. This phase difference can be expressed in fractions of the wavelength, e.g. $\lambda//4$, or in degrees, e.g. 90 degrees - with one full cycle of the wave being 360 degrees. A phase difference of 360 degrees is the same as that of 0 degrees: no phase difference.

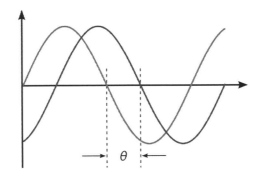

Figure RP 2: Phase Difference between Two Waves

Polarization

Another important quality of electromagnetic waves is *polarization*. Polarization describes the direction of the electrical field vector.

If you imagine a vertically aligned dipole antenna (the straight piece of wire), electrons can only move up and down, not sideways (because there is no room to move) and thus electrical fields only ever point up or down, vertically. The field leaving the wire and travelling as a wave has a strict linear (and in this case, vertical) polarization. If we put the antenna flat on the ground, we would find horizontal linear polarization.

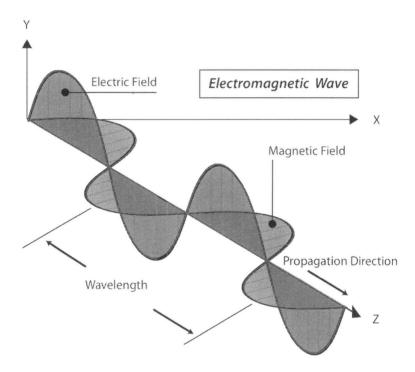

Figure RP 3: Vertically polarized electromagnetic wave

Linear polarization is just one special case, and is never quite so perfect: in general, we will always have some component of the field pointing in other directions too. If we combine two equal dipoles fed with the same signal, we can generate a circularly polarized wave, in which the electric field vector keeps rotating perpendicularly to the wave's trajectory.

The most general case is elliptical polarization, in which the electric field vector maximum value is not the same in the vertical and horizontal direction. As one can imagine, polarization becomes important when aligning antennas. If you ignore polarization, you might have very little signal even though you have the best antennas. We call this polarization mismatch.

Much in the same way, polarization may also be used in a smart way, to keep two wireless links independent and without interference, even though they might use the same end points (or even share a common reflector) and therefore the same trajectory: if one link is polarized vertically and the other horizontally, they will not "see" each other. This is a convenient way to double data rates over one link using a single frequency.

The antennas used in this kind of application must be carefully built in order to reject the "unwanted" polarization, i.e. an antenna meant for vertical polarization must not receive or transmit any horizontally polarized signal, and vice versa. We say they must have a high "cross polarization" rejection.

The electromagnetic spectrum

Electromagnetic waves span a wide range of frequencies (and, accordingly, wavelengths). This range of frequencies or wavelengths is called the *electromagnetic spectrum*. The part of the spectrum most familiar to humans is probably light, the visible portion of the electromagnetic spectrum. Light lies roughly between the frequencies of $7.5*10^{14}$ Hz and $3.8*10^{14}$ Hz, corresponding to wavelengths from circa 400 nm (violet/blue) to 800 nm (red).

We are also regularly exposed to other regions of the electromagnetic spectrum, including Alternating Current (*AC*) or grid electricity at 50/60 Hz, AM and FM radio, Ultraviolet (at frequencies higher than those of visible light), Infrared (at frequencies lower than those of visible light), X-Ray radiation, and many others.

Radio is the term used for the portion of the electromagnetic spectrum in which waves can be transmitted by applying alternating current to an antenna. This is true for the range from 30 kHz to 300 GHz, but in the more narrow sense of the term, the upper frequency limit would be about 1 GHz, above which we talk of microwaves and millimetric waves.

When talking about radio, many people think of FM radio, which uses a frequency around 100 MHz. In between radio and infrared we find the region of microwaves - with frequencies from about 1 GHz to 300 GHz, and wavelengths from 30 cm to 1 mm.

The most popular use of microwaves might be the microwave oven, which in fact works in exactly the same region as the wireless standards we are dealing with. These regions lie within the bands that are being kept open for general unlicensed use. This region is called the **ISM band**, which stands for Industrial, Scientific, and Medical.

Most other parts of the electromagnetic spectrum are tightly controlled by licensing legislation, with license values being a huge economic factor. In many countries the right to use portions of the spectrum have been sold to communications companies for millions of dollars. In most countries, the ISM bands have been reserved for unlicensed use and therefore do not have to be paid for when used.

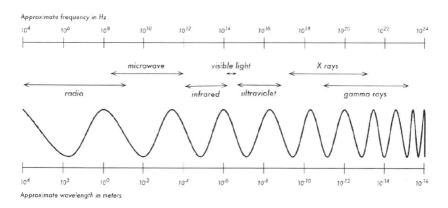

Figure RP 4: The electromagnetic spectrum.

The frequencies most interesting to us are 2.400 - 2.495 GHz, which is used by the 802.11b and 802.11g standards (corresponding to wavelengths of about 12.5 cm), and 5.150 - 5.850 GHz (corresponding to wavelengths of about 5 to 6 cm), used by 802.11a. The 802.11n standard can work in either of these bands.

See the Chapter called **WiFi Family** for an overview of standards and frequencies. In addition you can find out more about the Radio portion of the electromagnetic spectrum in the Chapter called **Radio Spectrum.**

Bandwidth

A term you will meet often in radio physics is **bandwidth**. Bandwidth is simply a measure of frequency range. If a range of 2.40 GHz to 2.48 GHz is used by a device, then the bandwidth would be 0.08 GHz (or more commonly stated as 80 MHz).

It is easy to see that the bandwidth we define here is closely related to the amount of data you can transmit within it - the more room in frequency space, the more data you can fit in at a given moment. The term bandwidth is often used for something we should rather call data rate, as in "my Internet connection has 1 Mbps of bandwidth", meaning it can transmit data at 1 megabit per second. How much exactly you can fit into a physical signal will depend on the modulation, encoding and other techniques. For example, 802.11g uses the same bandwidth as 802.11b, however it fits more data into those same frequency ranges transmitting up to 5 times more bits per second.

Another example we have mentioned: you may double your data rate by adding a second link at perpendicular polarization to an existing
radio link. Here, frequency and bandwidth have not changed, however the data rate is doubled.

Frequencies and channels

Let us look a bit closer at how the 2.4 GHz band is used in 802.11b. The spectrum is divided into evenly sized pieces distributed over the band as individual channels. Note that channels are 22 MHz wide, but are only separated by 5 MHz.

This means that adjacent channels overlap, and can interfere with each other. This is represented visually in Figure RP 5.

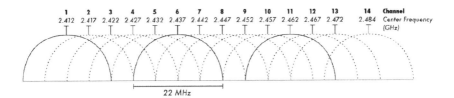

Figure RP 5: Channels and centre frequencies for 802.11b.
Note that channels 1, 6, and 11 do not overlap.

Behaviour of radio waves

There are a few simple rules of thumb that can prove extremely useful when making first plans for a wireless network:

- *the longer the wavelength, the further it goes;*
- *the longer the wavelength, the better it travels through and around things;*
- *the shorter the wavelength, the more data it can transport.*

All of these rules, simplified as they may be, are rather easy to understand by example.

Longer waves travel further

Waves with longer wavelengths tend to travel further than waves with shorter wavelengths. As an example, AM radio stations have a much greater range than FM stations, which use a frequency 100 times higher. Lower frequency transmitters tend to reach much greater distances than high frequency transmitters at the same power.

Longer waves pass around obstacles

A wave on water which is 5 metres long will not be affected by a 5 mm piece of wood floating on the water. If instead the piece of wood were 50 metres big (e.g. a ship), it would modify the behavior of the wave.

The distance a wave can travel depends on the relationship between the wavelength of the wave and the size of obstacles in its path of propagation. It is harder to visualize waves moving "through" solid objects, but this is the case with electromagnetic waves. Longer wavelength (and therefore lower frequency) waves tend to penetrate objects better than shorter wavelength (and therefore higher frequency) waves.

For example, FM radio (88-108 MHz) can travel through buildings and other obstacles easily, while shorter waves (such as GSM phones operating at 900 MHz or 1800 MHz) have a harder time penetrating buildings. This effect is partly due to the difference in power levels used for FM radio and GSM, but is also partly due to the shorter wavelength of GSM signals. At much higher frequencies, visible light does not go through a wall or even 1 mm of wood - as we all know, from practical experience. But metal will stop any kind of electromagnetic wave.

Shorter waves can carry more data

The faster the wave swings or beats, the more information it can carry - every beat or cycle could for example be used to transport a digital bit, a '0' or a '1', a 'yes' or a 'no'.

So the data rate scales with bandwidth, and can be further enhanced by advanced modulation and media access techniques such as OFDM, and MIMO (Multiple Input, Multiple Output).

The Huygens Principle

There is another principle that can be applied to all kinds of waves, and which is extremely useful for understanding radio wave propagation.

This principle is known as the *Huygens Principle*, named after Christiaan Huygens, Dutch mathematician, physicist and astronomer, 1629 - 1695.

Imagine you are taking a little stick and dipping it vertically into a still lake's surface, causing the water to swing and dance. Waves will leave the centre of the stick - the place where you dip in - in circles. Now, wherever water particles are swinging and dancing, they will cause their neighbor particles to do the same: from every point of disturbance, a new circular wave will start. This is, in simple form, the Huygens principle. In the words of wikipedia.org:

"The Huygens' principle is a method of analysis applied to problems of wave propagation in the far field limit. It recognizes that each point of an advancing wave front is in fact the centre of a fresh disturbance and the source of a new train of waves; and that the advancing wave as a whole may be regarded as the sum of all the secondary waves arising from points in the medium already traversed".

This view of wave propagation helps better understand a variety of wave phenomena, such as diffraction." This principle holds true for radio waves as well as waves on water, for sound as well as light, but for light the wavelength is far too short for human beings to actually see the effects directly.

This principle will help us to understand diffraction as well as Fresnel zones, and the fact that sometimes we seem to be able to transmit around corners, with no line of sight.

Let us now look into what happens to electromagnetic waves as they travel.

Absorption

When electromagnetic waves go through 'something' (some material), they generally get weakened or dampened.

How much they lose in power will depend on their frequency and of course the material.

Clear window glass is obviously transparent for light, while the glass used in sunglasses filters out quite a share of the light intensity and most of the ultraviolet radiation.

Often, an absorption coefficient is used to describe a material's impact on radiation.

For microwaves, the two main absorbent materials are:

Metal. Electrons can move freely in metals, and are readily able to swing and thus absorb the energy of a passing wave.

Water. Microwaves cause water molecules to jostle around, thus taking away some of the wave's energy.

For the purpose of practical wireless networking, we may well consider metal and water perfect absorbers: we will not be able to go through them (although thin layers of water will let some power pass). They are to microwave what a brick wall is to light.

When talking about water, we have to remember that it comes in different forms: rain, fog and mist, low clouds and so forth, all will be in the way of radio links. They have a strong influence, and in many circumstances a change in weather can bring a radio link down.

When talking about metal, keep in mind that it may be found in unexpected places: it may be hidden in walls (for example, as metal grids in concrete) or be a thin coat on modern types of glass (tinted glass, colored glass).

However thin the layer of metal, it might be enough to significantly absorb a radio wave.

There are other materials that have a more complex effect on radio absorption. For trees and wood, the amount of absorption depends on how much water they contain.

Old dead dry wood is more or less transparent, wet fresh wood will absorb a lot. Plastics and similar materials generally do not absorb a lot of radio energy, but this varies depending on the frequency and type of material.
Lastly, let us talk about ourselves: humans (as well as other animals) are largely made out of water.

As far as radio networking is concerned, we may well be described as big bags of water, with the same strong absorption.
Orienting an office access point in such a way that its signal must pass through many people is a key mistake when building office networks.
The same goes for hotspots, cafe installations, libraries, and outdoor installations.

Reflection

Just like visible light, radio waves are reflected when they come in contact with materials that are suited for that: for radio waves, the main sources of reflection are metal and water surfaces.
The rules for reflection are quite simple: the angle at which a wave hits a surface is the same angle at which it gets deflected.

Note that in the eyes of a radio wave, a dense grid of bars acts just the same as a solid surface, as long as the distance between bars is small compared to the wavelength.

At 2.4 GHz, a one cm metal grid will act much the same as a metal plate.

Although the rules of reflection are quite simple, things can become very complicated when you imagine an office interior with many many small metal objects of various complicated shapes.
The same goes for urban situations: look around you in city environment and try to spot all of the metal objects.

This explains why *multipath effects* (i.e. signal reaching their target along different paths, and therefore at different times) play such an important role in wireless networking.
Water surfaces, with waves and ripples changing all the time, effectively make for a very complicated reflection object which is more or less impossible to calculate and predict precisely.

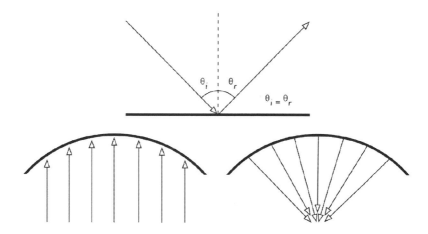

Figure RP 6: Reflection of radio waves. The angle of incidence is always equal to the angle of reflection. A metal parabolic surface uses this effect to concentrate radio waves spread out over it in a common direction.

We should also add that polarization has an impact: waves of different polarization in general will be reflected differently.

We use reflection to our advantage in antenna building: e.g. we put huge parabolas behind our radio transmitter/receiver to collect and bundle the radio signal into a single point, the focal point.

Diffraction

Diffraction is the apparent bending of waves when hitting an object.

It is the effect of "waves going around corners". Imagine a wave on water traveling in a straight wave front, just like a wave that we see rolling onto an ocean beach.

Now we put a solid barrier, say a wooden solid fence, in its way to block it. We cut a narrow slit opening into that wall, like a small door.

From this opening, a circular wave will start, and it will of course reach points that are not in a direct line behind this opening, but also on either side of it. If you look at this wavefront - and it might just as well be an electromagnetic wave - as a beam (a straight line), it would be hard to explain how it can reach points that should be hidden by a barrier.

When modelled as a wavefront, the phenomenon makes sense.

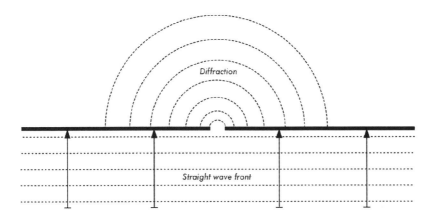

Figure RP 7: Diffraction through a narrow slit.

The Huygens Principle provides one model for understanding this behavior. Imagine that at any given instant, every point on a wavefront can be considered the starting point for a spherical "wavelet".

This idea was later extended by Fresnel, and whether it adequately describes the phenomenon is still a matter of debate. But for our purposes, the Huygens model describes the effect quite well.

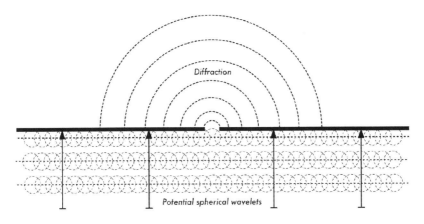

Figure RP 8: The Huygens Principle.

Through means of the effect of diffraction, waves will "bend" around corners or spread through an opening in a barrier.

The wavelengths of visible light are far too small for humans to observe this effect directly.

Microwaves, with a wavelength of several centimeters, will show the effects of diffraction when waves hit walls, mountain peaks, and other obstacles. It seems as if the obstruction causes the wave to change its direction and go around corners.

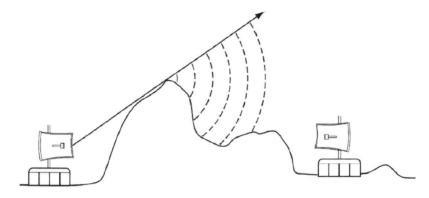

Figure RP 9: Diffraction over a mountain top.

Note that diffraction comes at the cost of power: the energy of the diffracted wave is significantly less than that of the wavefront that caused it. But in some very specific applications, you can take advantage of the diffraction effect to circumvent obstacles.

Interference

Interference is one of the most misunderstood terms and phenonema in wireless networking.

Interference often gets the blame when we are too lazy to find the real problem, or when a regulator wants to shut down someone else's network for business reasons. So, why all the misunderstandings?

It is mostly because different people mean different things though they are using the same word.

A physicist and a telecommunications engineer will use the word "Interference" in very different ways. The physicists' view will be concerned with the "behaviour of waves". The telecommunications engineer will talk about "... any noise that gets in the way".

Both views are relevant in wireless, and it is important to be able to know them both and know the difference. Let us start with the physicists' view: When working with waves, one plus one does not necessarily equal two. It can also result in zero.

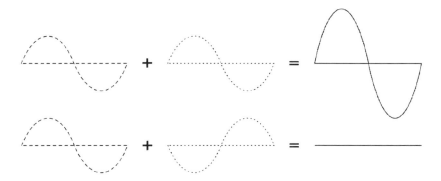

Figure RP 10: Constructive and destructive interference.

This is easy to understand when you draw two sine waves and add up the amplitudes. When the phase difference is zero, peak hits peak, and you will have maximum results (1 + 1 = 2).

This is called ***constructive interference***.

When the phase difference is 180 degrees, or $\lambda/2$, peak hits valley, and you will have complete annihilation ((1 + (-)1 = 0) - ***destructive interference***.

You can actually try this with waves on water and two little sticks to create circular waves - you will see that where two waves cross, there will be areas of higher wave peaks and others that remain almost flat and calm. In order for whole trains of waves to add up or cancel each other out perfectly, they have to have the exact same wavelength and a fixed phase relation.

You can see obvious examples of interference in action when you look at the way that antennas are arranged in what are called ***beamforming*** arrays, in order to give maximum constructive interference in the directions where you want the signal, and destructive interference (no signal) where you want no signal.

Technically, this is achieved by a combination of physical dimensioning and control of phase shifts.

Simplified, imagine that you have three antennas - and you don't want antenna 3 to pick up signal from antenna 1 and 2. You would then place antenna 3 at a position where the signals from antennas 1 and 2 cancel each other out.

Now let us have a look at the way the word interference is typically used: in a wider sense, for any disturbance through other RF sources, any noise that might get in our way, e.g. from neighboring channels or competing providers. So, when wireless networkers talk about interference they typically talk about all these kinds of disturbance by other networks, and any other sources of microwave, whether it has exactly the same frequency and a fixed phase relation or not. Interference of this kind is one of the main sources of difficulty in building wireless links, especially in urban environments or closed spaces (such as a conference space) where many networks may compete for use of the spectrum.

But, interference of this kind is also often overrated: for example, imagine you had to build a point to point link that has to cross a crowded inner city area, before reaching its target on the other side of the city. Such a highly directional beam will cross the "electric smog" of the urban centre without any problem. You may imagine this like a green and a red light beam crossing each other in a 90 degrees angle: while both beams will overlap in a certain area, the one will not have any impact on the other at all.

Generally, managing spectrum and coexistence has become a main issue especially in dense indoor environments and urban areas.

Line of sight

The term *line of sight*, often abbreviated as *LOS*, is quite easy to understand when talking about visible light: if we can see a point B from point A where we are, we have line of sight. Simply draw a line from A to B, and if nothing is in the way, we have line of sight.

Things get a bit more complicated when we are dealing with microwaves. Remember that most propagation characteristics of electromagnetic waves scale with their wavelength.

This is also the case for the widening of waves as they travel.

Light has a wavelength of about 0.5 micrometres, microwaves as used in wireless networking have a wavelength of a few centimetres.

Consequently, their beams are a lot wider - they need more space, so to speak.

Note that visible light beams widen just the same, and if you let them travel long enough, you can see the results despite their short wavelength. When pointing a well focussed laser at the moon, its beam will widen to well over 100 metres in radius by the time it reaches the surface. You can see this effect for yourself using an inexpensive laser pointer and a pair of binoculars on a clear night. Rather than pointing at the moon, point at a distant mountain or unoccupied structure (such as a water tower). The radius of your beam will increase as the distance increases. This is due to the diffraction.

The line of sight that we need in order to have an optimal wireless connection from A to B is more than just a thin line - its shape is more like that of a cigar, an ellipsoid. Its width can be described by the concept of Fresnel zones - see next section for an explanation. You will also find the abbreviation *NLOS*, for "non line of sight", which is mostly used to describe and advertise technologies that allow for dealing with waves that reach the receiver through multiple trajectories (multipath) or diffraction. It does not indicate that the single electromagnetic beam goes "around corners" (other than through diffraction) or "through obstacles" any better than that of other technologies. For example, you might call White Space technology NLOS, as its lower frequencies (longer wavelengths) allow it to permeate objects and utilize diffraction much better than comparable 2.4 GHz or 5 GHz transmissions.

Understanding the Fresnel zone

The exact theory of Fresnel (pronounced "Fray-nell") zones is quite complicated. However, the concept is quite easy to understand: we know from the Huygens principle that at each point of a wavefront new circular waves start, we know that microwave beams widen as they leave the antenna, we know that waves of one frequency can interfere with each other. Fresnel zone theory simply looks at a line from A to B, and then at the space around that line that contributes to what is arriving at point B. Some waves travel directly from A to B, while others travel on paths off axis and reach the receiver by reflection.
Consequently, their path is longer, introducing a phase shift between the direct and indirect beam.
Whenever the phase shift is one half wavelength, you get destructive interference: the signals cancel.

Taking this approach you find that when the reflected path is less than half a wavelength longer than the direct path, the reflections will add to the received signal. Conversely, when the reflected path length exceeds the direct path by more than one half wavelength, its contribution will decrease the received power.

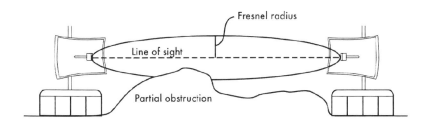

Figure RP 11: The Fresnel zone is partially blocked on this link, although the visual line of sight appears clear.

Note that there are many possible Fresnel zones, but we are chiefly concerned with the first zone, because the contributions from the second zone are negative. The contributions from the third zone are positive again, but there is no practical way to take advantage of those without the penalty incurred in going through the second Fresnel Zone.

If the first Fresnel zone is partially blocked by an obstruction, e.g. a tree or a building, the signal arriving at the far end would be diminished. When building wireless links, we therefore need to be sure that the first zone is kept free of obstructions. In practice, it is not strictly necessary that the whole of this zone is clear, in wireless networking we aim to clear about 60 percent of the radius of the first Fresnel zone.

Here is one formula for calculating the radius of the first Fresnel zone:

$$r = 17.31 \sqrt{\left(\frac{(d_1 * d_2)}{(f * d)}\right)}$$

...where r is the radius of the zone in metres, d_1 and d_2 are distances from the obstacle to the link end points in metres, d is the total link distance in metres, and f is the frequency in MHz.

The first Fresnel zone radius can also be calculated directly from the wavelength as:

$$r=\sqrt{(\frac{\lambda * d_1 * d_2}{d})}$$

with all the variables in metres

It is apparent that the maximum value of the first Fresnel zone happens exactly in the middle of the trajectory and its value can be found setting $d_1=d_2=d/2$ in the preceding formulas. Note that the formulae give you the radius of the zone, not the height above ground.

To calculate the height above ground, you need to subtract the result from a line drawn directly between the tops of the two towers.

For example, let's calculate the size of the first Fresnel zone in the middle of a 2 km link, transmitting at 2.437 GHz (802.11b channel 6):

$$r=17.31\sqrt{[\frac{(1000*1000)}{(2437*2000)}]}$$

$$r=17.31\sqrt{(\frac{1000000}{4874000})}$$

r = 7.84 metres

Assuming both of our towers were ten metres tall, the first Fresnel zone would pass just 2.16 metres above ground level in the middle of the link.

But how tall could a structure at that point be to block no more than 60% of the first zone?

*r = 0.6 * 7.84 metres*
r = 4.70 metres

Subtracting the result from 10 metres, we can see that a structure 5.3 metres tall at the centre of the link would block up to 40% of the first Fresnel zone.

This is normally acceptable, but to improve the situation we would need to position our antennas higher up, or change the direction of the link to avoid the obstacle.

Power

Any electromagnetic wave carries energy - we can feel that when we enjoy (or suffer from) the warmth of the sun.

The amount of energy divided by the time during which we measure it is called power. The power P is measured in W (watts) and is of key importance for a wireless links to work: you need a certain minimum power in order for a receiver to make sense of the signal.

We will come back to details of transmission power, losses, gains and radio sensitivity in the chapter called *Antennas/Transmission Lines.*

Here we will briefly discuss how the power P is defined and measured.
The electric field is measured in V/m (potential difference per metre), the power contained within it is proportional to the square of the electric field:

$$P \sim E^2$$

Practically, we measure the power in watts by means of some form of receiver, e.g. an antenna and a voltmetre, power metre, oscilloscope, spectrum analyser or even a radio card and laptop.
Looking at the signal's power directly means looking at the square of the signal in volts and dividing by the electrical resistance.

Calculating with dB

By far the most important technique when calculating power is calculating with decibels (dB). There is no new physics hidden in this - it is just a convenient method which makes calculations a lot simpler.

The decibel is a dimensionless unit, that is, it defines a relationship between two measurements of power. It is defined by:

$$dB = 10 * Log\ (P_1\ /\ P_0)$$

where P_1 and P_0 can be whatever two values you want to compare. Typically, in our case, this will be some amount of power.
Why are decibels so handy to use? Many phenomena in nature happen to behave in a way we call exponential.

For example, the human ear senses a sound to be twice as loud as another one if it has ten times the physical signal power.

Another example, quite close to our field of interest, is absorption. Suppose a wall is in the path of our wireless link, and each metre of wall takes away half of the available signal. The result would be:

0 metres = 1 (full signal)
1 metre = 1/2
2 metres = 1/4
3 metres = 1/8
4 metres = 1/16
n metres = $1/2^n$ = 2^{-n}

This is exponential behaviour.

But once we have used the trick of applying the logarithm (log), things become a lot easier: instead of taking a value to the n-th power, we just multiply by n. Instead of multiplying values, we just add.

Here are some commonly used values that are important to remember:

+3 dB = double power
-3 dB = half the power
+10 dB = order of magnitude (10 times power)
-10 dB = one tenth power

In addition to dimensionless dB, there are a number of definitions that are based on a certain base value P_0. The most relevant ones for us are:

dBm relative to P_0 = 1 mW
dBi relative to an ideal isotropic antenna

An isotropic antenna is a hypothetical antenna that evenly distributes power in all directions.

It is approximated by a dipole, but a perfect isotropic antenna cannot be built in reality. The isotropic model is useful for describing the relative power gain of a real world antenna.

Another common (although less convenient) convention for expressing power is in milliwatts. Here are equivalent power levels expressed in milliwatts and dBm:

$$1 \ mW = 0 \ dBm$$
$$2 \ mW = 3 \ dBm$$
$$100 \ mW = 20 \ dBm$$
$$1 \ W = 30 \ dBm$$

For more details on dB refer to the dB math lecture of the Wireless Training kit:
http://wtkit.org/sandbox/groups/wtkit/wiki/820cb/attachments/ebdac/02 -dB_Math-v1.12_with-notes.pdf

Physics in the real world

Don't worry if the concepts in this chapter seem challenging. Understanding how radio waves propagate and interact with the environment is a complex field of study in itself.

Most people find it difficult to understand phenomenon that they can't even see with their own eyes.

By now you should understand that radio waves don't travel only in a straight, predictable path.

To make reliable communication networks, you will need to be able to calculate how much power is needed to cross a given distance, and predict how the waves will travel along the way.

2. TELECOMMUNICATIONS BASICS

The purpose of any telecommunications system is to transfer *information* from the sender to the receiver by a means of a communication *channel*.

The information is carried by a *signal*, which is certain physical quantity that changes with time.

The signal can be a voltage proportional to the amplitude of the voice, like in a simple telephone, a sequence of pulses of light in an optical fibre, or a radio-electric wave irradiated by an antenna.

For analog signals, these variations are directly proportional to some physical variable like sound, light, temperature, wind speed, etc. The information can also be transmitted by digital binary signals, that will have only two values, a digital *one* and a digital *zero*. Any analog signal can be converted into a digital signal by appropriately *sampling* and then coding it. The sampling frequency must be at least twice the maximum frequency present in the signal in order to carry *all* the information contained therein. Random signals are the ones that are unpredictable and can be described only by statistical means.

Noise is a typical random signal, described by its mean power and frequency distribution. A signal can be characterised by its behaviour over time or by its frequency components, which constitute its spectrum. Some examples of signals are shown in Figure TB 1.

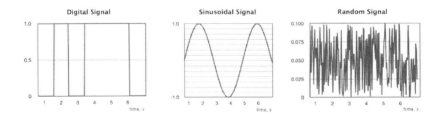

Figure TB 1: Examples of signals

Any periodic signal is composed of many sinusoidal components, all of them multiples of the fundamental frequency, which is the inverse of the period of the signal. So a signal can be characterised either by a graph of its amplitude over time, called a waveform, or a graph of of the amplitudes of its frequency components, called a spectrum.

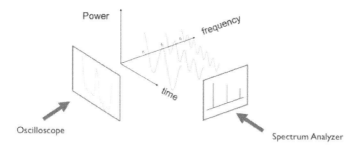

Figure TB 2: Waveforms, Spectrum and filters

Figure TB 2 shows how the same signal can be seen from two different perspectives. The waveform can be displayed by an instrument called an oscilloscope, while the spectrum can be displayed by what is called a Spectrum Analyzer. The spectrum distribution relays very important information about the signal and allows for the intuitive understanding of the concept of filtering of electrical signals. In the example shown, the signal is formed by the superposition of three sinusoidal components of frequency f_1, f_2 and f_3. If we pass this signal through a device that will remove f_2 and f_3, the output is a pure sinusoidal at frequency f_1.

We call this operation "**Low Pass filtering**" because it removes the higher frequencies. Conversely, we can apply the signal to a "**High Pass Filter**", a device that will remove f_1 and f_2 leaving only a sinusoidal signal at the f_3 frequency.

Other combinations are possible, giving rise to a variety of filters. No physical device can transmit all the infinite frequencies of the radio-electric spectrum, so every device will always perform some extent of filtering to the signal that goes through it.

The ***bandwidth*** of a signal is the difference between the highest and the lowest frequency that it contains and is expressed in Hz (number of cycles per second).

While travelling through the communication channel, the signal is subject to *interference* caused by other signals and is also affected by the electrical *noise* always present in any electrical or optical component. ***Intra-channel*** interference originates in the same channel as our signal. ***Co-channel*** interference is due to the imperfection of the filters that will let in signals from adjacent channels.

Consequently, the received signal will always be a distorted replica of the transmitted signal, from which the original information must be retrieved by appropriate means to combat the effect of interference and noise. Furthermore, the received signal will be subject to *attenuation* and *delay* that increase with the distance between the transmitter and the receiver.

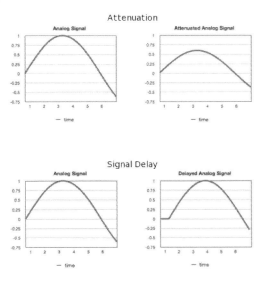

Figure TB 3: Attenuation and delay

Although it is relatively simple to restore the amplitude of signal by means of an electrical *amplifier*, the components of the amplifier will add additional noise to the signal, so at very long distances where the received signal is feeble, the amplifier will produce a signal so garbled with noise that the information originally transmitted will no longer be retrievable.

One way to address this problem consists in converting the continuous quantity carrying the information into a sequence of very simple *symbols* which can be easier to recognise even at great distance. For instance, the flag of a ship is a convenient way to distinguish the nationality of the ship even at distances at which the letters on the hull cannot be read.

This technique has been extended to carry generalised messages by assigning different position of flags to every letter of the alphabet, in an early form of long distance telecommunications by means of *digital* or *numeric* signals.

The limitation of this method is obvious; to be able to distinguish among, say, 26 symbols corresponding to each letters of the alphabet, one must be quite close to the communicating ship.

On the other hand, if we code each letter of the alphabet in a sequence of only two symbols, these symbols can be distinguished at much longer distance, for example the dot and dashes of the telegraph system.

The process of transforming a continuous analog signal into a discontinuous digital one is called Analog to Digital Conversion (**ADC**), and conversely we must have a Digital to Analog Converter (**DAC**) at the receiving end to retrieve the original information.

This is the reason why most modern telecommunication systems use digital binary signals to convey all sorts of information in a more robust way. The receiver must only distinguish between two possible symbols, or in other words between two possible values of the received *bit* (***binary digit***). For instance, the CD has replaced the vinyl record, and analogue television is being replaced by digital television. Digital signals can use less bandwidth, as exemplified by the "***digital dividend***" currently being harnessed in many countries which consists in bandwidth that has become available thanks to the transition from analog to digital transmission in TV broadcasting.

Although in the process of converting from an analog to a digital information system there is always some loss of information, we can engineer the system so as to make this loss negligible.

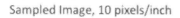

Normal, 72pixels/inch   Sampled Image, 10 pixels/inch

Figure TB 4: Undersampled Image

For example, in a digital camera we can choose the number of bits used to record the image.

The greater the number of bits (proportional to the amount of *megapixels*), the better the rendering, but more memory will be used and longer time to transmit the image will be needed.

So most modern communication systems deal with digital signals, although the original variable that we want to transmit might be analog, like the voice. It can be shown that any analog signal can be reconstructed from discrete samples if the sampling rate is at least twice as high as the highest frequency content of the signal.

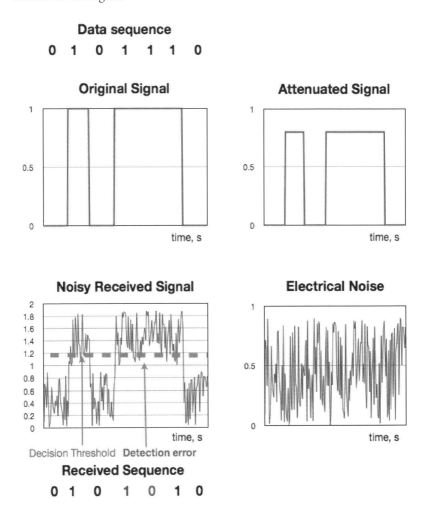

Figure TB 5: detection of a noisy signal

Then each sample is coded in as many bits as necessary to achieve the desired amount of precision.

These bits can now be efficiently stored or transmitted, since for the recovery of the information one needs to distinguish among only two states, and not among the infinite nuances of an analog signal.

This is shown in Figure TB 5, where the original data consists of the **0 1 0 1 1 1 0** sequence. The **0's** are represented as zero volts and the **1's** as 1 V. As the signal moves towards the receiver, its amplitude will diminish. This effect is called "attenuation" and is shown in the figure. Likewise, there will also be a delay as the signal moves from the transmitter to the receiver, the variability in the delay of the received signal is called *jitter*. Attenuation, noise or jitter (or their combination) if severe enough, can cause a detection error. An amplifier can be used to overcome the attenuation, but the electrical noise always present in the system will add to the received signal.

The noisy received signal is therefore quite different from the original signal, but in a digital system we can still recover the information contained by sampling the received signal at the correct time and comparing the value at the sampling time with a suitable threshold voltage. In this example the noise received signal has a peak of 1.8 V, so we might choose e threshold voltage of 1.1 V. If the received signal is above the threshold, the detector will output a digital **1**, otherwise, it will output a **0**. In this case we can see that because of the effect of the noise the fifth bit was erroneously detected as a zero.

Transmission errors can also occur if the sampling signal period is different from that of the original data (difference in the clock rates), or if the receiver clock is not stable enough (*jitter*).

Any physical system will have an upper limit in the frequencies that will transmit faithfully (the bandwidth of the system), higher frequencies will be blocked, so the abrupt rise and fall of the voltage will be smoothed out as the signal goes through the channel.

Therefore, we must make sure that each of the elements of the system has enough bandwidth to handle the signal. On the other hand, the greater the bandwidth of the receiver system, the greater the amount of the noise that will affect the received signal.

Modulation

The robustness of the digital signal is also exemplified by the fact that it was chosen for the first trials of radio transmission. Marconi showed the feasibility of long distance transmission, but pretty soon realised that there was a need to share the medium among different users.

This was achieved by assigning different *carrier* frequencies which were *modulated* by each user's *message*. *Modulation* is a scheme to modify the *amplitude*, *frequency* or *phase* of the carrier according with the information one wants to transmit. The original information is retrieved at destination by the corresponding demodulation of the received signal.

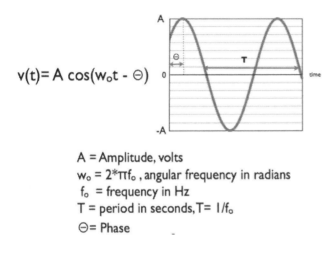

$$v(t) = A \cos(w_o t - \Theta)$$

A = Amplitude, volts
$w_o = 2*\pi f_o$, angular frequency in radians
f_o = frequency in Hz
T = period in seconds, T= $1/f_o$
Θ = Phase

Figure TB 6: Sinusoidal Carrier Signal

Figure TB 6 shows a carrier signal with Amplitude **A**, phase **θ**, and frequency f_o which is the reciprocal of the period **T**.

The combination of different modulation schemes has resulted in a plethora of modulation techniques depending on which aspect one wants to optimise: robustness against noise, amount of information transmitted per second (*capacity* of the link in bits/second) or *spectral efficiency* (number of bits/s per Hertz).

For instance, **BPSK** -Binary Phase Shift Keying- is a very robust modulation technique but transmits only one bit per symbol, while **256 QAM** -Quaternary Amplitude Modulation- will carry 8 bits per symbol, thus multiplying by a factor of eight the amount of information transmitted per second, but to correctly distinguish amongst the 256 symbols transmitted, the received signal must be very strong as compared with the noise (a very high **S/N** -Signal/Noise ratio- is required).

The ultimate measure of quality in digital transmission is the **BER** -Bit Error Rate- which corresponds to the fraction of erroneously decoded bits. Typical values of BER range between 10^{-3} and 10^{-9}.

The modulation also allows us to choose which range of frequency we want to use for a given transmission. All frequencies are not created equal and the choice of the carrier frequency is determined by legal, commercial and technical constraints.

Multiplexing and duplexing

In general, the sharing of a channel among different users is called *multiplexing*.
This is shown in Figure TB 7.

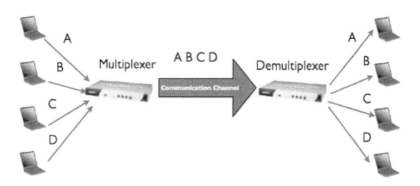

Figure TB 7: Multiplexing

Assigning different carrier frequencies to different users is called **FDMA** -Frequency Division Multiple Access-.

An alternative technique consists in assigning different *time slots* to different users, in what is known as **TDMA** -Time Division Multiple Access-, or even different codes in **CDMA** -Code Division Multiple Access- where the different users are recognised at the receiver by the particular mathematical code assigned to them. See Figure TB 8.

By using two or more antennas simultaneously, one can take advantage of the different amount of fading introduced in the different paths to the receiver establishing a difference among users in what is known as **SDMA** - Space Division Multiple Access-, a technique employed in the **MIMO** -Multiple Input,Multiple Output- systems that have gained popularity recently.

Medium sharing techniques

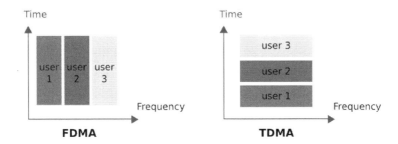

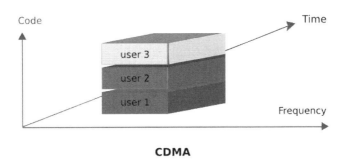

Figure TB 8: Medium Sharing techniques

Most communication systems transmit information in both directions, for instance from the Base Station to the subscriber in what is called the *downlink*, and from the subscriber to the base station in the ***uplink***.

To accomplish this, the channel must be shared between the two directions giving rise respectively to **FDD** -Frequency Division Duplexing- and **TDD** -Time Division Duplexing-.

Conclusions

The communication system must overcome the noise and interference to deliver a suitable replica of the signal to the receiver.

The capacity of the communication channel in bits/second is proportional to the bandwidth in Hz and to the logarithm of the S/N ratio.

Modulation is used to adapt the signal to the channel and to allow several signals to share the same channel. Higher order modulation schemes allow for a higher transmission rate, but require higher S/N ratio.

The channel can be shared by several users that occupy different frequencies, different time slots, different codes or by taking advantage of different propagation characteristics in what is called spatial multiplexing.

For more information and slides covering this topic please visit http://wtkit.org/groups/wtkit/wiki/820cb/download_page.html

3. LICENSING AND REGULATION

There are a number of areas where national and international laws and regulations can influence your ability to set up wireless networks.

Since these rules vary from country to country it is impossible to give an overview of which regulations may apply in your region.

It is also worth noting that there may be a huge difference in which laws exist, and how they are regulated in practice. In other words there may be countries where using the 2.4 GHz / 5 GHz spectrum for outdoor wireless is technically illegal, but where everyone does it anyway.

As a rule of thumb, if other people are building similar networks to what you intend, contact them and find out what legal issues they may have run into. If such networks are very widely deployed in your country, then you probably don't need to worry too much. On the other hand, it is always advisable to seek local advice, from hardware vendors, wireless experts or others who have come before you, before committing time and resources to building a wireless network. Whatever you do, it is important you take local laws and regulations into consideration.

Examples of relevant types of regulation

Each country may have different rules, and each scenario may come across different types of regulations. The areas where regulations may be relevant include licenses for using specific radio frequencies, rules regarding the right to install towers for antennas, the maximum power allowed and telecom licensing rules limiting your ability to provide Internet access to others.

The types of legal issues that may (or may not) be worth considering when planning a wireless network include:

- Spectrum Licensing
- ISP/Telecommunications Licenses
- Tower permits for antennas
- Transmission power and antenna gain limits
- Certification of equipment
- ISP Terms of Use

Spectrum Licensing

Most countries consider RF spectrum as an exclusive property of the state. The RF spectrum is a national resource, much like water, land, gas and minerals. Unlike these, however, RF is reusable. The purpose of spectrum management is to mitigate radio spectrum pollution and maximize the benefit of usable radio spectrum.

The first sentence of the International Telecommunications Union (ITU) constitution fully recognises "the sovereign right of each State to regulate its telecommunication". Effective spectrum management requires regulation at national, regional and global levels.

Licensing is an orderly way to manage who, when, where and how spectrum resource is used. The unlicensed wireless spectrum was set around the 2.4 GHz band.

In June 2003, the ITU made available the 5 GHz band for license-exempt technology deployment. The 900 MHz band, unlicensed in the United States, is presently used in Western Europe and in many developing countries for GSM phones. Each country has the sovereign right to regulate its telecommunication and to interpret the international Radio Regulations. Governments define the rules and conditions of the frequency use.

(*From: Wikipedia "Spectrum Management"*)

The technologies described in this book (mostly) use a license-exempt slice of the spectrum referred to as the ISM (Industrial, Scientific and Medical radio bands). Radio frequencies in the ISM bands have been used for communication purposes, although such devices may experience interference from non-communication sources.

The ISM bands are defined by the ITU-R (ITU's Radiocommunication Sector) at 2.4 and 5 GHz. Individual countries' use of the bands designated in these sections may differ due to variations in national radio regulations. Because communication devices using the ISM bands must tolerate any interference from ISM equipment, unlicensed operations are typically permitted to use these bands, since unlicensed operation typically needs to be tolerant of interference from other devices anyway.

In the US, the FCC (Federal Communications Commission) first made unlicensed spread spectrum available in the ISM bands in rules adopted on May 9, 1985. Many other countries later adopted these FCC regulations, enabling use of this technology in many countries.

(*From: Wikipedia "ISM Band"*)

ISP/Telecommunications Licenses

In some countries an ISP license would be required before deploying any network infrastructure for sharing networks over public spaces. In other countries this would only be required to run commercial networks.

Tower permits for antennas

When deploying long-range outdoor networks, it is often necessary to build a tower for the antenna. Many countries have regulations regarding the building of such antenna-towers if they are more than 5 or 10 metres above the roof or ground.

Transmission Power limits

When setting transmission power limits, regulatory agencies generally use the Equivalent Isotropically Radiated Power (EIRP), as this is the power actually radiated by the antenna element. Power limits can be imposed on output power of devices as well.

As an example, the FCC enforces certain rules regarding the power radiated by the antenna element, depending on whether the implementation is point-to-multipoint (PtMp) or point-to-point (PtP). It also enforces certain rules regarding the maximum power transmitted by the radio.

When an omnidirectional antenna is used, the FCC automatically considers the link a PtMP link. In the setup of a 2.4 GHz PtMP link, the FCC limits the EIRP to 4 Watts and the power limit set for the intentional radiator is 1 Watt.

Things are more complicated in the 5 Ghz band. The Unlicensed National Information Infrastructure (U-NII) radio band is part of the radio frequency spectrum used by IEEE-802.11a devices and by many wireless ISPs. It operates over three ranges:

U-NII Low (U-NII-1): 5.15-5.25 GHz. Regulations require use of an integrated antenna. Power limited to 50 mW.

U-NII Mid (U-NII-2): 5.25-5.35 GHz. Regulations allow for a user-installable antenna, subject to Dynamic Frequency Selection (DFS, or radar avoidance). Power limited to 250 mW.

U-NII Worldwide: 5.47-5.725 GHz. Both outdoor and indoor use, subject to Dynamic Frequency Selection (DFS, or radar avoidance). Power limited to 250 mW.

This spectrum was added by the FCC in 2003 to "align the frequency bands used by U-NII devices in the United States with bands in other parts of the world".
The FCC currently has an interim limitation on operations on channels which overlap the 5600 - 5650 MHz band.

U-NII Upper (U-NII-3): 5.725 to 5.825 GHz.
Sometimes referred to as U-NII / ISM due to overlap with the ISM band. Regulations allow for a user-installable antenna. Power limited to 1W. Wireless ISPs generally use 5.725-5.825 GHz.
(*From: Wikipedia "U-NII"*)

For PtP in the 5 GHz band the maximum EIRP allowed is considerable higher, since a high gain antenna produces a very narrow beam and therefore the interference caused to other users is considerably less than in PtMPt topology.

Certification of equipment

Governments may require a formal certification that a given radio equipment comply with specific technical standards and local regulations.
This is often referred to as *homologation*, and the process must be done by an independent laboratory authorised by the government of the country.
Certified equipment is allowed to operate without an individual license. It is worth noting that certification may only apply to the original factory state for radio equipment.
For example, changing the antenna on a wireless access point in the United States invalidates the FCC Certification.

ISP Terms of Use

Many ISP's include in their "Terms of Use" a clause that prohibits users from sharing an internet connection with other users.
There may also be commercial grade connections that do not have these limitations.
It is important to note that this is NOT a legal issue, but a clause of the contract with the ISP, and the repercussions for breaching these is usually a disconnection of the Internet connection.

4. RADIO SPECTRUM

What is the electromagnetic spectrum?

There is not a simple definition of the spectrum. From the technical viewpoint the spectrum is simply the range of electromagnetic waves that can be used to transmit information, but from the practical viewpoint the economic and political aspects, as well as the technology actually used to convey the information by means of these waves, play pivotal roles.

As an example, when Marconi in 1902 first spanned the Atlantic with his "wireless telegraph message", he used the whole spectrum available at the time to send a few bits/s over an area of thousands of square kilometres.

With the spark transmitter used for this achievement that occupied all the frequencies that the existing receivers were able to understand, nobody else could use radio for communications on a radius of some 3500 km from the transmitting station in England. So, if other users wanted to send messages in the same area, they would need to coordinate their transmissions in different "time slots" in order to share the medium. This technique is called "**TDMA**", Time Division Multiple Access.

Users located at distances much greater than 3500 km from Marconi's transmitter could use the spectrum again, since the power of the radio waves decreases as we move farther away from the transmitter. Reusing the spectrum in different geographical areas is called "**SDMA**", Space Division Multiple Access. Marconi was later able to build a transmitter that could restrict emissions to just a range of frequencies, and a receiver that could be "tuned" to a particular frequency range. Now, many users could transmit simultaneously in the same area (space) and at the same time. "**FDMA**", Frequency Division Multiple Access was borne. Radio then became a practical means of communications, and the only one that was available to reach a ship in the open seas. The coordination of the frequencies allocated to different users was done by national agencies created to this effect, but since radio waves are not stopped by national borders, international agreements were needed. The international organization that had been created to regulate the transmission of telegrams among different countries was commissioned to allocate the use of the electromagnetic spectrum.

Nowadays, ITU, International Telecommunications Union, is the oldest International Organization, tasked with issuing recommendations about which frequencies to use for which services to its 193 nation members.

The use of the spectrum for military applications raised a new issue; "jamming", the intentional interference introduced by the enemy to impede communication. To avoid jamming, a new technique was developed in which the information to be transmitted was combined with a special mathematical code; only receivers with the knowledge of that particular code could interpret the information. The coded signal was transmitted at low power but using a very wide interval of frequencies to make jamming more difficult.

This technique was later adapted to civilian applications in what is called **"CDMA"**, Code Division Multiple Access, one of the flavours of **spread spectrum communication**, extensively used in modern communications systems. In summary, the spectrum can be shared among many users by assigning different *time slots*, different *frequency intervals*, different *regions of space*, or different *codes*. A combination of these methods is used in the latest cellular systems. Besides issues of sovereignty and its defence, very strong economic and political interests play a determinant role in the management of the spectrum, which also needs to be constantly updated to take advantage of the advances in the communications technology.

Telecommunications engineers keep finding more efficient ways to transmit information using time, frequency and space diversity by means of ever advancing modulation and coding techniques. The goal is to increase the "spectrum efficiency", defined as the amount of bits per second (bit/s) that can be transmitted in each Hz of bandwidth per square kilometre of area. For example, the first attempts to provide mobile telephone services were done by using a powerful transmitter, conveniently located to give coverage to a whole city.

This transmitter (called a **Base Station** in this context), divided the allocated frequency band into say, 30 channels. So only 30 conversations could be held simultaneously in the whole city.

As a consequence, the service was very expensive and only the extremely wealthy could afford it. This situation prevailed for many years, until the advances in electronic technology allowed the implementation of a scheme to take advantage of "Space Diversity".

Instead of using a single powerful transmitter to cover the whole city, the area to be serviced was divided into many "cells", each one served by a low power transmitter. Now cells that are sufficiently apart can utilize the same channels without interference, in what is known as "frequency reuse". With the cellular scheme, the first 10 channels are to use frequency band 1, the second 10 channels frequency band 2 and the remaining 10 channels frequency band 3. This is shown in Figure RS 1, in which the colours correspond to different frequency bands. Notice that the colours repeat only at distances far enough to avoid interference. If we divide the city in say, 50 cells, we can now have 10X50 = 500 simultaneous users in the same city instead of 30. Therefore, by adding cells of smaller dimensions (specified by lower transmission power) we can increase the number of available channels until we reach a limit imposed by the interference.

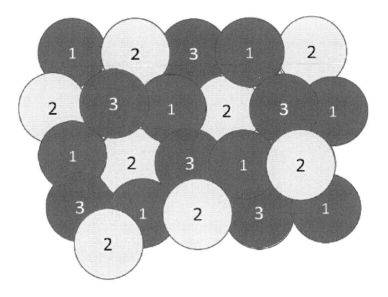

Figure RS 1: Cellular sharing of spectrum

This example shows that a clever use of existing resources can dramatically increase its usefulness. Although the main use of the spectrum is for communication purposes, there are also other uses, like cooking food in microwave ovens, medical applications, garage door openers and so on.

So some frequency bands are allocated for these purposes in what is known as the **ISM** (Industrial, Scientific and Medical) bands.

This spectrum usage is normally for short distance applications.

A breakthrough occurred in 1985 when the FCC (Federal Commission of Communications), the agency that oversees the spectrum in the U.S., allowed the use of this spectrum for communications applications as well, provided that the transmission power was kept to a very low level to minimize interference.

People could freely use these "Unlicensed" bands without previously applying for a permit, provided that the equipment used had been certified by an authorized laboratory that ensured compliance with interference mitigation measures.

Several manufacturers began taking advantage of this opportunity by offering equipment that could be used to communicate between computers without the need for cables, and some wireless data networks covering significant geographic areas were built with them, but the turning point happened after the 1997 approval of the IEEE (Institute of Electrical and Electronics Engineers) 802.11 Standard, the basis of what is known as WiFi.

The existence of a standard that guaranteed the interoperability of equipment produced by different manufacturers fuelled an impressive growth of the market, which in turn drove the competition that fostered a dramatic decrease in the cost of the devices.

In particular, the portion of the ISM band between 2400 and 2483 MHz is nowadays available in most of the world without the need for previously applying for a license and is widely used by laptops, tablets, smart phones and even photographic cameras.

It is important to stress the role of the unlicensed spectrum in the enormous success of WiFi high speed Internet access.

Many airports, hotels and cafes all over the world offer free WiFi Internet access on their premises, and low cost wireless community networks have been built both in rural area and in cities covering considerable geographic areas, thanks to the availability of free spectrum.

Mobile phone operators, who have to pay dearly for frequency licenses to use the spectrum, were quite hostile to this apparently unfair competition.

But when they started offering smart phones, which make very intensive use of the Internet, they pretty soon realized that off-loading the traffic to WiFi was in their best interest, because it relieved the traffic in their distribution network (known as the *backhaul*).

So now they encourage their customers to use WiFi wherever it is available and use the more expensive cellular service only when out of range of any WiFi Access Point.

This is a remarkable example of the usefulness of the unlicensed spectrum even to traditional telecommunications operators who often have lobbied against it.

How is the spectrum adjudicated?

Currently the main methods to gain access to a given spectrum band are auctions and the so called "beauty contest".

The auction method is straightforward; interested parties bid for a given spectrum chunk; whoever commits the higher sum gets the right to use the frequencies.

In theory this method guarantees that the adjudication will be transparent, in practice this has often been circumvented and there have been instances of powerful commercial interests that acquire frequencies only to avoid their use by the competition, with the result of highly valuable spectrum not being used.

Also there is the temptation on the part of governments to use this method as a means to generate revenues and not necessarily in the best public interest.

As an example, in the year 2000 there were auctions in several countries of Europe to adjudicate spectrum for mobile phones, which resulted in a total income of 100 billion (100 000 000 000) euros to the government coffers.

The "beauty contest" method is for the interested parties to submit proposals about how they intend to use the spectrum.

A committee of the spectrum regulating agency then decides which of the proposals better serves the public goals.

This method relies on the objectivity, technical proficiency and honesty of the members of the deciding committee, which is not always guaranteed.

In many countries there are rules for spectrum adjudication that call for the relinquishing of spectrum bands that have been acquired but are not being used; however their enforcement is often lacking due to the strong economic interests affected.

Figure RS 2: A special vehicle for spectrum monitoring in Montevideo, Uruguay.

Figure RS 2 shows a photograph of a spectrum monitoring vehicle in Montevideo, Uruguay and Figure RS 3 that of the same kind of equipment being used in Jakarta, Indonesia.

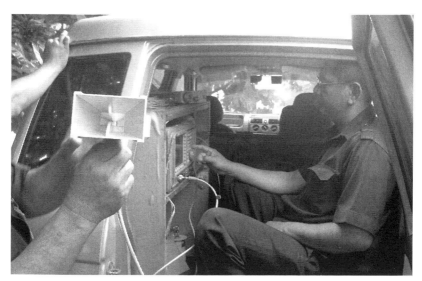

Figure RS 3: The "spectrum Police" at work in Jakarta

Note that the open spectrum used in the unlicensed bands cannot prevent interference issues, especially in very crowded areas, but nevertheless it has proved a fantastic success for short distance applications in cities and also for long distance applications in rural areas.

It is therefore advisable to investigate new forms of spectrum allocation, taking into consideration the needs of many stakeholders and striking a balance among them.

A dynamic spectrum allocation mechanism seems to be the best choice given the advances in technology that make this viable nowadays.

As an example, the current method of spectrum allocation is similar to the railway system, the railroads can be idle a considerable amount of time, whereas the dynamic spectrum allocation is akin to the freeway system that can be used at all times by different users.

Political issues

The importance of the spectrum as a communications enabler cannot be overstated. Television and radio broadcasting have a strong influence in shaping public perception of any issue, and have been used overtly for political propaganda (It has been said that the election of Kennedy as president of the U.S. was due mainly to his television campaign).

During the cold war, *The Voice of America, Moscow Radio* and *Radio Havana Cuba* were very effective ways to sway a global audience.

More recent examples include the influence of CNN and Al Jazeera in the public interpretation of the Arab Spring. Spectrum used for two way communications has also been subject to government interventions, especially in cases of political unrest. On the other hand, economic interests also play a vital role in broadcasting; the consumer society relies heavily on radio and television to create artificial needs or to veer the consumer towards a particular brand. We can conclude that the electromagnetic spectrum is a natural resource whose usefulness is heavily conditioned by technological, economic and political factors.

Explosion in spectrum demand

As the number of tablets and smart phones grows, telecom operators are trying to get access to new frequency bands, but the traditional way of adjudicating the spectrum is facing a dead end.

Keep in mind that the spectrum is used for radio and television broadcasts, for satellite communications, for airplane traffic control, for geolocalisation (Global Postioning Systems-GPS), as well as for military, police and other governmental purposes. Traditionally, the demand for additional spectrum has been met thanks to the advances of electronics that have permitted the use of higher frequencies at an affordable cost. Higher frequencies are well suited for high speed transmissions, but they have a limited range and are highly attenuated by walls and other obstacles as well as by rain.

This is exemplified by comparing the coverage of an AM radio broadcasting station to that of an FM one: the greatest range of the AM station is due to its use of lower frequencies. On the other hand, FM stations can make use of higher bandwidths and as consequence can offer greater audio quality at the expense of a more limited range.

Current cellular operators use even higher frequencies, usually above 800 MHz. Accordingly, the TV broadcasting frequencies are coveted by the cellular telephone providers, because by using lower frequencies they will need less base stations, with huge savings in deployment, operation and maintenance costs. This is why these frequencies are commonly referred to as "beach front property".

Techniques for more efficient spectrum usage by means of advanced modulation and coding methods have had the greatest impact in allowing more bits/s per Hz of bandwidth availability. This, in turn, was made possible by the great strides in electronics (fabrication of ever advanced integrated circuits) that now make it economically feasible to implement the required sophisticated modulation and coding techniques.

According to the calculations performed in 1948 by Claude Shannon - the father of modern telecommunications, a typical telephone line could carry up to 30 kbit/s. But this was only achieved in the 90's when integrated circuits implementing the required techniques were actually built. In particular, the transition to digital terrestrial television broadcasting, which is more efficient in spectrum usage compared with analogue transmission, has freed some spectrum in the so called "White Spaces", the frequencies that had to be left fallow in between analogue television channels to prevent interference.

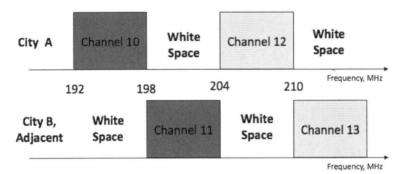

Figure RS 4: Example of TV Channels adjudication in two cities that are close enough that transmissions from one can reach the other. White spaces are kept fallow to minimize interference.

In traditional analogue TV broadcasting, adjacent channels cannot be used at the same time, because the signal from one channel will "spill" over to the two adjacent channels and would cause interference. This is similar to the central reservation used in freeways to separate the two directions of traffic in order to prevent collisions. So a "White Space" must be left between two contiguous analogue TV channels to prevent interference. Digital TV broadcasting is much more efficient in spectrum utilization, and several digital TV channels can be accommodated in the same frequency band formerly used by a single analogue channel without "spillover" into adjacent channels. So, in places where Analog TV is replaced by Digital TV a "digital dividend" is being harvested.

In conclusion, the concept of white spaces can be applied to three different frequency chunks:

- The spectrum that has been assigned to TV broadcasting but it is not currently being used. This applies particularly to developing countries, in which there has been no economic incentive for broadcaster to use every available TV channel.

- The spectrum that must be left free in between two analogous TV channels to prevent interference.

- The spectrum that has been reclaimed as a consequence of the transition to digital terrestrial TV, which is more spectrum efficient. This currently applies to developed countries, but will soon apply to developing countries as well.

In the last 20 years there has been a tremendous growth in the demand for more spectrum for mobile communication services, in which data services are consuming much more bandwidth than voice and the growing use of video is presenting an additional challenge.

Not surprisingly, telecom operators everywhere are trying to get a portion of these "White Spaces" allocated to them to fulfill their needs. Broadcasters, on the other hand, are very reluctant to concede any spectrum at all to what are now their direct competitors.

Spectrum scarcity or spectrum hoarding?

Although the available spectrum is currently totally adjudicated in developed countries, many independent studies have found that the actual simultaneous usage of the spectrum is a tiny fraction of the total. This is caused by the way spectrum was originally adjudicated and also because often spectrum is used intermittently; for instance some TV broadcasting stations do not transmit 24 hours a day.

As a consequence, a radically new way to use the spectrum has been suggested; instead of leasing spectrum to a given organization in an exclusive basis, the new dynamic spectrum management paradigm proposes to use whatever spectrum is available in a certain place at a certain time and switch to another frequency whenever interference is detected in a given band.

An analogy can be made to explain this concept: the current way to allocate the spectrum is similar to a railroad system; the railroads are never used 100% of the time, a more efficient use of the same amount of terrain can be done with a highway in which many different users can share the same path according to their current needs.

Of course to implement dynamic spectrum access requires new technologies and new legislation; many vested interests are fighting it alleging the possibility of interference. The key issue is how to determine when a particular chunk of spectrum is really being used in a particular place and how to move quickly to a new frequency band when an existing user with higher priority is detected. The technology to accomplish this feat has already being demonstrated and implemented in the new IEEE 802.22 standard recently approved, as well as in others currently being considered.

IEE 802.22

Stimulated by the impressive success of WiFi (due mostly to the use of unlicensed, open spectrum), the IEEE created a working group to address the requirements of a Wireless Regional Area Network. The challenge was to develop a technology suitable for long distance transmission that could be deployed in different countries (with quite different spectrum allocations), so they focused on the spectrum currently allocated to TV broadcasting which spans approximately from 50 to 800 MHz. Nowhere is this spectrum being used in its entirety all the time, so there are "White Spaces", fallow regions that could be "re-farmed" and put to use for bidirectional communications. In rural areas all over the world, but specially in developing countries there are large portions of spectrum currently under utilised. It is expected that IEEE 802.22 will enable dynamic spectrum access in a similar way to IEEE 802.11 (WiFi), allowing access to open spectrum. Of course not all the spectrum can be liberated at once, a gradual process is required as the many technical, legal, economic and political hurdles are solved, but there is no doubt that this is the trend and that IEE 802.22 paves the way to the future of spectrum allocation. In order to assess the availability of a given frequency channel at a given time two methods are being considered: channel sensing and a database of primary users in a given geographic location at a given time.

Channel sensing means that prior to an attempt to use a channel, the base stations will listen to the channel; if it is being used it will try another one, repeating the procedure until a free channel is found. This procedure is repeated at regular intervals to account for the possibility of stations coming alive at any time. This method should suffice, nevertheless current spectrum holders have successfully lobbied the regulators to enforce the implementation of the second method, which is much more complicated and imposes additional complexity and costs in the consumer equipment.

The second method consists in the building of a database of all the existing incumbent transmission stations, with their position and respective coverage area in order to establish an "off limit" zone in a given channel.

A new station wishing to transmit must first determine its exact position (so it must have a GPS receiver or other means to determine its geographic location) and then interrogate the database to ascertain that its present location is not in the forbidden zone of the channel it is attempting to use.

To interrogate the database, it must have Internet access by some other means (ADSL-Asymmetrical Digital Subscriber Loop, Cable, Satellite, or Cellular), besides the 802.22 radio (which cannot be used until the channel is confirmed as usable), so this adds a considerable additional burden to the station hardware which translates into additional cost, beside the cost of building and maintaining the database.

In the US the FCC (Federal Communications Commission, the spectrum regulatory agency) has been promoting the building of the database of registered users and have authorised 10 different private enterprises to build, operate and maintain such repositories.

Furthermore, field trials of the standard are been conducted. In the UK, OFCOM (the spectrum regulator) is also conducting IEE 802.22 trials and concentrating on the database method having ruled out the spectrum sensing method for interference mitigation. Although IEEE 802.22 is the formally approved standard that has received the most publicity, there are several competing candidates that are being explored to leverage the TV White Spaces to provide two-way communication services, among them:

IEEE 802.11af

This amendment takes advantage of the enormous success of IEEE 802.11 by adapting the same technology to work in the frequency bands allocated to TV transmission, thus relieving the spectrum crowding of the 2.4 GHz band and offering greater range due to the use of lower transmission frequencies. Its details are still being discussed by the corresponding IEEE 802.11 working group.

IEEE 802.16h

This amendment of the 802.16 standard was ratified in 2010 and describes the mechanism for implementing the protocol in uncoordinated operation, licensed or license exempt applications. Although most deployments have been in the 5 GHz band, it can also be applied to the TV band frequencies and can profit from the significant deployments of WiMAX (Wireless Microwave Access) systems in many countries.

Developing countries advantage

It is noteworthy that in developing countries the spectrum allocated to broadcast television is only partially used. This presents a magnificent opportunity to introduce wireless data networking services in the channels that are not currently allocated, and to start reaping the benefits of 802.22 in a more benign environment, where the spectrum sensing and agile frequency changing required to share the crowded spectrum in developed countries can be dispensed with. The usefulness of the lower frequencies for two-way data communications has been proved by the the successful deployment of CDMA (Code Division Modulation Access) cellular systems in the 450 MHz band, right in the middle of the TV allocated frequencies, in rural areas like the Argentinian Patagonian, currently served by "Cooperativa Telefonica de Calafate-COTECAL". COTECAL offer voice and data services to customers at distances up to 50 km from the Base Station, in the beautiful area shown in the figure below:

Figure RS 5: Region served with voice and data services by COTECAL, in Calafate and El Chalten, Argentina.

So there is an opportunity for stakeholders to lobby for the introduction of TV Band Device based solutions at an early stage, while the issues of the digital transition are being considered. This will help ensure that commercial interests of a few do not prevail over the interests of society at large. Activists/lobbyists should emphasize the need for transparency in the frequency allocation process and for accountability of the administration of spectrum in their country or region.

Also it is important that those who wish to deploy networks gain an understanding of the real spectrum usage by spectrum holders in each region of their country. The monitoring of the spectrum requires expensive instruments with a steep learning curve to use them, but recently an affordable and easy to use device has become available that permits analysis of the frequency band between 240 MHz and 960 MHz, which encompasses the higher part of the TV band. Details of this open hardware based RF Explorer Spectrum Analyzer for the upper TV band are at:

http://www.seeedstudio.com/depot/rf-explorer-model-wsub1g-p-922.html

Figure RS 6 shows the RF Explorer for the 2.4 GHz band being used to test an antenna built by participants of the 2012 ICTP Wireless training workshop in Trieste, Italy.

Figure RS 6: Participants from Albania, Nepal, Malawi and Italy testing an antenna with the RF Explorer Spectrum Analyzer in Trieste, February 2012.

This low cost instrument paves the way for a wide involvement of people in the measurement of the real spectrum usage on their own country which hopefully can lead to a better spectrum management.

For additional information see:

http://www.apc.org/en/faq/citizens -guide-airwaves

5. ANTENNAS / TRANSMISSION LINES

The transmitter that generates the RF power to drive the antenna is usually located at some distance from the antenna terminals. The connecting link between the two is the *RF transmission line*. Its purpose is to carry RF power from one place to another, and to do this as efficiently as possible. From the receiver side, the antenna is responsible for picking up any radio signals in the air and passing them to the receiver with the minimum amount of distortion and maximum efficiency, so that the radio has its best chance to decode the signal. For these reasons, the RF cable has a very important role in radio systems: it must maintain the integrity of the signals in both directions.

Wireless system connections

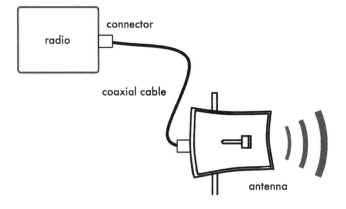

Figure ATL 1: Radio, transmission line and antenna

The simplest transmission line one can envisage is the bifilar or twin lead, consisting of two conductors separated by a dielectric or insulator. The dielectric can be air or a plastic like the one used for flat transmission lines used in TV antennas. A bifilar transmission line open at one end will not radiate because the current in each wire has the same value but opposite direction, so that the fields created on a given point at some distance from the line cancel.

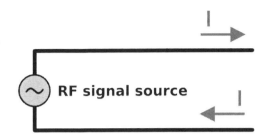

Figure ATL 2: Bifilar transmission line

If we bend the open ends of the transmission line in opposite directions, the currents will now generate electric fields that are in phase and reinforce each other and will therefore radiate and propagate at a distance. We now have an antenna at the end of the transmission line.

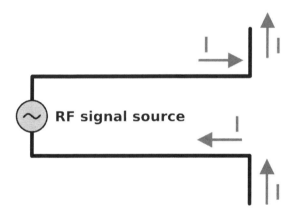

Figure ATL 3: Antenna from transmission line

The length of the bent portion of the transmission line will determine the antenna feature. If this length corresponds to a quarter of a wavelength we will have a half wave dipole antenna with a gain of 2.15 dBi.

The functioning of the bifilar transmission line just described is strongly affected by any metal in its proximity, so a better solution is to confine the electrical fields by means of an external conductor that shields the internal one. This constitutes a *coaxial* cable. Alternatively, a hollow metallic pipe of the proper dimensions will also effectively carry RF energy in what is known as a *waveguide*.

Cables

For frequencies higher than HF the coaxial cables (or **coax** for short, derived from the words "of common axis") are used almost exclusively. Coax cables have a core conductor wire surrounded by a non-conductive material called **dielectric**, or simply **insulation**.

The dielectric is then surrounded by an encompassing shielding which is often made of braided wires. The dielectric prevents an electrical connection between the core and the shielding. Finally, the coax is protected by an outer casing which is generally made from a PVC material.

The inner conductor carries the RF signal, and the outer shield prevents the RF signal from radiating to the atmosphere, and also prevents outside signals from interfering with the signal carried by the core. Another interesting fact is that high frequency electrical signal travels only along the outer layer of a conductor, the inside material does not contribute to the conduction, hence the larger the central conductor, the better the signal will flow. This is called the "skin effect".

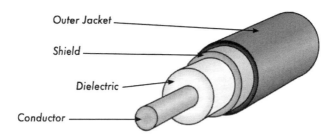

Figure ATL 4: Coaxial cable with jacket, shield, dielectric, and core conductor.

Even though the coaxial construction is good at transporting the signal, there is always resistance to the electrical flow: as the signal travels along, it will fade away.

This fading is known as attenuation, and for transmission lines it is measured in decibels per metre (dB/m).

The rate of attenuation is a function of the signal frequency and the physical construction of the cable itself. As the signal frequency increases, so does its attenuation.

Obviously, we need to minimise the cable attenuation as much as possible by keeping the cable very short and using high quality cables.

Here are some points to consider when choosing a cable for use with microwave devices:

1. "The shorter the better!" The first rule when you install a piece of cable is to try to keep it as short as possible. The power loss is not linear, so doubling the cable length means that you are going to lose much more than twice the power. In the same way, reducing the cable length by half gives you more than twice the power at the antenna. The best solution is to place the transmitter as close as possible to the antenna, even when this means placing it on a tower.

2. "The cheaper the worse!" The second golden rule is that any money you invest in buying a good quality cable is a bargain. Cheap cables can be used at low frequencies, such as VHF. Microwaves require the highest quality cables available.

3. Avoid RG-58. It is intended for thin Ethernet networking, CB or VHF radio, not for microwave.

4. Avoid RG-213 or RG-8. They are intended for CB and HF radio. In this case even if the diameter is large the attenuation is significant due to the cheap insulator used.

5. Whenever possible, use the best rated LMR cable or equivalent you can find. LMR is a brand of coax cable available in various diameters that works well at microwave frequencies. The most commonly used are LMR-400 and LMR-600. Heliax cables are also very good, but expensive and difficult to use.

6. Whenever possible, use cables that are pre-crimped and tested in a proper lab. Installing connectors to cable is a tricky business, and is difficult to do properly even with the specific tools. Never step over a cable, bend it too much, or try to unplug a connector by pulling the cable directly. All of these behaviours may change the mechanical characteristic of the cable and therefore its impedance, short the inner conductor to the shield, or even break the line.

7. Those problems are difficult to track and recognise and can lead to unpredictable behaviour on the radio link.

8. For very short distances, a thin cable of good quality maybe adequate since it will not introduce too much attenuation.

Waveguides

Above 2 GHz, the wavelength is short enough to allow practical, efficient energy transfer by different means. A waveguide is a conducting tube through which energy is transmitted in the form of electromagnetic waves. The tube acts as a boundary that confines the waves in the enclosed space. The Faraday cage phenomenon prevents electromagnetic effects from being evident outside the guide. The electromagnetic fields are propagated through the waveguide by means of reflections against its inner walls, which are considered perfect conductors. The intensity of the fields is greatest at the center along the X dimension, and must diminish to zero at the end walls because the existence of any field parallel to the walls at the surface would cause an infinite current to flow in a perfect conductor.

The X, Y and Z axis of a rectangular waveguide can be seen in the following figure:

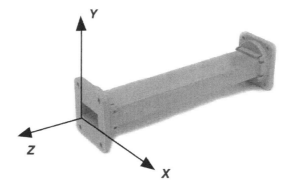

Figure ATL 5: The X, Y, and Z axis of a rectangular waveguide.

There are an infinite number of ways in which the electric and magnetic fields can arrange themselves in a waveguide for frequencies above the low cutoff. Each of these field configurations is called a mode. The modes may be separated into two general groups. One group, designated TM (Transverse Magnetic), has the magnetic field entirely transverse to the direction of propagation, but has a component of the electric field in the direction of propagation. The other type, designated TE (Transverse Electric) has the electric field entirely transverse, but has a component of magnetic field in the direction of propagation.

The mode of propagation is identified by the group letters followed by two subscript numerals. For example, TE 10, TM 11, etc.

The number of possible modes increases with the frequency for a given size of guide, and there is only one possible mode, called the *dominant mode*, for the lowest frequency that can be transmitted. In a rectangular guide, the critical dimension is X. This dimension must be more than 0.5 λ at the lowest frequency to be transmitted. In practice, the Y dimension is usually about 0.5 X to avoid the possibility of operation in other than the dominant mode. Cross-sectional shapes other than the rectangle can be used, the most important being the circular pipe. Much the same considerations apply as in the rectangular case. Wavelength dimensions for rectangular and circular guides are given in the following table, where X is the width of a rectangular guide and r is the radius of a circular guide. All figures apply to the dominant mode.

Type of guide	Rectangular	Circular
Cutoff wavelength	2X	3.41r
Longest wavelength transmitted with little attenuation	1.6X	3.2r
Shortest wavelength before next mode becomes possible	1.1X	2.8r

Energy may be introduced into or extracted from a waveguide by means of either the electric or magnetic field. The energy transfer typically happens through a coaxial line. Two possible methods for coupling to a coaxial line are using the inner conductor of the coaxial line, or through a loop. A probe which is simply a short extension of the inner conductor of the coaxial line can be oriented so that it is parallel to the electric lines of force. A loop can be arranged so that it encloses some of the magnetic lines of force. The point at which maximum coupling is obtained depends upon the mode of propagation in the guide or cavity. Coupling is maximum when the coupling device is in the most intense field.

If a waveguide is left open at one end, it will radiate energy (that is, it can be used as an antenna rather than a transmission line).

This radiation can be enhanced by flaring the waveguide to form a pyramidal horn antenna.

There are examples of practical waveguide antennas for WiFi shown in **Appendix A** called Antenna Construction.

Connectors and adapters

Connectors allow a cable to be connected to another cable or to a component in the RF chain. There are a wide variety of fittings and connectors designed to go with various sizes and types of coaxial lines.
We will describe some of the most popular ones.

BNC connectors were developed in the late 40s. BNC stands for Bayonet Neill Concelman, named after the men who invented it: Paul Neill and Carl Concelman.
The BNC product line is a miniature quick connect/disconnect connector. It features two bayonet lugs on the female connector, and mating is achieved with only a quarter turn of the coupling nut. BNCs are ideally suited for cable termination for miniature to subminiature coaxial cable (RG-58 to RG-179, RG-316, etc.). They are most commonly found on test equipment and 10base2 coaxial Ethernet cables.
TNC connectors were also invented by Neill and Concelman, and are a threaded variation of the BNC. Due to the better interconnect provided by the threaded connector, TNC connectors work well through about 12 GHz. TNC stands for Threaded Neill Concelman.
Type N (again for Neill, although sometimes attributed to "Navy") connectors were originally developed during the Second World War. They are usable up to 18 GHz, and very commonly used for microwave applications. They are available for almost all types of cable. Both the plug / cable and plug / socket joints are supposedly waterproof, providing an effective cable clamp. Nevertheless for outdoor use they should be wrapped in self agglomerating tape to prevent water from seeping in.
SMA is an acronym for Sub Miniature version A, and was developed in the 60s. SMA connectors are precision, subminiature units that provide excellent electrical performance up to 18 GHz. These threaded high-performance connectors are compact in size and mechanically have outstanding durability.
The *SMB* name derives from Sub Miniature B, and it is the second subminiature design. The SMB is a smaller version of the SMA with snap-on coupling. It provides broadband capability through 4 GHz with a snap-on connector design.
MCX connectors were introduced in the 80s.
While the MCX uses identical inner contact and insulator dimensions as the SMB, the outer diameter of the plug is 30% smaller than the SMB.

This series provides designers with options where weight and physical space are limited. MCX provides broadband capability though 6 GHz with a snap-on connector design. In addition to these standard connectors, most WiFi devices use a variety of proprietary connectors. Often, these are simply standard microwave connectors with the centre conductor parts reversed, or the thread cut in the opposite direction. These parts are often integrated into a microwave system using a short, flexible jumper called a *pigtail* that converts the non-standard connector into something more robust and commonly available.

Some of these connectors include:

RP-TNC. This is a TNC connector with the genders reversed.

U.FL (also known as *MHF*). This is possibly the smallest microwave connector currently in wide use. The U.FL/MHF is typically used to connect a mini-PCI radio card to an antenna or larger connector (such as an N or TNC) using a thin cable in waht is known as a *pigtail*.

The *MMCX* series, which is also called a MicroMate, is one of the smallest RF connector line and was developed in the 90s. MMCX is a micro-miniature connector series with a lock-snap mechanism allowing for 360 degrees rotation enabling flexibility.

MC-Card connectors are even smaller and more fragile than MMCX. They have a split outer connector that breaks easily after just a few interconnects. Adapters are short, two-sided devices which are used to join two cables or components which cannot be connected directly. For example, an adapter can be used to connect an SMA connector to a BNC.

Adapters may also be used to fit together connectors of the same type, but of different gender.

Figure ATL 6: An N female barrel adapter.

For example a very useful adapter is the one which enables to join two Type N connectors, having socket (female) connectors on both sides.

Choosing the proper connector

"The gender question." Most connectors have a well defined gender. Male connectors have an external housing or sleeve (frequently with an inner thread) that is meant to surround the body of the female connector. They normally have a pin that inserts in the corresponding socket of the female connector, which has a housing threaded on the outer surface or two bayonet struds protruding from a cylinder. Beware of reverse polarity connectors, in which the male has an inner socket and the female an inner pin. Usually cables have male connectors on both ends, while RF devices (i.e. transmitters and antennas) have female connectors. Lightning arrestors, directional couplers and line-through measuring devices may have both male and female connectors. Be sure that every male connector in your system mates with a female connector.

"Less is best!" Try to minimise the number of connectors and adapters in the RF chain. Each connector introduces some additional loss (up to a dB for each connection, depending on the connector!)

"Buy, don't build!" As mentioned earlier, buy cables that are already terminated with the connectors you need whenever possible. Soldering connectors is not an easy task, and to do this job properly is almost impossible for small connectors as U.FL and MMCX. Even terminating "Foam" cables is not an easy task. Don't use BNC for 2.4 GHz or higher. Use N type connectors (or SMA, SMB, TNC, etc.)

Microwave connectors are precision-made parts, and can be easily damaged by mistreatment. As a general rule, you should rotate the outer sleeve to tighten the connector, leaving the rest of the connector (and cable) stationary. If other parts of the connector are twisted while tightening or loosening, damage can easily occur.

Never step over connectors, or drop connectors on the floor when disconnecting cables (this happens more often than you may imagine, especially when working on a mast over a roof).

Never use tools like pliers to tighten connectors. Always use your hands. When working outside, remember that metals expand at high temperatures and contract at low temperatures: connector too tight in the summer can bind or even break in winter.

Antennas radiation patterns

Antennas are a very important component of communication systems. By definition, an antenna is a device used to transform an RF signal traveling on a transmission line into an electromagnetic wave in free space. Antennas have a property known as reciprocity, which means that an antenna will maintain the same characteristics regardless if whether it is transmitting or receiving. All antennas operate efficiently over a relatively narrow frequency band. An antenna must be tuned to the same frequency band of the radio system to which it is connected, otherwise the reception and the transmission will be impaired. In broadcasting, we can make do with inefficient *receiving* antennas, because the transmitters are very powerful, but in two-way communications we must have properly sized antennas. When a signal is fed into an antenna, the antenna will emit radiation distributed in space in a certain way. A graphical representation of the relative distribution of the radiated power in space is called a radiation pattern.

Antenna term glossary

Before we talk about specific antennas, there are a few common terms that must be defined and explained:

Input Impedance

For an efficient transfer of energy, the impedance of the radio, antenna, and transmission cable connecting them must be the same. Transceivers and their transmission lines are typically designed for 50 Ω impedance. If the antenna has an impedance different from 50 Ω there will be a mismatch and reflections will occur unless an impedance matching circuit is inserted. When any of these components are mismatched, transmission efficiency suffers.

Return loss

Return loss is another way of expressing mismatch. It is a logarithmic ratio measured in dB that compares the power reflected by the antenna Pr to the power that is fed into the antenna from the transmission line Pi:

$$Return\ Loss\ (in\ dB) = 10\ log_{10} Pi/Pr$$

While some energy will always be reflected back into the system, a high return loss will yield unacceptable antenna performance.

The interaction between the wave travelling from the transmitter to the antenna and the wave reflected by the antenna towards the transmitter creates what is known as a stationary wave, therefore an alternative way to measure the impedance mismatch is by means of the *Voltage Standing Wave Ratio (VSWR):*

$$Return\ Loss\ (in\ dB) = 20\ log_{10}\ (VSWR+1\,/\,VSWR-1)$$

In a perfectly matched transmission line, VSWR = 1.

In practice, we strive to maintain a VSWR below 2.

Bandwidth

The bandwidth of an antenna refers to the range of frequencies F_H - F_L over which the antenna can operate correctly. The antenna's bandwidth is the number of Hz for which the antenna meets certain requirements, like exhibiting a gain within 3 dB of the maximum gain or a VSWR less than 1.5. The bandwidth can also be described in terms of percentage of the centre frequency of the band.

$$Bandwidth = 100\ (F_H - F_L\,)/F_C$$

...where F_H is the highest frequency in the band, F_L is the lowest frequency in the band, and F_C is the centre frequency in the band.

In this way, bandwidth is constant relative to frequency. If bandwidth was expressed in absolute units of frequency, it would be different depending upon the center frequency.

Different types of antennas have different bandwidth limitations.

Directivity and Gain

Directivity is the ability of an antenna to focus energy in a particular direction when transmitting, or to receive energy from a particular direction when receiving.

If a wireless link uses fixed locations for both ends, it is possible to use antenna directivity to concentrate the radiation beam in the wanted direction.

In a mobile application where the transceiver is not fixed, it may be impossible to predict where the transceiver will be, and so the antenna should ideally radiate as well as possible in all directions. An omnidirectional antenna is used in these applications. Gain cannot be defined in terms of a physical quantity such as the watt or the ohm, but it is a dimensionless ratio. Gain is given in reference to a standard antenna.

The two most common reference antennas are the isotropic antenna and the half-wave dipole antenna.

Directivity and Gain

Directivity is the ability of an antenna to focus energy in a particular direction when transmitting, or to receive energy from a particular direction when receiving. If a wireless link uses fixed locations for both ends, it is possible to use antenna directivity to concentrate the radiation beam in the wanted direction. In a mobile application where the transceiver is not fixed, it may be impossible to predict where the transceiver will be, and so the antenna should ideally radiate as well as possible in all directions. An omnidirectional antenna is used in these applications. Gain cannot be defined in terms of a physical quantity such as the watt or the ohm, but it is a dimensionless ratio. Gain is given in reference to a standard antenna.

The two most common reference antennas are the isotropic antenna and the half-wave dipole antenna.

The isotropic antenna radiates equally well in all directions. Real isotropic antennas do not exist, but they provide useful and simple theoretical antenna patterns with which to compare real antennas. Any real antenna will radiate more energy in some directions than in others. Since antennas cannot create energy, the total power radiated is the same as an isotropic antenna. Any additional energy radiated in the direction it favours is offset by equally less energy radiated in some other direction.The gain of an antenna in a given direction is the amount of energy radiated in that direction compared to the energy an isotropic antenna would radiate in the same direction when driven with the same input power. Usually we are only interested in the maximum gain, which is the gain in the direction in which the antenna is radiating most of the power, the so called *boresight*. An antenna gain of 3 dB compared to an isotropic antenna would be written as 3 dBi.

The half-wave dipole can be a useful standard for comparing to other antennas at one frequency or over a very narrow band of frequencies.

Unlike the isotropic, is very easy to build and sometimes manufacturers will express the gain with reference to the half-wave dipole instead of the isotropic. An antenna gain of 3 dB compared to a dipole antenna would be written as 3 dBd. Since a half-wave dipole has a gain of 2.15 dBi, we can find the dBi gain of any antenna by adding 2.15 to its dBd gain.

The method of measuring gain by comparing the antenna under test against a known standard antenna, which has a calibrated gain, is technically known as a gain transfer technique.

Radiation Pattern

The radiation pattern or antenna pattern describes the relative strength of the radiated field in various directions from the antenna, at a constant distance. The radiation pattern is a reception pattern as well, since it also describes the receiving properties of the antenna, as a consequence of reciprocity. The radiation pattern is three-dimensional, but usually the published radiation patterns are a two-dimensional slice of the three-dimensional pattern, in the horizontal and vertical planes.

These pattern measurements are presented in either a rectangular or a polar format.

The following figure shows a rectangular plot presentation of a typical ten-element Yagi antenna radiation pattern.

The detail is good but it is difficult to visualize the antenna behaviour in different directions.

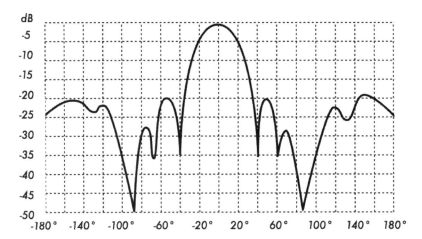

Figure ATL 7: A rectangular plot of the radiation pattern of a Yagi antenna.

Polar coordinate systems are used almost universally.

In the polar-coordinate graph, points are located by projection along a rotating axis (radius) to an intersection with one of several concentric circles that represent the correspong gain in dB, referenced to 0 dB at the outer edge of the plot.

This representation makes it easier to grasp the radial distribution of the antenna power.

Figure ATL 8 is a polar plot of the same 10 element Yagi antenna.

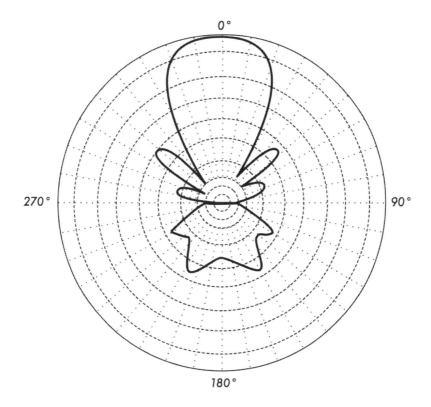

Figure ATL 8: The polar radiation pattern plot of the same antenna

The field pattern that exists close to the antenna is different from the one at a distance, which is the one of interest.

The far-field is also called the radiation field.

For radiation pattern measurement it is important to choose a distance sufficiently large.

The minimum permissible distance depends on the dimensions of the antenna in relation to the wavelength.

The accepted formula for this distance is:

$$r_{min} = 2d^2 / \lambda$$

where r_{min} is the minimum distance from the antenna, d is the largest dimension of the antenna, and λ is the wavelength.

Beamwidth

An antenna's beamwidth is usually understood to mean the half-power beamwidth. The peak radiation intensity is found, and then the points on either side of the peak at which the power has reduced by half are located. The angular distance between the half power points is defined as the beamwidth. Half the power expressed in decibels is -3 dB, so the half power beamwidth is sometimes referred to as the 3 dB beamwidth. Both horizontal and vertical beamwidth are usually considered.

Assuming that most of the radiated power is not divided into sidelobes, the directive and hence the gain is inversely proportional to the beamwidth: as the beamwidth decreases, the gain increases. A very high gain antenna can have a beamwidth of a few degrees and will have to be pointed very carefully in order not to miss the target. The beamwidth is defined by the half power points and in turn determines the coverage area. Coverage area refers to geographical space "illuminated" by the antenna and it is roughly defined by the intersection of the beamwidth with the earth surface. On a base station, it is normally desired to maximise the coverage area, but sometimes one must resort to "downtilting" the antenna, either mechanically or electrically, in order to provide services to customers very close to the base station and therefore below the beamwidth of a non tilted antenna. This down tilting could be achieved by mechanically inclining the antenna, but often the beam can be steered by changing the phase of the signal applied to the different elements of the antenna in what is known as *electrically* downtilting.

Sidelobes

No antenna is able to radiate all the energy in one preferred direction. Some is inevitably radiated in other directions. These smaller peaks are referred to as sidelobes, commonly specified in dB down from the main lobe.

In antenna design, a balance must be struck between gain and sidelobes.

Nulls

In an antenna radiation pattern, a null is a zone in which the effective radiated power is at a minimum. A null often has a narrow directivity angle compared to that of the main beam. Thus, the null is useful for several purposes, such as suppression of interfering signals in a given direction

Polarization

Polarization is defined as the orientation of the electric field of an electromagnetic wave. The initial polarization of a radio wave is determined by the antenna. Most antennas are either vertically or horizontally polarized.

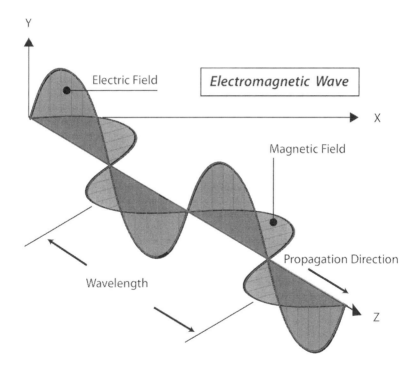

Figure ATL 9: The electric field is perpendicular to magnetic field, both of which are perpendicular to the direction of propagation.

The polarization of the transmitting and the receiving antenna must match, or a very big loss will be incurred.

Some modern systems take advantage of polarization by sending two independent signals at the same frequency, separated by the polarization. Polarization is in general described by an ellipse. Two special cases of elliptical polarization are linear polarization and circular polarization.
With linear polarization, the electric field vector stays in the same plane all the time.
The electric field may leave the antenna in a vertical orientation, a horizontal orientation, or at some angle between the two.
Vertically polarized radiation is somewhat less affected by reflections over the transmission path.
Omnidirectional antennas normally have vertical polarization.
Horizontal antennas are less likely to pick up man- made interference, which ordinarily is vertically polarized.

In circular polarization the electric field vector appears to be rotating with circular motion about the direction of propagation, making one full turn for each RF cycle. This rotation may be right-hand or left-hand.
Choice of polarization is one of the design choices available to the RF system designer.

Polarization Mismatch

In order to transfer maximum power between a transmit and a receive antenna, both antennas must have the same spatial orientation, and the same polarization sense.
When the antennas are not aligned or do not have the same polarization, there will be a reduction in power transfer between the two antennas. This reduction in power transfer will reduce the overall system efficiency and performance.
When the transmit and receive antennas are both linearly polarized, physical antenna misalignment will result in a polarization mismatch loss, which can be determined using the following formula:

$$Loss\ (dB) = 20\ log_{10}(cos\ \theta)$$

...where θ is the difference in the polarization angle between the two antennas.

For 15° the loss is approximately 0.3 dB, for 30° we lose 1.25 dB, for 45° we lose 3 dB and for 90° we have an infinite loss.

In short, the greater the mismatch in polarization between a transmitting and receiving antenna, the greater the loss.

In the real world, a 90° mismatch in polarization is quite large but not infinite. Some antennas, such as Yagis or can antennas, can be simply rotated 90° to match the polarization of the other end of the link.

You can use the polarization effect to your advantage on a point-to-point link.

Use a monitoring tool to observe interference from adjacent networks, and rotate one antenna until you see the lowest received signal. Then bring your link online and orientate the other end to match polarization.

This technique can sometimes be used to build stable links, even in noisy radio environments.

Polarization mismatch can be exploited to send two different signals on the same frequency at the same time, thus doubling the throughput of the link. Special antennas that have dual feeds can be used for this purpose. They have two RF connectors that attach to two independent radios. The real life throughput is somewhat lower than twice the single antenna throughput because of the inevitable cross polarization interference.

Front-to-back ratio

It is often useful to compare the front-to-back ratio of directional antennas. This is the ratio of the maximum directivity of an antenna to its directivity in the opposite direction.

For example, when the radiation pattern is plotted on a relative dB scale, the front-to-back ratio is the difference in dB between the level of the maximum radiation in the forward direction and the level of radiation at 180 degrees from it.

This number is meaningless for an omnidirectional antenna, but it is quite relevant when building a system with repeaters, in which the signal sent backward will interfere with the useful signal and must be minimised.

Antenna Aperture

The electrical "aperture" of a receiving antenna is defined as the cross section of a parabolic antenna that would deliver the same power to a matched load.

It is easy to see that a parabolic grid has an aperture very similar to a solid paraboloid.

The aperture of an antenna is proportional to the gain.

By reciprocity, the aperture is the same for a transmitting antenna.

Notice that the concept of aperture is not easily visualised in the case of a wire antenna in which the physical area is negligible. In this case the antenna aperture must be derived from the formula of the gain.

Types of antennas

A classification of antennas can be based on:

Frequency and size.

Antennas used for HF are different from antennas used for VHF, which in turn are different from antennas for microwave. The wavelength is different at different frequencies, so the antennas must be different in size to radiate signals at the correct wavelength.

We are particularly interested in antennas working in the microwave range, especially in the 2.4 GHz and 5 GHz frequencies.

At 2.4 GHz the wavelength is 12.5 cm, while at 5 GHz it is 6 cm.

Directivity.

Antennas can be omnidirectional, sectorial or directive. Omnidirectional antennas radiate roughly the same signal all around the antenna in a complete 360° pattern.

The most popular types of omnidirectional antennas are the dipole and the ground plane. Sectorial antennas radiate primarily in a specific area. The beam can be as wide as 180 degrees, or as narrow as 60 degrees.

Directional or directive antennas are antennas in which the beamwidth is much narrower than in sectorial antennas. They have the highest gain and are therefore used for long distance links.

Types of directive antennas are the **Yagi**, the **biquad**, the **horn**, the **helicoidal**, the **patch antenna**, the **parabolic dish**, and many others.

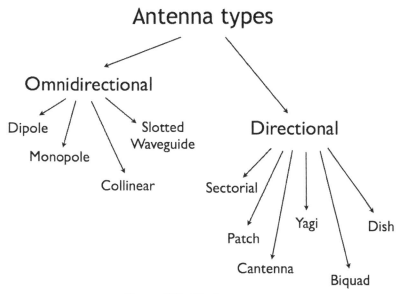

Figure ATL 10: Antenna types

Physical construction.

Antennas can be constructed in many different ways, ranging from simple wires, to parabolic dishes, to coffee cans.

When considering antennas suitable for 2.4 GHz WLAN use, another classification can be used:

Application.

Access points tend to make point-to-multipoint networks, while remote links or backbones are point-to-point. Each of these suggest different types of antennas for their purpose. Nodes that are used for multipoint access will likely use omni antennas which radiate equally in all directions, or several sectorial antennas each focusing into a small area. In the point-to-point case, antennas are used to connect two single locations together. Directive antennas are the primary choice for this application.

A brief list of common type of antennas for the 2.4 GHz frequency is presented now, with a short description and basic information about their characteristics.

1/4 wavelength ground plane.

The 1/4 wavelength ground plane antenna is very simple in its construction and is useful for communications when size, cost and ease of construction are important. This antenna is designed to transmit a vertically polarized signal. It consists of a 1/4 wavelength element as active element and three or four 1/4 wavelength ground elements bent 30 to 45 degrees down. This set of elements, called radials, is known as a ground plane.

Figure ATL 11: Quarter wavelength ground plane antenna.

This is a simple and effective antenna that can capture a signal equally from all directions. The gain of this antenna is in the order of 2 - 4 dBi.

Yagi-Uda antenna

A basic Yagi or more properly Yagi-Uda antenna consists of a certain number of straight elements, each measuring approximately half wavelength. The driven or active element of a Yagi is the equivalent of a centre-fed, half-wave dipole antenna.

Parallel to the driven element, and approximately 0.2 to 0.5 wavelength on either side of it, are straight rods or wires called reflectors and directors, or simply passive elements.

A reflector is placed behind the driven element and is slightly longer than half wavelength; directors are placed in front of the driven element and are slightly shorter than half wavelength. A typical Yagi has one reflector and one or more directors.

The antenna propagates electromagnetic field energy in the direction running from the driven element toward the directors, and is most sensitive to incoming electromagnetic field energy in this same direction. The more directors a Yagi has, the greater the gain.

Following is the photo of a Yagi antenna with 5 directors and one reflector. Yagi antennas are often enclosed in a cylindrical radome to afford protection from the weather.

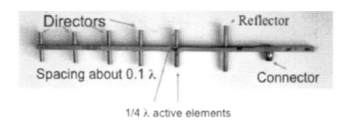

Figure ATL 12: Yagi-Uda

Yagi antennas are used primarily for Point-to-Point links, have a gain from 10 to 20 dBi and a horizontal beamwidth of 10 to 20 degrees.

Horn

The horn antenna derives its name from the characteristic flared appearance.

The flared portion can be square, rectangular, cylindrical or conical.

The direction of maximum radiation corresponds with the axis of the horn.

It is easily fed with a waveguide, but can be fed with a coaxial cable and a proper transition.

While it is cumbersome to make a horn antenna at home, a cylindrical can with proper dimensions will have similar characteristics.

Figure ATL 13: Feed horn made from a food can.

Horn antennas are commonly used as the active element in a dish antenna. The horn is pointed toward the centre of the dish reflector.

The use of a horn, rather than a dipole antenna or any other type of antenna, at the focal point of the dish minimizes loss of energy around the edges of the dish reflector.

At 2.4 GHz, a simple horn antenna made with a tin can has a gain in the order of 10 dBi.

Parabolic Dish

Antennas based on parabolic reflectors are the most common type of directive antennas when a high gain is required.

The main advantage is that they can be made to have gain and directivity as large as required. The main disadvantage is that big dishes are difficult to mount and are likely to have a large wind load.

Randomes can be used to reduce the wind load or windage, as well as for weather protection.

Figure ATL 14: A solid dish antenna.

Dishes up to one metre are usually made from solid material.

Aluminum is frequently used for its weight advantage, its durability and good electrical characteristics.

Windage increases rapidly with dish size and soon becomes a severe problem. Dishes which have a reflecting surface that uses an open mesh are frequently used.

These have a poorer front-to-back ratio, but are safer to use and easier to build.

Copper, aluminum, brass, galvanized steel and steel are suitable mesh materials.

BiQuad

The BiQuad antenna is simple to build and offers good directivity and gain for Point-to-Point communications. It consists of a two squares of the same size of 1/4 wavelength as a radiating element and of a metallic plate or grid as reflector. This antenna has a beamwidth of about 70 degrees and a gain in the order of 10-12 dBi. It can be used as stand-alone antenna or as feeder for a Parabolic Dish.

The polarization is such that looking at the antenna from the front, if the squares are placed side by side the polarization is vertical.

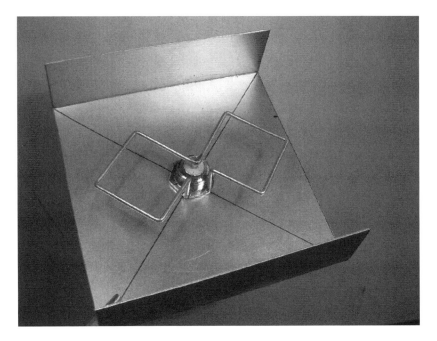

Figure ATL 15: The BiQuad.

Log Periodic Antennas

Log periodic antennas have moderate gain over a wide frequency band, They are often used in spectrum analysers for testing purposes and are also popular as TV receiving antennas since they can efficiently cover from channel 2 up to channel 14. These antennas are used in White space devices that require the ability to work in widely different channels.

Figure ATL 16: Log periodic antenna

Other Antennas

Many other types of antennas exist and new ones are created following advances in technology.

Sector or Sectorial antennas: they are widely used in cellular telephony infrastructure and are usually built adding a reflective plate to one or more phased dipoles.
Their horizontal beamwidth can be as wide as 180 degrees, or as narrow as 60 degrees, while the vertical is usually much narrower.
Composite antennas can be built with many Sectors to cover a wider horizontal range (multisectorial antenna).
Panel or Patch antennas: they are solid flat panels used for indoor coverage, with a gain up to 23 dBi.

Reflector theory

The basic property of a perfect parabolic reflector is that it converts a spherical wave irradiating from a point source placed at the focus into a plane wave. Conversely, all the energy received by the dish from a distant source is reflected to a single point at the focus of the dish. The position of the focus, or focal length, is given by:

$$f = D^2 / 16\,c$$

...where D is the dish diameter and c is the depth of the parabola at its centre.

The size of the dish is the most important factor since it determines the maximum gain that can be achieved at the given frequency and the resulting beamwidth. The gain and beamwidth obtained are given by:

$$Gain = ((3.14\,D)^2 / \lambda^2)\,\eta$$

$$Beamwidth = 70\,\lambda / D$$

...where $\mathbf{D}$ is the dish diameter and $\boldsymbol{\eta}$ is the efficiency. The efficiency is determined mainly by the effectiveness of illumination of the dish by the feed, but also by other factors. Each time the diameter of a dish is doubled, the gain is four times or 6 dB greater. If both stations double the size of their dishes, signal strength can be increased by 12 dB, a very substantial gain. An efficiency of 50% can be assumed when hand-building antennas.

The ratio f / $\mathbf{D}$ (focal length/diameter of the dish) is the fundamental factor governing the design of the feed for a dish. The ratio is directly related to the beamwidth of the feed necessary to illuminate the dish effectively. Two dishes of the same diameter but different focal lengths require different design of feed if both are to be illuminated efficiently. The value of 0.25 corresponds to the common focal-plane dish in which the focus is in the same plane as the rim of the dish.

The optimum illumination of a dish is a compromise between maximising the gain and minimising the sidelobes.

Amplifiers

As mentioned earlier, antennas do not actually create power. They simply direct all available power into a particular pattern. By using a power amplifier, you can use DC power to augment your available signal. An amplifier connects between the radio transmitter and the antenna, and has an additional cable that connects to a power source.

Amplifiers are available that work at 2.4 GHz, and can add several Watts of power to your transmission. These devices sense when an attached radio is transmitting, and quickly power up and amplify the signal. They then switch off again when transmission ends. When receiving, they also add amplification to the signal before sending it to the radio.

Unfortunately, simply adding amplifiers will not magically solve all of your networking problems.

We do not discuss power amplifiers at length in this book because there are a number of significant drawbacks to using them:

- They are expensive. Amplifiers must work at relatively wide bandwidths at 2.4 GHz, and must switch quickly enough to work for Wi- Fi applications.
- They provide no additional directionality. High gain antennas not only improve the available amount of signal, but tend to reject noise from other directions. Amplifiers blindly amplify both desired and interfering signals, and can make interference problems worse.
- Amplifiers generate noise for other users of the band. By increasing your output power, you are creating a louder source of noise for other users of the unlicensed band. Conversely, adding antenna gain will improve your link and can actually decrease the noise level for your neighbours.
- Using amplifiers is often illegal. Every country imposes power limits on use of unlicensed spectrum. Adding an antenna to a highly amplified signal will likely cause the link to exceed legal limits.

Antennas cost far less than amps, and can improve a link simply by changing the antenna on one end.

Using more sensitive radios and good quality cable also helps significantly on long distance wireless links.

These techniques are unlikely to cause problems for other users of the band, and so we recommend pursuing them before adding amplifiers.

Many manufacturers offer high power versions of their WiFi radios at both 2 and 5 GHz, which have a built in amplifiers.

These are better than external amplifiers, but do not assume that it is always smart to use the high power version, for many application the standard power coupled with a high gain antenna is actually better.

Practical antenna designs

The cost of 2.4 GHz antennas has fallen dramatically with the increased popularity of WiFi. Innovative designs use simpler parts and fewer materials to achieve impressive gain with relatively little machining. Unfortunately, availability of good antennas is still limited in some areas of the world, and importing them can be expensive.

While actually designing an antenna can be a complex and error-prone process, constructing antennas from locally available components is very straightforward, and can be a lot of fun.

In **Appendix A** called Antenna Construction we present some practical antenna designs that can be built for very little money.

Antenna measurements

Precise antenna instruments require expensive instruments and installations. It is therefore advisable to obtain the antenna parameters values directly from a reputable manufacturer.

An anechoic chamber is needed to perform accurate antenna measurements, otherwise the reflections will cause false readings.

Ice affects the performance of all antennas to some degree and the problem gets more serious at higher frequencies. The impedance of free space is 377 ohms. If the air immediately surrounding the dipole elements is replaced by ice which has a lower impedance than air, then the impedance match and radiation patterns of the antenna will change.

These changes become progressively worse as the ice loading increases.

Antenna elements are usually encased in a plastic protective housing (radome). This provides an air space between the elements and ice casing so that the lower impedance of the ice layer has only a small effect on the radiators.

Detuning is greatly reduced but radiation pattern distortion may still be encountered (detuning reduces usable antenna bandwidth). For a given ice thickness, deviation from nominal performance values become worse as frequency increases.

In areas where severe icing and wet snow are common, it is prudent to install a full radome over solid parabolic antennas, to use panel antennas instead of corner reflectors, and to stay away from grid parabolics.

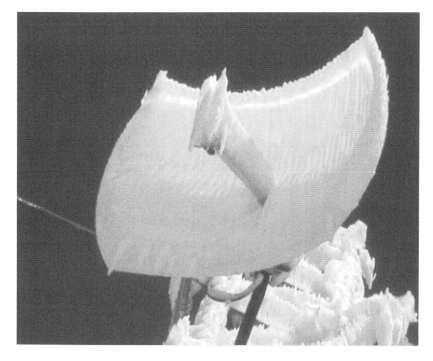

Figure ATL 17: Effect of ice on a parabolic grid antenna

NETWORKING

6. NETWORKING

Before purchasing equipment or deciding on a hardware platform, you should have a clear idea of the nature of your communications problem. Most likely, you are reading this book because you need to connect computer networks together in order to share resources and ultimately reach the larger global Internet.
The network design you choose to implement should fit the communications problem you are trying to solve.

Do you need to connect a remote site to an Internet connection in the centre of your campus? Will your network likely grow to include several remote sites? Will most of your network components be installed in fixed locations, or will your network expand to include hundreds of roaming laptops and other devices?

In this chapter, we will review the networking concepts that define TCP/IP, the primary family of networking protocols currently used on the Internet.
We will also look at the hardware options that are likely to form the underlying physical layer of your TCP/IP network and end with some examples of wireless configurations. This will prepare you very well for the chapter called **Deployment Planning** later in this book.

TCP/IP refers to the suite of protocols that allow conversations to happen on the global Internet.
By understanding TCP/IP, you can build networks that will scale to virtually any size, and will ultimately become part of the global Internet.

This edition of the book now includes an introduction to IPv6 which is the new numbering system of the Internet.
As it is very likely you will be deploying networks using IPv6, it is highly recommended you become familiar with how this works and also how it can work alongside the older IPv4 networks that will continue to operate on the Internet for some while yet.

Introduction

Venice, Italy is a fantastic city to get lost in. The roads are mere foot paths that cross water in hundreds of places, and never go in a simple straight line. Postal carriers in Venice are some of the most highly trained in the world, specialising in delivery to only one or two of the six sestieri (districts) of Venice. This is necessary due to the intricate layout of that ancient city. Many people find that knowing the location of the water and the sun is far more useful than trying to find a street name on a map.

Imagine a tourist who happens to find papier-mâché mask as a souvenir, and wants to have it shipped from the studio in S. Polo, Venezia to their home in London, United Kingdom. This may sound like an ordinary (or even trivial) task, but let's look at what actually happens.

Figure NG 1: Another kind of network mask.

The artist first packs the mask into a shipping box and addresses it to the home of the tourist.

They then hand this to a postal employee in Venice, who attaches some official forms and sends it to a central package processing hub for international destinations.

After several days, the package clears Italian customs and finds its way onto a flight to the UK, arriving at a central import processing depot at Heathrow airport. Once it clears through customs, the package is sent to a distribution point in the city of London, then on to the local district postal processing centre of Camden where the tourist lives.
The package eventually makes its way onto a delivery van which has a route that brings it to the correct house on the correct street in Camden. A member of the family accepts and signs for the package from the delivery van driver and then leaves it in the home studio of the tourist who enjoys unpacking it some time later.

The sorting clerk at the office in Camden neither knows nor cares about how to get to the sestiere of S. Polo, Venezia.
His job is simply to accept packages as they arrive, and deliver them to the correct person in Camden.
Similarly, the postal employee in Venice has no need to worry about how to get to the correct address in London. His job is to accept packages from his local neighborhood and forward them to the next closest hub in the delivery chain.

This is very similar to how Internet routing works. A message is split up into many individual packets, and are labelled with their source and destination.
The computer then sends these packets to a router, which decides where to send them next.

The router needs only to keep track of a handful of routes (for example, how to get to the local network, the best route to a few other local networks, and one route to a gateway to the rest of the Internet). This list of possible routes is called the routing table.
As packets arrive at the router, the destination address is examined and compared against its internal routing table.

If the router has no explicit route to the destination in question, it sends the packet to the closest match it can find, which is often its own Internet gateway (via the default route).

And the next router does the same, and so forth, until the packet eventually arrives at its destination.

Packages can only make their way through the international postal system because we have established a standardised addressing scheme for packages.

For example, the destination address must be written legibly on the front of the package, and include all critical information (such as the recipient's name, street address, city, country, and postal code). Without this information, packages are either returned to the sender or are lost in the system. Packets can only flow through the global Internet because we have agreed on a common addressing scheme and protocol for forwarding packets.

These standard communication protocols make it possible to exchange information on a global scale.

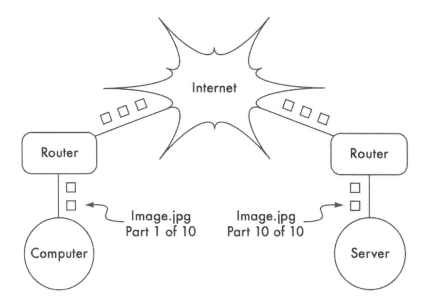

Figure NG 2: Internet networking. Packets are forwarded between routers until they reach their ultimate destination.

Cooperative communications

Communication is only possible when the participants speak a common language. But once the communication becomes more complex than a simple conversation between two people, protocol becomes just as important as language.

All of the people in an auditorium may speak English, but without a set of rules in place to establish who has the right to use the microphone, the communication of an individual's ideas to the entire room is nearly impossible. Now imagine an auditorium as big as the world, full of all of the computers that exist.

Without a common set of communication protocols to regulate when and how each computer can speak, the Internet would be a chaotic mess where every machine tries to speak at once. People have developed a number of communications frameworks to address this problem. The most well-known of these is the OSI model.

The OSI model

The international standard for Open Systems Interconnection (OSI) is defined by the document ISO/IEC 7498-1, as outlined by the International Standards Organization and the International Electrotechnical Commission. The full standard is available as publication "ISO/IEC 7498-1:1994," available from http://standards.iso.org/ittf/PubliclyAvailableStandards/.

The OSI model divides network traffic into a number of layers. Each layer is independent of the layers around it, and each builds on the services provided by the layer below while providing new services to the layer above. The abstraction between layers makes it easy to design elaborate and highly reliable protocol stacks, such as the ubiquitous TCP/IP stack. A protocol stack is an actual implementation of a layered communications framework. The OSI model doesn't define the protocols to be used in a particular network, but simply delegates each communications "job" to a single layer within a well-defined hierarchy.

While the ISO/IEC 7498-1 specification details how layers should interact with each other, it leaves the actual implementation details up to the manufacturer. Each layer can be implemented in hardware (more common for lower layers) or software.

As long as the interface between layers adheres to the standard, implementers are free to use whatever means are available to build their protocol stack.

This means that any given layer from manufacturer A can operate with the same layer from manufacturer B (assuming the relevant specifications are implemented and interpreted correctly).

Here is a brief outline of the seven-layer OSI networking model:

Layer	Name	Description
7	Application	The Application Layer is the layer that most network users are exposed to; it is the level at which human communication happens. HTTP, FTP, and SMTP are all application layer protocols. The human sits above this layer, interacting with the application.
6	Presentation	The Presentation Layer deals with data representation, before it reaches the application. This would include HTML, MIME encoding, data compression, formatting checks, byte ordering, etc.
5	Session	The Session Layer manages the logical communications session between applications. RPC is an example of a layer five protocol.
4	Transport	The Transport Layer provides a method of reaching a particular service on a given network node. Examples of protocols that operate at this layer are TCP, UDP and SCTP. Some protocols at the transport layer (such as TCP) ensure that all of the data has arrived at the destination, and is reassembled and delivered to the next layer in the proper order. UDP is a "connectionless" protocol commonly used for video and audio streaming and doesnt check arrival of data packets.
3	Network	IP (the Internet Protocol) is the most common Network Layer protocol. This is the layer where routing occurs. Packets can leave the link local network and be retransmitted on other networks. Routers perform this function on a network by having at least two network interfaces, one on each of the networks to be interconnected. Nodes on the Internet are reached by their globally unique IP address. Another critical Network Layer protocol is ICMP, which is a special protocol which provides various management messages needed for correct operation of IP. This layer is also sometimes referred to as the Internet Layer.
2	Data Link	Whenever two or more nodes share the same physical medium (for example, several computers plugged into a hub, or a room full of wireless devices all using the same radio channel) they use the Data Link Layer to communicate. Common examples of data link protocols are Ethernet, Token Ring, ATM, and the wireless networking protocols (IEEE 802.11A/B/G). Communication on this layer is said to be link-local, since all nodes connected at this layer communicate with each other directly. This layer is sometimes known as the Media Access Control (MAC) layer. On Ethernet networks, nodes are referred to by their MAC address. This is a unique 48-bit number assigned to every networking device when it is manufactured.
1	Physical	The Physical Layer is the lowest layer in the OSI model, and refers to the actual physical medium over which communicationstake place. This can be a copper CAT5 cable, a fibre optic bundle, radio waves, or just about any other medium capable of transmitting signals. Cut wires, broken fibre, and RF interference are all physical layer problems.

The layers in this model are numbered one through seven, with seven at the top. This is meant to reinforce the idea that each layer builds upon, and depends upon, the layers below. Imagine the OSI model as a building, with the foundation at layer one, the next layers as successive floors, and the roof at layer seven. If you remove any single layer, the building will not stand. Similarly, if the fourth floor is on fire, then nobody can pass through it in either direction.

The first three layers (Physical, Data Link, and Network) all happen "on the network." That is, activity at these layers is determined by the configuration of cables, switches, routers, and similar devices. A network switch can only distribute packets by using MAC addresses, so it need only implement layers one and two. A simple router can route packets using only their IP addresses, so it needs to implement only layers one through three. A web server or a laptop computer runs applications, so it must implement all seven layers. Some advanced routers may implement layer four and above, to allow them to make decisions based on the higher-level information content in a packet, such as the name of a website, or the attachments of an email.

The OSI model is internationally recognised, and is widely regarded as the complete and definitive network model. It provides a framework for manufacturers and network protocol implementers that can be used to build networking devices which interoperate in just about any part of the world. From the perspective of a network engineer or troubleshooter, the OSI model can seem needlessly complex. In particular, people who build and troubleshoot TCP/IP networks rarely need to deal with problems at the Session or Presentation layers. For the majority of Internet network implementations, the OSI model can be simplified into a smaller collection of five layers.

The TCP/IP model

Unlike the OSI model, the TCP/IP model is not an international standard and its definitions vary. Nevertheless, it is often used as a pragmatic model for understanding and troubleshooting Internet networks.

The vast majority of the Internet uses TCP/IP, and so we can make some assumptions about networks that make them easier to understand.

The TCP/IP model of networking describes the following five layers:

Layer	Name
5	Application
4	Transport
3	Internet
2	Data Link
1	Physical

In terms of the OSI model, layers five through seven are rolled into the topmost layer (the Application layer). The first four layers in both models are identical.

Many network engineers think of everything above layer four as "just data" that varies from application to application.

Since the first three layers are interoperable between virtually all manufacturers' equipment, and layer four works between all hosts using TCP/IP, and everything above layer four tends to apply to specific applications, this simplified model works well when building and troubleshooting TCP/IP networks.

We will use the TCP/IP model when discussing networks in this book.

The TCP/IP model can be compared to a person delivering a letter to a city office building.

The person first needs to interact with the road itself (the Physical layer), pay attention to other traffic on the road (the Data Link layer), turn at the correct junction to join another road and arrive at the correct address (the Internet layer), go to the correct floor and room number (the Transport layer), and finally give it to a receptionist who can take the letter from there (the Application layer).

Once they have delivered the message to the receptionist, the delivery person is free to go on their way. The five layers can be easily remembered by using the mnemonic "Please Don't Look In The Attic," which of course stands for "Physical / Data Link / Internet / Transport / Application."

The Internet Protocols

TCP/IP is the protocol stack most commonly used on the global Internet. The acronym stands for Transmission Control Protocol (TCP) and Internet Protocol (IP), but actually refers to a whole family of related communications protocols. TCP/IP is also called the Internet protocol suite, and it operates at layers three and four of the TCP/IP model.

In this discussion, we will focus on version six of the IP protocol (IPv6) as since 2012 this is the version to deploy in parallel with the previous version four (IPv4). In 2012, about half of the Internet content is available with a better user experience by using IPv6.

The previous version is explained in this chapter too because some old content or old applications (Skype in 2012) still require IPv4. And indeed many networks that you might have to interconnect to will still have the legacy IPv4 technology deployed for some years to come.

Besides the length of the address, IPv4 and IPv6 are quite similar: they are connectionless network protocols running on the same data-link layer (WiFi, Ethernet...) and serving the same transport protocols (TCP, SCTP, UDP...) In this book, when IP is written without any version, then it means that it applies to both versions. A dual-stack network is a network that runs IPv6 and IPv4 and the same time. It is expected that dual-networks will be the norm at least until 2020 when IPv6-only will become the norm.

IPv6 Addressing

The IPv6 address is a 128-bit number usually written as multiple hexadecimal numbers. In order to make this address human-readable, it is written in chunks of 32 bits or 4 hexadecimal numbers separated by a colon ':'. The hexadecimal number should be written in lowercase but can also be written in uppercase. An example of an IPv6 address is:

2001:0db8:1234:babe:0000:0000:0000:0001

This address corresponds to:

2001	0db8	1234	babe	0	0	0	1

As these addresses are quite long, it is common to remove the leading 0 in each chunk, so, the same address can also be written as:

2001:db8:1234:babe:0:0:0:1

This address can further be simplified by grouping one block of consecutive chunks of '0' into the abbreviated form of '::', the same address becomes then:

2001:db8:1234:babe::1

There are some specific IPv6 addresses:

- ::1 (or 0000:00000:0000:0000:0000:0000:0000:0001) represents the loopback address, this is the node itself when the node wants to send packets to itself;
- :: (all zero) is the undetermined address, to be used by a node when it does not know its global address, for instance when it boots.

IPv6 Prefixes

IPv6 nodes on the same link or network share the same IPv6 prefix, which is defined as the most-significant part of the IPv6 address. The prefix length is usually 64 bit on a LAN. So, our usual address of 2001:db8:babe::1 can be written as 2001:db8:1234:babe::1/64 (the prefix length is added at the end of the address after a '/'). Defining a prefix length on an address actually splits the address in two parts: the prefix itself and the interface identifier (IID).

2001	0DB8	1234	babe	0	0	0	0
Prefix				Interface Identifier			

On a LAN or WLAN, the prefix length must be 64 bits else some protocols will not work correctly. All nodes on the same LAN or WLAN usually share the same prefix but their IID must be unique to avoid confusion.

The analogy with a postal address in big cities is that the street name is the prefix and the house number is the IID.

The prefix length can be different on links that are neither LAN or WLAN. The network itself is identified by the prefix without any IID but with the prefix length, for example: 2001:db8:1234:babe::/64

IPv4 Addressing

In an IPv4 network, the address is a 32-bit number, normally written as four 8-bit numbers expressed in decimal form and separated by periods. Examples of IPv4 addresses are 10.0.17.1, 192.168.1.1, or 172.16.5.23.

If you enumerated every possible IPv4 address, they would range from 0.0.0.0 to 255.255.255.255.
This yields a total of more than four billion possible IPv4 addresses (255 x 255 x 255 x 255 = 4,228,250,625); although many of these are reserved for special purposes and should not be assigned to hosts.

Some IPv4 addresses that are special:

- 127.0.0.1 represents the loopback address (similar to ::1 for IPv6);
- 0.0.0.0 represents the unspecified address (similar to :: for IPv6).

IPv4 Subnets

By applying a subnet mask (also called a network mask, or simply netmask or even prefix) to an IPv4 address, you can logically define both a host and the network to which it belongs.

Traditionally, subnet masks are expressed using dotted decimal form, much like an IPv4 address. For example, 255.255.255.0 is one common netmask. You will find this notation used when configuring network interfaces, creating routes, etc. However, subnet masks are more succinctly expressed using CIDR notation, which simply enumerates the number of bits in the mask after a forward slash (/).
Thus, 255.255.255.0 can be simplified as /24. CIDR is short for Classless Inter-Domain Routing, and is defined in RFC1518. A subnet mask determines the size of a given network. Using a /24 netmask, 8 bits are reserved for hosts (32 bits total - 24 bits of netmask = 8 bits for hosts). This yields up to 256 possible host addresses (28 = 256). By convention, the first value is taken as the network address (.0 or 00000000), and the last value is taken as the 'broadcast address (.255 or 11111111).

This leaves 254 addresses available for hosts on this network.

Subnet masks work by applying AND logic to the 32 bit IPv4 number.

In binary notation, the "1" bits in the mask indicate the network address portion, and "0" bits indicate the host address portion. A logical AND is performed by comparing two bits. The result is "1" if both of the bits being compared are also "1". Otherwise the result is "0".

Here are all of the possible outcomes of a binary AND comparison between two bits.

Bit 1	Bit 2	Result
0	0	0
0	1	0
1	0	0
1	1	1

To understand how a netmask is applied to an IPv4 address, first convert everything to binary. The netmask 255.255.255.0 in binary contains twenty-four "1" bits:

$$255 \quad 255 \quad 255 \quad 0$$

$$11111111.11111111.11111111.00000000$$

When this netmask is combined with the IPv4 address 10.10.10.10, we can apply a logical AND to each of the bits to determine the network address.

10.10.10.10: 00001010.00001010.00001010.00001010

255.255.255.0: 11111111.11111111.11111111.00000000

10.10.10.0: 00001010.00001010.00001010.00000000

This results in the network 10.10.10.0/24.

This network consists of the hosts 10.10.10.1 through 10.10.10.254, with 10.10.10.0 as the network address and 10.10.10.255 as the broadcast address.

Subnet masks are not limited to entire octets. One can also specify subnet masks like 255.254.0.0 (or /15 CIDR). This is a large block, containing 131,072 addresses, from 10.0.0.0 to 10.1.255.255.

It could be further subdivided, for example into 512 subnets of 256 addresses each. The first one would be 10.0.0.0-10.0.0.255, then 10.0.1.0-10.0.1.255, and so on up to 10.1.255.0-10.1.255.255. Alternatively, it could be subdivided into 2 blocks of 65,536 addresses, or 8192 blocks of 16 addresses, or in many other ways. It could even be subdivided into a mixture of different block sizes, as long as none of them overlap, and each is a valid subnet whose size is a power of two.

While many netmasks are possible, common netmasks include:

CIDR	Decimal	# of Hosts
/30	255.255.255.252	4
/29	255.255.255.248	8
/28	255.255.255.240	16
/27	255.255.255.224	32
/26	255.255.255.192	64
/25	255.255.255.128	128
/24	255.255.255.0	256
/16	255.255.0.0	65 536
/8	255.0.0.0	16 777 216

With each reduction in the CIDR value the IPv4 space is doubled. Remember that two IPv4 addresses within each network are always reserved for the network and broadcast addresses.

There are three common netmasks that have special names. A /8 network (with a netmask of 255.0.0.0) defines a Class A network. A /16 (255.255.0.0) is a Class B, and a /24 (255.255.255.0) is called a Class C. These names were around long before CIDR notation, but are still often used for historical reasons.

In many ways as you can already see IPv6 is easier to plan for than IPv4.

Global IP Addresses

Interconnected networks must agree on an IP addressing plan for IPv6 and IPv4 addresses.

IP addresses must be unique and generally cannot be used in different places on the Internet at the same time; otherwise, routers would not know how best to route packets to them.

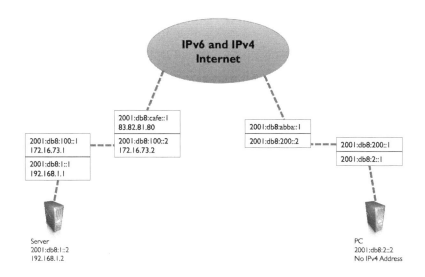

Figure NG 3: With unique IP addresses, ambiguous global routing is impossible. If the PC requests a web page from 2001:db8:1::2, it will reach the correct server.

In order to keep IP addresses unique and globally routable, they are allocated by a central numbering authority that provides a consistent and coherent numbering method. This ensures that duplicate addresses are not used by different networks.

The authority assigns large blocks of consecutive addresses to smaller authorities, who in turn assign smaller consecutive blocks within these ranges to other authorities, or to their customers. The groups of addresses are called subnets or prefixes as we have already mentioned.

A group of related addresses is referred to as an address space.

Both IPv4 and IPv6 addresses are administered by the Internet Assigned Numbers Authority (IANA, http://www.iana.org/).

IANA has divided these address spaces into large subnets, and these subnets are delegated to one of the five regional Internet registries (RIRs), who have been given authority over large geographic areas.

IP addresses are assigned and distributed by Regional Internet Registrars (RIRs) to ISPs. The ISP then allocates smaller IP blocks to their clients as required. Virtually all Internet users obtain their IP addresses from an ISP.

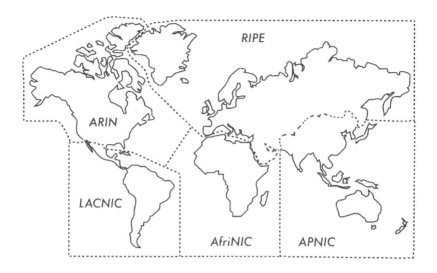

Figure NG 4: Authority for Internet IP address assignments is delegated to the five Regional Internet Registrars.

The five RIRs are:

1) African Network Information Centre
   (AfriNIC, http://www.afrinic.net)
2) Asia Pacific Network Information Centre
   (APNIC, http://www.apnic.net)
3) American Registry for Internet Numbers
   (ARIN, http://www.arin.net)
4) Regional Latin-American and Caribbean IP Address Registry
   (LACNIC, http://www.lacnic.net)
5) Réseaux IP Européens
   (RIPE NCC, http://www.ripe.net)

Your ISP will assign globally routable IP address space to you from the pool allocated to it by your RIR.

The registry system assures that IP addresses are not reused in any part of the network anywhere in the world.

Once IP address assignments have been agreed upon, it is possible to pass packets between networks and participate in the global Internet.

The process of moving packets between networks is called routing.

Static IP Addresses

A static IP address is an address assignment that never changes.

Static IP addresses are important because servers using these addresses may have DNS mappings pointing towards them, and typically serve information to other machines (such as email services, web servers, etc.).

Blocks of static IP addresses may be assigned by your ISP, either by request or automatically depending on your means of connection to the Internet.

Dynamic IP Addresses

Dynamic IP addresses are assigned by an ISP for non-permanent nodes connecting to the Internet, such as a home computer which is on a dial-up connection or a laptop connecting to a wireless hotspot.

Dynamic IP addresses can be assigned automatically using the Dynamic Host Configuration Protocol (DHCP), or the Point-to-Point Protocol (PPP), depending on the type of Internet connection.

A node using DHCP first requests an IP address assignment from the network, and automatically configures its network interface. IP addresses can be assigned randomly from a pool by your ISP, or might be assigned according to a policy. IP addresses assigned by DHCP are valid for a specified time (called the lease time).

The node must renew the DHCP lease before the lease time expires. Upon renewal, the node may receive the same IP address or a different one from the pool of available addresses.

While DHCP works for IPv6 and IPv4, IPv6 has another primary mechanism which is more commonly used for address assignment - it is called Stateless Address Auto-Configuration (SLAAC) which is the default on routers and hosts running IPv6.

It does not require a DHCP server; the router sends periodically Router Advertisement (RA) messages on all connected (W)LAN's which contain the 64-bit prefix to be used on that (W)LAN; hosts then generate their 64-bit interface identifier (usually a random number or a number based on their MAC address – see further) and build their 128-bit address by concatenating the 64-bit prefix from the RA and the newly created 64-bit IID.

Dynamic addresses are popular with Internet Service Providers, because it enables them to use fewer IP addresses than their total number of customers.
They only need an address for each customer who is active at any one time.
Globally routable IP addresses cost money, and there is now a shortage of IPv4 addresses.
Assigning addresses dynamically allows ISPs to save money, and they will often charge extra to provide a static IP address to their customers.

Private IPv4 addresses

Around 2000, it became clear that there would not be enough IPv4 addresses for everyone; this is the reason that IPv6 was specified and developed.
But there was also a temporary trick as most private networks do not require the allocation of globally routable, public IPv4 addresses for every computer in the organisation.
In particular, computers which are not public servers do not need to be addressable from the public Internet.
Organisations typically use IPv4 addresses from the private address space for machines on the internal network.

There are currently three blocks of private address space reserved by IANA: 10.0.0.0/8, 172.16.0.0/12, and 192.168.0.0/16.

These are defined in RFC1918.
These addresses are not intended to be routed on the Internet, and are typically unique only within an organisation or group of organisations that choose to follow the same numbering scheme.

This means that several distinct organisations can use the same addresses as long as they never interconnect their networks directly.

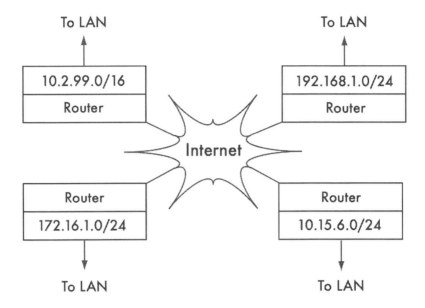

Figure NG 5: RFC1918 private addresses may be used within an organisation, and are not routed on the global Internet.

If you ever intend to link together private networks that use RFC1918 address space, be sure to use unique addresses throughout all of the networks.

For example, you might break the 10.0.0.0/8 address space into multiple Class B networks (10.1.0.0/16, 10.2.0.0/16, etc.).
One block could be assigned to each network according to its physical location (the campus main branch, field office one, field office two, dormitories, and so forth).

The network administrators at each location can then break the network down further into multiple Class C networks (10.1.1.0/24, 10.1.2.0/24, etc.) or into blocks of any other logical size.

In the future, should the networks ever be linked (either by a physical connection, wireless link, or VPN), then all of the machines will be reachable from any point in the network without having to renumber network devices.

Some Internet providers may allocate private addresses like these instead of public addresses to their customers, although this has serious disadvantages.

Since these addresses cannot be routed over the Internet, computers which use them are not really "part" of the Internet, and are not directly reachable from it. In order to allow them to communicate with the Internet, their private addresses must be translated to public addresses.

This translation process is known as Network Address Translation (NAT), and is normally performed at the gateway between the private network and the Internet.

We will look at NAT in more detail later on in this chapter.

As there are huge numbers of IPv6 addresses, there is no need for private IPv6 addresses, although there are Unique Local Addresses (ULA) that are suitable for non connected networks such as labs.

Discovering Neighbours

Imagine a network with three hosts: H_A, H_B, and H_C. They use the corresponding IP addresses A, B and C.

These hosts are part of the same subnet/prefix.

For two hosts to communicate on a local network, they must determine each others' MAC addresses. It is possible to manually configure each host with a mapping table from IP address to MAC address, but it is easier to dynamically discover the neighbour's MAC address through Neighbor Discovery Protocol (NDP) in IPv6 and Address Resolution Protocol (ARP) in IPv4. NDP and ARP work in a very similar way.

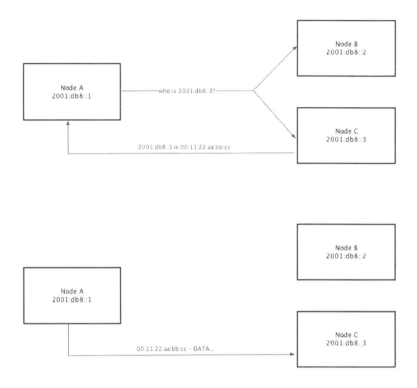

Figure NG 6: IPv6 node A, 2001:db8::1 needs to send data to 2001:db8::3 in the same network (2001:db8::/64 prefix). But it must first ask for the MAC address that corresponds to 2001:db8::3.

When using NDP, node A multicasts to some hosts the question, "Who has the MAC address for the IPv6 2001:db8::3?"

When node C sees a Neighbor Solicitation (NS) for an IPv6 address of its own, it replies with its MAC address with a Neighbor Advertisement (NA) message.

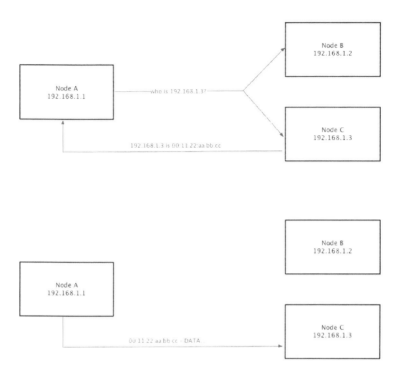

Figure NG 7: IPv4 node A, 192.168.1.1, needs to send data to 192.168.1.3
in the same subnet (192.168.1.0/24). But it must first ask the whole network
for the MAC address that corresponds to 192.168.1.3.

When using ARP, node A broadcasts to all hosts the question,
"Who has the MAC address for the IPv4 192.168.1.3?"

When node C sees an ARP request for its own IPv4 address, it replies
with its MAC address. Node B will also see the ARP request but will not
reply as 192.168.1.3 is not an address of it. This is very similar to NDP
for IPv6 except that an IPv4 node has only a single IPv4 address.

Also ARP broadcasts the request, this means that it is received by all IPv4
nodes in the network causing more host CPU utilisation than IPv6 NDP
which only multicasts to some hosts.

IP Routing to non-Neighbours

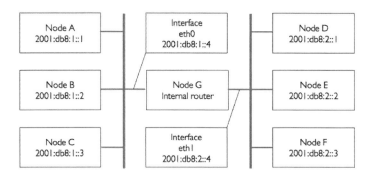

Figure NG 8: Two separate IPv6 networks.

Consider now another network with 3 nodes, D, E, and F, with the corresponding IPv6 addresses 2001:db8:2::1, 2001:db8:2::2, and 2001:db8:2::3.

This is another /64 network, but it is not in the same range as the network on the left hand side.

All three hosts can reach each other directly (first using NDP to resolve the IPv6 address into a MAC address, and then sending packets to that MAC address).

Now we will add node G. This node has two network cards (also called interfaces), with one plugged into each network. The first network card uses the IPv6 address 2001:db8:1::4, on interface eth0 and the other, eth1, uses 2001:db8:2::4.

Node G is now link-local to both networks, and can forward packets between them: node G can route packets between the two networks, it is therefore called a router or sometimes a gateway.

But what if hosts A, B, and C want to reach hosts D, E, and F? They need to know that they should use node G and so they will need to add a route to the other network via host G. For example, hosts A-C would add a static route via 2001:db8:1::4.

In Linux, this can be accomplished with the following command:

 # ip -6 route add 2001:db8:2::/64 via 2001:db8:1::4

...and hosts D-F would add the following:

 # ip -6 route add 2001:db8:1::/64 via 2001:db8:2::4

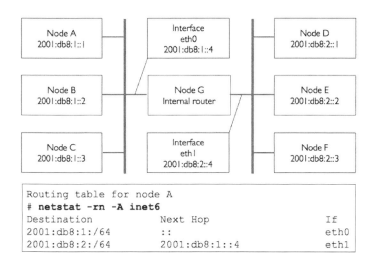

```
Routing table for node A
# netstat -rn -A inet6
Destination            Next Hop                    If
2001:db8:1:/64         ::                          eth0
2001:db8:2:/64         2001:db8:1::4               eth1
```

Figure NG 9: Node G acts as a router between the two networks, other hosts use static routes.

The result for node A is shown in Figure NG 9.

Notice that the route is added via the IPv6 address on host G that is link-local to the respective network.

Host A could not add a route via 2001:db8:2::4, even though it is the same physical machine as 2001:db8:1::4 (node G), since that IPv6 is not link-local.
The address of the next hop can be entered either as a global address (2001:db8:2::4) or as a link-local address (fe80::...); it is usually easier to configure a static route with a global address.

In IPv6, the router G also sends a solicitation and periodically router advertisements that contain its own link-local address, hence, all nodes using stateless auto-configuration or DHCP automatically add a default route via the router link-local address as shown in Figure NG 10.

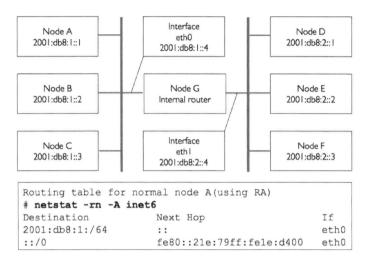

Figure NG 10: Node G acts as a router between the two networks, hosts use stateless address autoconfiguration.

This is a very simple routing example, where the destination is only a single hop away from the source. As networks get more complex, many hops may need to be traversed to reach the ultimate destination. Since it isn't practical for every machine on the Internet to know the route to every other, we make use of a routing entry known as the default route (also known as the default gateway).

When a router receives a packet destined for a network for which it has no explicit route, the packet is forwarded to its default gateway.
The default gateway is typically the best route out of your network, usually in the direction of your ISP.

An example of a router that uses a default gateway is shown in Figure NG 11. Figure NG 11 shows the routing table (which is the set of all routes) on the internal router G which includes the two directly connected networks 2001:db8:1::/64 and 2001:db8:2::/64 as well as a route to all other hosts on the Internet ::/0.

A node always uses the most specific route; that is the route with the longest match to the destination, in Figure NG 11 eth0 will be used for destination 2001:db8:1::1 (match length /64) rather than the less specific ::/0 (match length of 0).

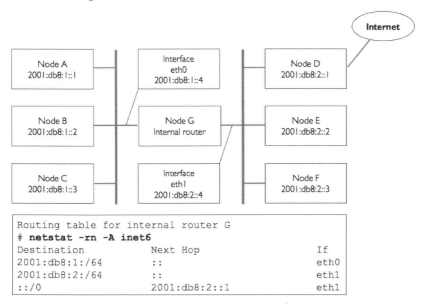

Figure NG 11: Node G is the internal router and uses the Internet router.

A route tells the OS that the desired network doesn't lie on the immediate link-local network, and it must forward the traffic through the specified router.

If host A wants to send a packet to host F, it would first send it to node G. Node G would then look up host F in its routing table, and see that it has a direct connection to host F's network.
Finally, host G would resolve the hardware (MAC) address of host F and forward the packet to it.

Routes can be updated manually, or can dynamically react to network outages and other events.
Some examples of popular dynamic routing protocols are RIP, OSPF, BGP.

Configuring dynamic routing is beyond the scope of this book, but for further reading on the subject, see the resources in **Appendix F**.
IPv4 behaves exactly the same way as depicted in figure NG 12.

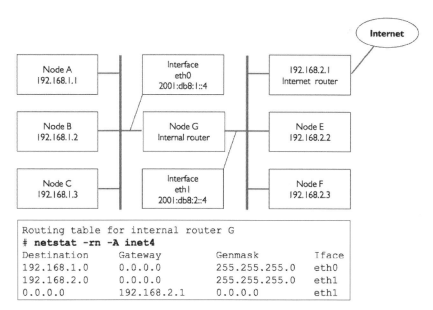

Figure NG 12: Node G is the Internet router on this IPv4 network.

As noted before, most networks and the Internet are dual-stack and all hosts and routers have both IPv4 and IPv6 addresses, this also means that the nodes will have routes for IPv4 and routes for IPv6. For instance, the set of all routes on node G of the previous figures will be:

 # netstat -rn -A inet6

Destination	Next Hop	If
2001:db8:1:/64	::	eth0
2001:db8:2:/64	::	eth1
::/0	2001:db8:2::1	eth1

 # netstat -rn -A inet4

Destination	Gateway	Genmask	Iface
192.168.1.0	0.0.0.0	255.255.255.0	eth0
192.168.2.0	0.0.0.0	255.255.255.0	eth1
0.0.0.0	192.168.2.1	0.0.0.0	eth1

Network Address Translation (NAT) for IPv4

In order to reach hosts on the Internet, private addresses must be converted to global, publicly routable IPv4 addresses.

This is achieved using a technique known as Network Address Translation, or NAT.

A NAT device is a router that manipulates the addresses of packets instead of simply forwarding them.

On a NAT router, the Internet connection uses one (or more) globally routed IPv4 addresses, while the private network uses an IPv4 address from the RFC1918 private address range.

The NAT router allows the global address(es) to be shared with all of the inside users, who all use private addresses.

It converts the packets from one form of addressing to the other as the packets pass through it. As far as the network users can tell, they are directly connected to the Internet and require no special software or drivers.

They simply use the NAT router as their default gateway, and address packets as they normally would.

The NAT router translates outbound packets to use the global IPv4 address as they leave the network, and translates them back again as they are received from the Internet.

The major consequence of using NAT is that machines from the Internet cannot easily reach servers within the organisation without setting up explicit forwarding rules on the router.

Connections initiated from within the private address space generally have no trouble, although some applications (such as Voice over IPv4 and some VPN software) can have difficulty dealing with NAT.

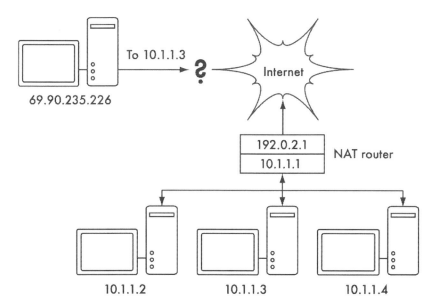

Figure NG 13: Network Address Translation allows you to share a single IPv4 address with many internal hosts, but can make it difficult for some services to work properly.

Depending on your point of view, this can be considered a bug (since it makes it harder to set up two-way communication) or a feature. RFC1918 addresses should be filtered on the edge of your network to prevent accidental or malicious RFC1918 traffic entering or leaving your network.

While NAT performs some firewall-like functions, it is not a replacement for a real firewall as most of the attacks happen now when an internal user visits some web sites with hostile content (called malware for malevolent software).

Internet Protocol Suite

Machines on the Internet use the Internet Protocol (IP) to reach each other, even when separated by many intermediary machines.

There are a number of protocols that are run in conjunction with IP that provide features which are as critical to normal operations as IP itself. Every packet specifies a protocol number that identifies the packet as one of these protocols.

The most commonly used protocols are the Transmission Control Protocol (TCP, number 6), User Datagram Protocol (UDP, number 17), and the Internet Control Message Protocol (ICMP, number 1 for IPv4 and number 58 for IPv6). Taken as a group, these protocols (and others) are known as the Internet Protocol Suite, or simply TCP/IP for short.

The TCP and UDP protocols introduce the concept of port numbers. Port numbers allow multiple services to be run on the same IP address, and still be distinguished from each other. Every packet has a source and destination port number. Some port numbers are well-defined standards, used to reach well-known services such as email and web servers. For example, web servers normally listen on TCP port 80 for insecure traffic and on TCP port 443 for encrypted/secure traffic, NTP time servers listen on UDP port 123, DNS domain name servers listen on UDP port 53, and SMTP email servers listen on TCP port 25.

When we say that a service "listens" on a port (such as port 80), we mean that it will accept packets that use its IP as the destination IP address, and 80 as the destination port.

Servers usually do not care about the source IP or source port, although sometimes they will use them to establish the identity of the other side.
When sending a response to such packets, the server will use its own IP as the source IP, and 80 as the source port.
When a client connects to a service, it may use any source port number on its side that is not already in use, but it must connect to the proper port on the server (e.g. 80 for web, 25 for email).

TCP is a session-oriented protocol with guaranteed and ordered delivery and transmission control features (such as detection and mitigation of network congestion, retries, packet reordering and reassembly, etc.).

UDP is designed for connectionless streams of information, and does not guarantee delivery at all, or in any particular order but can be faster so it is often used for real-time protocols such as for timing, voice or video.

The ICMP protocol is designed for debugging and maintenance on the Internet.
Rather than port numbers, it has message types, which are also numbers.

Different message types are used to request a simple response from another computer (echo request), notify the sender of another packet of a possible routing loop (time exceeded), or inform the sender that a packet that could not be delivered due to firewall rules or other problems (destination unreachable).

By now you should have a solid understanding of how computers on the network are addressed, and how information flows on the network between them.

Now let's take a brief look at the physical hardware that implements these network protocols.

Physical hardware

Ethernet

Ethernet is the name of the most popular standard for connecting together computers on a Local Area Network (LAN). It is sometimes used to connect individual computers to the Internet, via a router, ADSL modem, or wireless device.

However, if you connect a single computer to the Internet, you may not use Ethernet at all.

The name comes from the physical concept of the ether, the medium which was once supposed to carry light waves through free space. The official standard is called IEEE 802.3.

One widely deployed Ethernet standard is called 100baseT also known as Fast Ethernet.

This defines a data rate of 100 Megabits per second (hence the 100), running over twisted (hence the T) pair wires, with modular RJ-45 connectors on the end.

The network topology is a star, with switches or hubs at the centre of each star, and end nodes (devices and additional switches) at the edges. Servers are also connected using Gigabit Ethernet with a rate of 1 Gigabit per second.

Increasingly Gigabit Ethernet is replacing Fast Ethernet in many networks these days as demand for high volume video and other high data rate applications become more prevalent.

Medium Access Control (MAC) addresses

Every device connected to an Ethernet or WiFi network has a unique MAC address, assigned by the manufacturer of the network card. It serves as a unique identifier that enables devices to talk to each other. However, the scope of a MAC address is limited to a broadcast domain, which is defined as all the computers connected together by wires, hubs, switches, and bridges, but not crossing routers or Internet gateways.

MAC addresses are never used directly on the Internet, and are not transmitted across routers.

MAC addresses for Ethernet and IEEE 802.11 WiFi networks are 48 bits long and look like this - 00:1c:c0:17:78:8c or 40:6c:8f:52:59:41; for the latter MAC address, the first 24 bits 40:6c:8f indicates that Apple assigned this address.

Hubs

Ethernet hubs connect multiple twisted-pair Ethernet devices together. They work at the physical layer (the lowest or first layer).

They repeat the signals received by each port out to all of the other ports. Hubs can therefore be considered to be simple repeaters.

Due to this design, only one port can successfully transmit at a time. If two devices transmit at the same time, they corrupt each other's transmissions, and both must back off and retransmit their packets later. This is known as a collision, and each host remains responsible for detecting and avoiding collisions before transmitting, and retransmitting its own packets when needed.

When problems such as excessive collisions are detected on a port, some hubs can disconnect (partition) that port for a while to limit its impact on the rest of the network.

While a port is partitioned, devices attached to it cannot communicate with the rest of the network.

Hubs are limited in their usefulness, since they can easily become points of congestion on busy networks so they are no longer normally deployed in networks nowadays. Its only important to note that a WiFi access point acts as a hub on the radio side.

Switches

A switch is a device which operates much like a hub, but provides a dedicated (or switched) connection between ports.

Rather than repeating all traffic on every port, the switch determines which ports are communicating directly and temporarily connects them together. There can be several such temporary port connections at the same time.

Switches generally provide much better performance than hubs, especially on busy networks with many computers. They are not much more expensive than hubs, and are replacing them in most situations.

Switches work at the data link layer (the second layer), since they interpret and act upon the MAC address in the packets they receive. When a packet arrives at a port on a switch, it makes a note of the source MAC address, which it associates with that port. It stores this information in an internal MAC table often known as Content Addressable Memory (CAM) table. The switch then looks up the destination MAC address in its MAC table, and transmits the packet only on the matching port. If the destination MAC address is not found in the MAC table, the packet is then sent to all of the connected interfaces hoping to reach the right MAC.

Hubs vs. Switches

Hubs are considered to be fairly unsophisticated devices, since they inefficiently rebroadcast all traffic on every port. This simplicity introduces both a performance penalty and a security issue. Overall performance is slower, since the available bandwidth must be shared between all ports. Since all traffic is seen by all ports, any host on the network can easily monitor all of the network traffic.

Switches create temporary virtual connections between receiving and transmitting ports. This yields better performance because many virtual connections can be made simultaneously. More expensive switches can switch traffic by inspecting packets at higher levels (at the transport or application layer), allowing the creation of VLANs, and implementing other advanced features.

A hub can be used when repetition of traffic on all ports is desirable; for example, when you want to explicitly allow a monitoring machine to see all of the traffic on the network. Most switches provide monitor port functionality that enables repeating on an assigned port specifically for this purpose.

Hubs were once cheaper than switches. However, the price of switches has reduced dramatically over the years. Therefore, old network hubs should be replaced whenever possible with new switches.

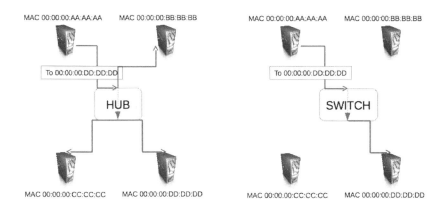

Figure NG 14: A hub simply repeats all traffic on every port, while a switch makes a temporary, dedicated connection between the ports that need to communicate.

Both hubs and switches may offer managed services.
Some of these services include the ability to set the link speed (10baseT, 100baseT, 1000baseT, full or half-duplex) per port, enable triggers to watch for network events (such as changes in MAC address or malformed packets), and usually include port counters for easy bandwidth accounting.

A managed switch that provides upload and download byte counts for every physical port can greatly simplify network monitoring.
These services are typically available via SNMP, or they may be accessed via telnet, ssh, a web interface, or a custom configuration tool.

Routers and Firewalls

While hubs and switches provide connectivity on a local network segment, a router's job is to forward packets between different network segments.
A router typically has two or more physical network interfaces.

It may include support for different types of network media, such as Ethernet, WiFi, optical fibre, DSL, or dial-up.

Routers can be dedicated hardware devices or they can be made from a standard PC with multiple network cards and appropriate software.

Routers sit at the edge of two or more networks. By definition, they have one connection to each network, and as border machines they may take on other responsibilities as well as routing. Many routers have firewall capabilities that provide a mechanism to filter or redirect packets that do not fit security or access policy requirements.

They may also provide Network Address Translation (NAT) services for IPv4.

Routers vary widely in cost and capabilities.

The lowest cost and least flexible are simple, dedicated hardware devices, often with NAT functionality, used to share an Internet connection between a few computers; well known brands include Linksys, D-Link, Netgear.

The next step up is a software router, which consists of an operating system running on a standard PC with multiple network interfaces. Standard operating systems such as Microsoft Windows, Linux, and BSD are all capable of routing, and are much more flexible than the low-cost hardware devices; it is often called Internet Connection Sharing.

However, they suffer from the same problems as conventional PCs, with high power consumption, a large number of complex and potentially unreliable parts, and more involved configuration.

The most expensive devices are high-end dedicated hardware routers, made by companies like Cisco and Juniper.

They tend to have much better performance, more features, and higher reliability than software routers on PCs.

It is also possible to purchase technical support and maintenance contracts for them.

Most modern routers offer mechanisms to monitor and record performance remotely, usually via the Simple Network Management Protocol (SNMP), although the least expensive devices often omit this feature.

Other equipment

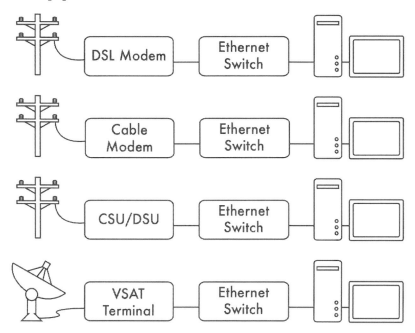

Figure NG 15: Many DSL modems, cable modems, wireless access points, and VSAT terminals terminate at an Ethernet jack.

Each physical network has an associated piece of terminal equipment. For example, VSAT connections consist of a satellite dish connected to a terminal that either plugs into a card inside a PC, or ends at a standard Ethernet connection. DSL lines use a DSL modem that bridges the telephone line to a local device, either an Ethernet network or a single computer via USB. Cable modems bridge the television cable to Ethernet, or to an internal PC card bus.

Standard dialup lines use modems to connect a computer to the telephone, usually via a plug-in card or serial port. And there are many different kinds of wireless networking equipment that connect to a variety of radios and antennas, but nearly always end at an Ethernet jack. The functionality of these devices can vary significantly between manufacturers. Some provide mechanisms for monitoring performance, while others may not.

Since your Internet connection ultimately comes from your ISP, you should follow their recommendations when choosing equipment that bridges their network to your Ethernet network.

Putting it all together

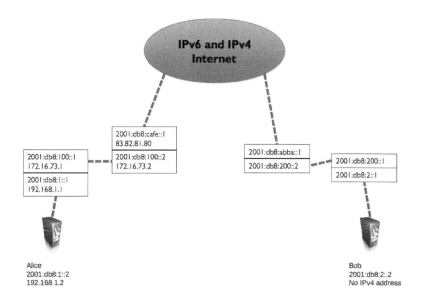

Figure NG 16: Internet networking. Each network segment has a router with two IP addresses, making it "link local" to two different networks. Packets are forwarded between routers until they reach their ultimate destination.

Once all network nodes have an IP address, they can send data packets to the IP address of any other node. Through the use of routing and forwarding, these packets can reach nodes on networks that are not physically connected to the originating node. This process describes much of what "happens" on the Internet.

In this example, you can see the path that the packets take as Alice chats with Bob using an instant messaging service.
Each dotted line represents an Ethernet cable, a wireless link, or any other kind of physical network.
The cloud symbol is commonly used to stand in for "The Internet", and represents any number of intervening IP networks.

Neither Alice nor Bob need to be concerned with how those networks operate, as long as the routers forward IP traffic towards the ultimate destination.

If it weren't for Internet protocols and the cooperation of everyone on the net, this kind of communication would be impossible.

In Figure NG 16, Alice is dual-stack and has IPv4 and IPv6 addresses, and as Bob has only IPv6 addresses, they will communicate by using IPv6 which is the common IP version between them.

Designing the physical network

It may seem odd to talk about the "physical" network when building wireless networks.

After all, where is the physical part of the network? In wireless networks, the physical medium we use for communication is obviously electromagnetic energy.

But in the context of this chapter, the physical network refers to the mundane topic of where to put things. How do you arrange the equipment so that you can reach your wireless clients?

Whether they fill an office building or stretch across many miles, wireless networks are naturally arranged in these three logical configurations: point-to-point links, point-to-multipoint links, and multipoint-to-multipoint clouds. While different parts of your network can take advantage of all three of these configurations, any individual link will fall into one of these topologies.

Point-to-point

Point-to-point links typically provide an Internet connection where such access isn't otherwise available. One side of a point-to-point link will have an Internet connection, while the other uses the link to reach the Internet.

For example, a university may have a fast frame relay or VSAT connection in the middle of campus, but cannot afford such a connection for an important building just off campus. If the main building has an unobstructed view of the remote site, a point-to-point connection can be used to link the two together. This can augment or even replace existing dial-up links. With proper antennas and clear line of sight, reliable point-to-point links in excess of thirty kilometres are possible.

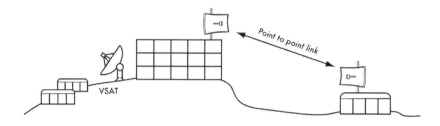

Figure NG 17: A point-to-point link allows a remote site to share a central Internet connection.

Of course, once a single point-to-point connection has been made, more can be used to extend the network even further. If the remote building in our example is at the top of a tall hill, it may be able to see other important locations that can't be seen directly from the central campus. By installing another point-to-point link at the remote site, another node can join the network and make use of the central Internet connection.

Point-to-point links don't necessarily have to involve Internet access. Suppose you have to physically drive to a remote weather monitoring station, high in the hills, in order to collect the data which it records over time. You could connect the site with a point-to-point link, allowing data collection and monitoring to happen in realtime, without the need to actually travel to the site.

Wireless networks can provide enough bandwidth to carry large amounts of data (including audio and video) between any two points that have a connection to each other, even if there is no direct connection to the Internet.

Point-to-multipoint

The next most commonly encountered network layout is point-to-multipoint. Whenever several nodes are talking to a central point of access, this is a point-to-multipoint application. The typical example of a point-to-multipoint layout is the use of a wireless access point that provides a connection to several laptops. The laptops do not communicate with each other directly, but must be in range of the access point in order to use the network.

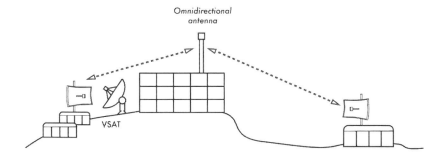

Figure NG 18: The central VSAT is now shared by multiple remote sites. All three sites can also communicate directly at speeds much faster than VSAT.

Point-to-multipoint networking can also apply to our earlier example at the university. Suppose the remote building on top of the hill is connected to the central campus with a point-to-point link.

Rather than setting up several point-to-point links to distribute the Internet connection, a single antenna could be used that is visible from several remote buildings. This is a classic example of a wide area point (remote site on the hill) to multipoint (many buildings in the valley below) connection.

Note that there are a number of performance issues with using point-to-multipoint over very long distance, which will be addressed in the chapter called **Deployment Planning**. Such links are possible and useful in many circumstances, but don't make the classic mistake of installing a single high powered radio tower in the middle of town and expecting to be able to serve thousands of clients, as you would with an FM radio station. As we will see, two-way data networks behave very differently than broadcast radio.

Multipoint-to-multipoint

The third type of network layout is multipoint-to-multipoint, which is also referred to as an ad-hoc or mesh network. In a multipoint-to-multipoint network, there is no central authority. Every node on the network carries the traffic of every other as needed, and all nodes communicate with each other directly.

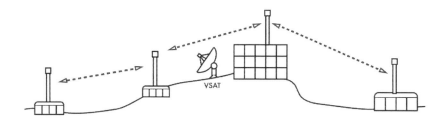

Figure NG 19: A multipoint-to-multipoint mesh. Every point can reach every other at very high speed, or any of them can use the central access point for a VSAT connection to the Internet.

The benefit of this network layout is that even if none of the nodes are in range of a central access point, they can still communicate with each other.

Good mesh network implementations are self-healing, which means that they automatically detect routing problems and fix them as needed. Extending a mesh network is as simple as adding more nodes.

If one of the nodes in the "cloud" happens to be an Internet gateway, then that connection can be shared among all of the clients.

Several disadvantages of this topology include increased complexity and lower performance.

Security in such a network is also a concern, since every participant potentially carries the traffic of every other.

Multipoint-to-multipoint networks tend to be difficult to troubleshoot, due to the large number of changing variables as nodes join and leave the network.

Multipoint-to-multipoint clouds typically have reduced capacity compared to point-to-point or point-to-multipoint networks, due to the additional overhead of managing the network routing and increased contention in the radio spectrum.

Nevertheless, mesh networks are useful in many circumstances.

For more information about them please read the chapter called **Mesh Networking.**

Use the technology that fits

All of these network designs can be used to complement each other in a large network, and additionally they can make use of traditional wired networking techniques whenever possible. Wired networks still often have higher bandwidth capacity than wireless so should be used whenever appropriate or affordable.

But looking at the wireless, it is a common practice, for example, to use a long distance wireless link to provide Internet access to a remote location, and then set up an access point on the remote side to provide local wireless access. One of the clients of this access point may also act as a mesh node, allowing the network to spread organically between laptop users who all ultimately use the original point-to-point link to access the Internet.

This is just one common scenario for wireless deployments, there are many others.

Now that we have a clear idea of how wireless networks are typically arranged, we can begin to understand how communication is possible over such networks.

7. WIFI FAMILY

IEEE 802: What is it, and why should I care?

The early days of networking only saw wireline networks (if you don't count the old AT&T Long lines microwave backbone that at one time crossed the United States) Now many networks are built using both wireline and wireless solutions. Typically networks based on wires or more usually these days, fibres have greater capacity than wireless.
But laying fibre is much more expensive and takes time. So often networks begin as wireless networks and as they grow in use, fibre based networks start to be deployed. In access networks (those near the consumers) or in dense urban environments, often wireless is more practical too. So very importantly as you begin to think about deploying wireless networks in your local area or community, your network could form the basis of the future growth of networking in your region.

An aspect of wired and wireless networks important to understand is the various standards that exist today as well as those new standards that are being developed. Wireless standards are the basis for many wireless products, ensuring interoperability and useability by those who design, deploy and manage wireless networks. We touched on this subject already in the chapter called **Radio Spectrum**. The Standards used in the vast majority of the networks come from the IEEE Standard Association's IEEE 802 Working Group. **IEEE 802** refers to a family of IEEE standards dealing with local area networks and metropolitan area networks.

More specifically, the IEEE 802 standards are restricted to networks carrying variable-size packets. (By contrast, in cell relay networks data is transmitted in short, uniformly sized units called cells). The number 802 was simply the next free number IEEE could assign, though "802" is sometimes associated with the date the first meeting was held – February 1980. The services and protocols specified in IEEE 802 map to the lower two layers (Data Link and Physical) of the seven-layer OSI networking reference model. In fact, IEEE 802 splits the OSI Data Link Layer into two sub-layers named Logical Link Control (LLC) and Media Access Control (MAC).

The IEEE 802 family of standards is maintained by the IEEE 802 LAN/MAN Standards Committee (LMSC).

The most widely used standards are for the Ethernet family, Token Ring, Wireless LAN, Bridging and Virtual Bridged LANs.

An individual Working group provides the focus for each area and they are listed in the table below.

Name	Description
IEEE 802.1	Bridging and Network Management
IEEE 802.3	Ethernet
IEEE 802.11 a/b/g/n	Wireless LAN (WLAN)
IEEE 802.15	Wireless PAN
IEEE 802.15.1	Bluetooth certification
IEEE 802.15.2	IEEE 802.15 and IEEE 802.11 coexistence
IEEE 802.15.3	High-Rate wireless PAN
IEEE 802.15.4	Low-Rate wireless PAN eg. Zigbee
IEEE 802.15.5	Mesh networking for WPAN
IEEE 802.15.6	Body Area Network
IEEE 802.16	Broadband Wireless Access (basis of WiMAX)
IEEE 802.16.1	Local Multipoint Distribution Service
IEEE 802.18	Radio Regulatory TAG
IEEE 802.19	Coexistence TAG
IEEE 802.20	Mobile Broadband Wireless Access
IEEE 802.21	Media Independent Handoff
IEEE 802.22	Wireless Regional Area Network
IEEE 802.23	Emergency Services Working Group
IEEE 802.24	Smart Grid TAG
IEEE 802.25	Omni-Range Area Network

The 802.11 standard

The standard we are most interested in is 802.11 as it defines the protocol for Wireless LAN.
The 802.11 Amendments are so numerous they have in the last few years started using 2 letters instead of 1. (802.11z - the DLS amendment - gave way to 802.11aa, ab, ac..., etc)

Below is a table of the variants of 802.11, their frequencies and approximate ranges.

802.11 protocol	Release	Freq.	Bandwidth	Data Rate per stream	Approximate indoor range		Approximate outdoor range	
		(GHz)	(MHz)	(Mbit/s)	(m)	(ft)	(m)	(ft)
-	Jun 1997	2.4	20	1, 2	20	66	100	330
a	Sep 1999	5	20	6,9,12, 18, 24, 36, 48, 54	35	115	120	390
b	Sep 1999	2.4	20	1, 2, 5.5, 11	35	115	140	460
g	Jun 2003	2.4	20	6,9,12, 18, 24, 36, 48, 54	38	125	140	460
n	Oct 2009	2.4/5	20	7.2, 14.4, 21.7, 28.9, 43.3, 57.8, 65, 72.2	70	230	250	820
			40	15, 30, 45, 60 , 90, 120, 135, 150				
ac	Nov.2011	5	20	Up to 87.6				
			40	Up to 200				
			80	Up to 433.3				
			160	Up to 866.7				

In 2012 IEEE issued the 802.11-2012 Standard that consolidates all the previous amendments.

The document is freely downloadable from:
http://standards.ieee.org/findstds/standard/802.11-2012.html

Deployment planning for 802.11 wireless networks

Before packets can be forwarded and routed to the Internet, layers one (the physical) and two (the data link) need to be connected. Without link local connectivity, network nodes cannot talk to each other and route packets.

To provide physical connectivity, wireless network devices must operate in the same part of the radio spectrum.

This means that 802.11a radios will talk to 802.11a radios at around 5 GHz, and 802.11b/g radios will talk to other 802.11b/g radios at around 2.4 GHz.

But an 802.11a device cannot interoperate with an 802.11b/g device, since they use completely different parts of the electromagnetic spectrum.

More specifically, wireless interfaces must agree on a common channel. If one 802.11b radio card is set to channel 2 while another is set to channel 11, then the radios cannot communicate with each other.

The centre frequencies of each channel for 802.11a and 802b/g are given in **Appendix B: Channel Allocations.**

When two wireless interfaces are configured to use the same protocol on the same radio channel, then they are ready to negotiate data link layer connectivity. Each 802.11a/b/g device can operate in one of four possible modes:

1. Master mode (also called AP or infrastructure mode) is used to create a service that looks like a traditional access point. The wireless interface creates a network with a specified name (called the SSID) and channel, and offers network services on it. While in master mode, wireless interfaces manage all communications related to the network (authenticating wireless clients, handling channel contention, repeating packets, etc.) Wireless interfaces in master mode can only communicate with interfaces that are associated with them in managed mode.

2. Managed mode is sometimes also referred to as client mode. Wireless interfaces in managed mode will join a network created by a master, and will automatically change their channel to match it. They then present any necessary credentials to the master, and if those credentials are accepted, they are said to be associated with the master. Managed mode interfaces do not communicate with each other directly, and will only communicate with an associated master.

3. Ad-hoc mode creates a multipoint-to-multipoint network where there is no single master node or AP. In ad-hoc mode, each wireless interface communicates directly with its neighbours.

   Nodes must be in range of each other to communicate, and must agree on a network name and channel. Ad-hoc mode is often also called Mesh Networking and you can find details of this type of networking in the chapter called **Mesh Networking.**

4. Monitor mode is used by some tools (such as Kismet) to passively listen to all radio traffic on a given channel. When in monitor mode, wireless interfaces transmit no data. This is useful for analysing problems on a wireless link or observing spectrum usage in the local area. Monitor mode is not used for normal communications.

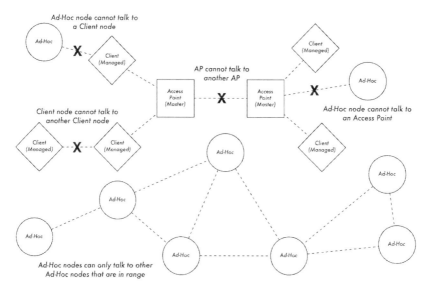

Figure WF 1: APs, Clients, and Ad-Hoc nodes.

When implementing a point-to-point or point-to-multipoint link, one radio will typically operate in master mode, while the other(s) operate in managed mode. In a multipoint-to-multipoint mesh, the radios all operate in ad-hoc mode so that they can communicate with each other directly.

It is important to keep these modes in mind when designing your network layout.

Remember that managed mode clients cannot communicate with each other directly, so it is likely that you will want to run a high repeater site in master or ad-hoc mode.

Ad-hoc is more flexible but has a number of performance issues as compared to using the master / managed modes.

The 802.22 Standard

Did you ever wonder why one of the biggest users of wireless spectrum in almost any country on earth, never got into the 2 way communications business? No? Well, you ask, why did the Television Broadcast Industry never want to do two-way communications.

The simple answer is that this was not the business they were in.

What they did instead was get access to and use of the best "beach front" spectrum between DC and day light.. and almost for free to boot.

As analogue TV gets replaced by digital TV, some of that beach front spectrum is being made available for wireless networking. And in parts of the world where TV has been deployed less extensively, these same parts of the radio spectrum are available for wireless networking as well.

The new wireless technology is commonly called TVWS (TV White Spaces) and although relatively new at the time of writing, this technology is in many trials for rural broadband wireless.

From *wikipedia* -

IEEE 802.22, known informally as Super Wi-Fi, is a standard for Wireless Regional Area Networks (WRAN) using white spaces in the TV frequency spectrum.

The development of the IEEE 802.22 WRAN standard is aimed at using cognitive radio (CR) techniques to allow sharing of geographically unused spectrum allocated to the Television Broadcast Service, on a non-interfering basis, to bring broadband access to hard-to-reach, low population density areas, typical of rural environments, and is therefore timely and has the potential for a wide applicability worldwide.

IEEE 802.22 WRANs are designed to operate in the TV broadcast bands while assuring that no harmful interference is caused to the incumbent operation, i.e., digital TV and analog TV broadcasting, and low power licensed devices such as wireless microphones. The standard was finally published in July 2011.

Technology of 802.22 or TVWS

The initial drafts of the 802.22 standard specify that the network should operate in a point to multipoint topology (p2m). The system will be formed by Base Stations (BS) and Customer Premise Equipment (CPE). The CPEs will be attached to a BS via a wireless link.

One key feature of the 802.22 WRAN Base Stations is that they will be capable of performing sensing of available spectrum.
The Institute of Electrical and Electronic Engineers (IEEE), together with the FCC, in the US, pursued a centralised approach for available spectrum discovery.
Specifically each Base Station (BS) would be armed with a GPS receiver which would allow its position to be reported.
This information would be sent back to centralised servers which would respond with the information about available free TV channels and guard bands in the area of the Base Station.

Other proposals would allow local spectrum sensing only, where the BS would decide by itself which channels are available for communication. This is called *distributed sensing*.

The CPEs will be sensing the spectrum and sending periodic reports to the BS informing it about what they sense. The BS, with the information gathered, will evaluate whether a change is necessary in the channel used, or on the contrary, if it should stay transmitting and receiving in the same one. A combination of these two approaches is also envisioned.

These sensing mechanisms are primarily used to identify if there is an incumbent transmitting, and if there is a need to avoid interfering with it. This means that the physical layer must be able to adapt to different conditions and be flexible in jumping from channel to channel without errors in transmission or losing clients (CPEs).

This flexibility is also required for dynamically adjusting the bandwidth, modulation and coding schemes. OFDMA *(Orthogonal Frequency-Division Multiple Access)* is the modulation scheme for transmission in up and downlinks. With OFDMA it will be possible to achieve this fast adaptation needed for the BSs and CPEs.

By using just one TV channel (a TV channel has a bandwidth of 6 MHz, in some countries they can be 7 or 8 MHz) the approximate maximum bit rate is 19 Mbit/s at a 30 km distance.

The speed and distance achieved is not enough to fulfill the requirements of the standard.

There is a feature called *Channel Bonding* which deals with this problem. Channel Bonding uses more than one channel for transmit/receive. This allows the system to have higher bandwidth which will be reflected in a better system performance.

Summary

As we can see the standards for wireline and wireless are mostly incorporated into the IEEE 802 Working Group.

At the moment the 802.11 family of WiFi standards and equipment are by far and away the most widely manufactured and deployed in indoor and outdoor wireless links.

The chapter called **Hardware Selection and Configuration** looks at equipment in much more detail.

The new 802.22 Standard is expected to play an increasing role in many rural (and urban) wireless networks. The free-ing up of unlicensed spectrum currently used by broadcast TV will enable this to happen. As yet the standards and the various groups involved in this standard are in their infancy, as are the bodies around the world involved in spectrum re-allocation. Available equipment is still very new and deployments are few and far between. In the next 2-3 years it is anticipated that this will change significantly and the next revision of this book may well have case studies and deployment information to share with respect to 802.22 based networks. Meanwhile there is an interesting project underway in Scotland, United Kingdom which is deploying TVWS 802.22 networks.

You can read more about the project here:
http://www.wirelesswhitespace.org/projects.aspx

8. MESH NETWORKING

Introduction

Mesh networks are based on multipoint-to-multipoint (m2m) networking. In the nomenclature of IEEE 802.11, m2m networking is referred to as 'ad-hoc' or 'IBSS' mode.

Most wireless networks today are based on point-to-point (p2p) or point-to-multipoint (p2m) communication.

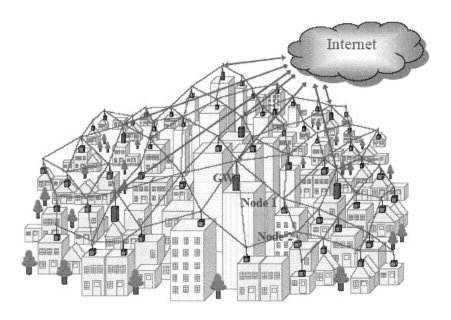

Figure MN 1: A metropolitan area mesh network, providing local connectivity and Internet access via multiple Internet gateways

A typical wireless hotspot operates in p2m infrastructure mode - it consists of an access point (with a radio operating in master mode), attached to a DSL line or other large scale wired network.
In a hotspot the access point acts as a master station that is distributing Internet access to its clients.

This hub-and-spoke topology is also typically used for the mobile phone (2G or 3G) service.

Mobile phones connect to a base station p2m - without the presence of a base station mobiles can't communicate with each other.

If you make a joke call to a friend that is sitting on the other side of the table, your phone sends data to the base station of your provider that may be a mile away - the base station then sends data back to the phone of your friend.

In a remote area without base stations a GSM phone is useless as a communication device, simply because GSM radios are designed in such a way that they can not communicate directly with each other.

This is unlike analog radio sets that can communicate m2m with each other as long as they are within range.

Wireless radio is by default a broadcast medium, and any station which can transmit and receive could communicate m2m.

With regards to the technological challenge, implementing m2m networking is much more demanding then p2m and p2p.

Strategies to implement channel access coordination are more complex, for example, there is no central authority to assign transmit time slots.

Because there is no central management, m2m stations need to mutually agree on cell coordination parameters such as the MAC like cell-id of the wireless cell.

The fact that 802.11 names the m2m mode of WiFi "ad-hoc" suggests that the IEEE board thought of a m2m network as a spontaneous, provisional, sub-optimal solution.

Multipoint-to-multipoint communication is actually more versatile and can be much more efficient than point-to-point or point-to-multipoint communication: m2m communication includes the ability to communicate p2p and p2m, because p2p and p2m are just subsets of m2m.

A network consisting of just two multipoint-to-multipoint capable devices simply communicates p2p:

A--B

A network of three mesh-capable devices A, B, C can form a topology like this, for example:

A--B--C

Here A can communicate only with B. C can only communicate with B, while B can communicate with A and C. B actually does communicate p2m. Without routing, A and C can not communicate with each other in 802.11 ad-hoc mode.

By adding a routing protocol A can automatically learn that behind B there is C and vice versa, and that B can be utilised as a communication relay so all nodes can communicate with each other. In this case B will act similarly to an access point in 802.11 infrastructure mode. WiFi cards configured as infrastructure clients can't communicate directly, so they would always need the access point B as a relay.

If the three devices move around, the topology might become a full mesh, where every node can communicate with every other node directly:

In this case relaying traffic is not necessary, given that the links are all good enough. In infrastructure mode, communicating directly is not possible. All the traffic between clients has to be relayed by the access point. If we now add D to the little chain topology example, all devices can communicate with each other if this is a mesh.

A--B--C--D

On the other hand, this would not be possible if the network is an infrastructure mode network and B is an access point. C and D would both be infrastructure clients, and as already mentioned before, infrastructure clients can not communicate directly with each other.

So client D could not join the infrastructure network because it is out of range of the access point B, while it would be still in the radio range of client C.

Bandwidth impact of multi-hop relaying routes

Mesh networks consisting of devices that feature only one radio are a low cost way to establish a ubiquitous wireless network, but this comes at a tradeoff. With only one wireless interface in each device, the radios have to operate on the same channel. Simply sending data through a routing path going from node A through B to C halves the available bandwidth. While A is sending data to B, B and C have to remain silent.
While B is forwarding data to C, A has to remain silent as well - and so on. Note that the same is true if two clients connected to an access point in infrastructure want to communicate with each other.

If we assume that all wireless links in the A-B-C-D mesh chain operate at the same speed, the communication between A and D would be roughly 1/3 of the speed of a single link, given that A and D can use the network capacity exclusively. The bandwidth impact can be mitigated or avoided by using devices with multiple radios, given that they are operating on different frequencies that don't interfere with each other.
Despite the bandwidth tradeoff, single radio mesh devices still have their merits. They are cheaper, less complex and consume less power than multi-radio devices. This can be important if the systems are solar or wind powered or require a battery backup. If the wireless links in a three hop network (a chain with 4 nodes like above) operate at 12 Mbit each, the total end-to-end bandwidth would still offer plenty bandwidth to saturate a 2 Mbit Internet uplink.

Summary

Mesh networking extends the range of wireless devices by multi-hop relaying traffic. By means of dynamic routing, meshes can be self- healing in case of node failure and grow organically if more nodes are added. If the mesh nodes have only one radio, the benefit in coverage comes at the tradeoff of reduced bandwidth. Here is an example of a currently deployed mesh network. You can find more information about this deployment at the url http://code.google.com/p/afrimesh/

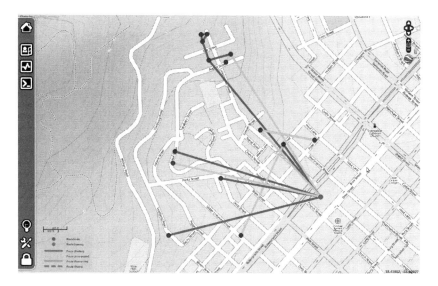

Figure MN 2: Screenshot of the Villagetelco mesh at Bo-Kaap in Capetown, South Africa

Routing protocols for mesh networks

Routing protocols for wireless mesh networks have to be designed with the challenges of radio communication in mind.

Wireless links and the topology of a mesh network are inherently unstable, devices can go up and down, the available bandwidth varies, and links are often flawed with packet loss.
A mesh routing protocol should be resilient against routing errors even if routing protocol messages are delayed or lost.
At the same time the available communication bandwidth and computing performance of mesh nodes is limited and shouldn't be wasted on protocol decisions and traffic overhead.

In 2005, when the first edition of the WNDW book was written, there were only few practically usable routing protocols for mesh networks. In previous editions, this chapter has been focused on OLSR. Back then, the OLSR daemon wasn't shipped with a working default configuration, so it was necessary to wade into the depths of the olsrd.conf configuration file to find out what the best configuration of the routing algorithm was.

The situation has changed quite a bit since 2005.

There are now a number of mesh protocols and implementations, and all the implementations which are mentioned in this chapter are readily available as installation packages for OpenWRT.

Mesh protocol developers are competing in a challenge to deliver the best mesh routing protocol. There is now a annual competition event for mesh protocol developers, taking place once a year. It is called "Battlemesh", www.battlemesh.org

Most mesh routing protocols (BABEL, BATMAN, OLSR, BMX, BMX6) take care of maintaining the IPv4 and IPv6 routing tables in a mesh node by adding, updating and deleting routes. These mesh protocols are using IP-based routing. They are layer 3 mesh protocols, since IP represents the third layer of the OSI networking layer model.

Batman-adv(anced) is a relatively new protocol that operates on the second layer of the networking model, hence it is a layer 2 mesh protocol. To the higher layers (including IP), Batman-adv makes the whole mesh appear as a switch, where all connections are link-local. A Batman-adv mesh is transparent for the higher layers of the networking model.

This simplifies the setup of a mesh network a great deal, since it is possible to use DHCP, mDNS or MAC bridging with Batman-adv. Batman-adv is a Linux kernel module, which is shipped with the official Linux kernel sources. Mesh routing protocols should also manage the announcement and selection of gateways to external networks like the Internet. A common problem with gateway selection mechanisms is that the routing protocol might decide to switch between gateways too often - for example, because one routing path to one gateway just got slightly better than the other. This is annoying because it can cause gateway flapping and results in stateful connection sessions breaking down frequently. If there is more than one Internet gateway in the mesh, using an advanced method for gateway selection is strongly recommended.

How about 802.11s?

The roadmap of 802.11s is to scale up to 32 nodes. According to Wikipedia it uses HWMP (Hybrid wireless mesh protocol) as the default routing protocol, with the option to use other routing protocols. Quote: "HWMP is inspired by a combination of AODV (RFC 3561[20] and tree-based routing". Since 802.11s is relatively new there is not much practical experience so far.

Devices and firmware for embedded devices

Not all WiFi devices on the market are suitable for mesh networking.
In 2005, when the first edition of this book was written, one of the clear hardware recommendations for mesh networking was the Linksys WRT54G router in combination with the Freifunk firmware.
While the WRT54G(L) is still on the market, it is no longer worth a recommendation.[1]

OpenWRT is a very versatile firmware development environment for developers and a firmware for advanced Linux users.
The old Freifunk firmware is based on the outdated OpenWRT version "White Russian". "White Russian" supported only devices with Broadcom chipsets with a proprietary binary wireless driver and is based on Linux 2.4. It has been ousted by the OpenWRT releases 'Kamikaze' and 'Backfire' (latest release). With 'Kamikaze' and 'Backfire' OpenWRT has gained support for many different wireless chipsets, CPU architectures and devices. The next stable release of OpenWRT will be named 'Attitude Adjustment', which, at the time of writing, is at 'release candidate' stage. The 'AR7xxx' and 'Atheros' platform ports of 'Attitude Adjustment' can be considered stable.
The mesh routing protocols mentioned before are all available as installation packages for OpenWRT.
A few open networking communities have developed their own customised firmware images with OpenWRT Kamikaze and Backfire.
However they are mostly geared to meet their local preferences and requirements and support a limited number of devices. Since they are often localised they might be of limited use for the general public.
OpenWRT has a package management system, which comes to the rescue. It is typical for OpenWRT to install software packages into the router after it has been flashed. There is now a meta package named 'luci-freifunk-community', which will automatically convert a stock OpenWRT image into a community mesh networking firmware.

---

1. While the WRT54GL is still available on the market, at prices of 60 US-$ it is overpriced. The WRT54G revision 4.0 was sold as WRT54GL revision 1.0 by Linksys in 2005, after Linksys had introduced the WRT54G revision 5.0, which wasn't Linux compatible anymore. The WRT54G revision 5.0 has only half the flash storage and RAM capacity. The WRT54G model is the WiFi router with the longest production lifetime. For the money you need to spend on a Linksys WRT54GL in 2011 you can buy two or three similar devices from other brands that are faster in terms of CPU and data rate.

The number of devices that can be converted to a mesh router has increased dramatically. On the other hand the process of converting a stock OpenWRT firmware into a mesh firmware via the package management system is often more error-prone unfortunately.

Some manufacturers of WiFi routers are shipping devices that come with OpenWRT as their factory firmware: Mesh-Potato, Dragino MS- 12, Allnet 0305.

Figure MN 3: Mesh-Potato outdoor mesh WiFi router with VoIP (with one FXS port to connect analog telephone handsets). www.villagetelco.org

The Mesh-Potato is a low-power outdoor device designed for mesh networking with an FXS (analog telephone) port, so you can plug an analog telephone handset in and make telephone calls via the mesh.
The Mesh-Potato is shipped with a mesh firmware that uses the BATMAN Layer 3 mesh protocol.

A second firmware named SECN (Small enterprise/campus network) is also available for the Mesh- Potato, which uses the BATMAN-ADV Layer 2 mesh.
But these are not the only devices that should be considered for purchase.
OpenWRT supports a very broad selection of wireless routers.
Replacing the factory firmware with OpenWRT is often quite easy.

Again, the selection of available devices is so diverse that there isn't a single update method that works for the whole variety of hardware, which could be described here.

The OpenWRT table of supported hardware is huge and it keeps expanding:
http://wiki.openwrt.org/toh/start

This site should be the place to start before you go and buy devices.

At the moment, if you are looking for a router based on a chipset that supports 802.11n in ad-hoc mode, my recommendation are devices that are supported by the AR7XXX port of OpenWRT.

Note that hardware manufacturers may change the chipsets of the devices without explicitly stating it. Newer hardware revisions are not guaranteed to work unless someone has tested and reported it in the OpenWRT wiki.

Notable devices which can be flashed with OpenWRT are the outdoor units made by Ubiquiti and the SOHO devices made by TP-Link.

TP- Link produces several low-cost SOHO devices with Atheros ar71xx chipsets (802.11n).
The TP-Link MR3220 (802.11n single stream) and MR3420 (802.11n dual stream) feature a 400 MHz Mips 24kc CPU, one USB 2.0 port, a 4 port 100 Mbit switch, WAN port, 4MB flash and 32MB RAM.
Prices start at around 30US $.

Since the TP-Link devices have a USB 2.0 port it is possible to add another WiFi interface via USB WiFi dongle.
Actually USB 2.0 adds many opportunities, like adding additional storage space, audio support, webcams and so on.

Figure MN 4:DIY outdoor routers based on PCBs taken from SOHO routers (Picture showing two devices based on TP-Link WR741 and WR941 and one based on a Fonera router)

Another firmware distribution which initially orginated as an alternative firmware of the WRT54G is DD-WRT. DD-WRT is a firmware distribution designed for end-users. It only supports the OLSR routing protocol.

Frequently observed problems

The typical problems of multipoint-to-multipoint communication are either on the physical radio or the MAC layer. The IEEE 802.11 suggestions about multipoint-to-multipoint mode were not up to the task.

The main challenges are:

Channel access coordination, namely the hidden node and exposed node problem

Going back to our little mesh with the topology A B C it can happen that A and C start to send data to B at the same time because they don't receive each other, resulting in a collision at the location of B. 802.11 has a mechanism to mitigate this: RTS/CTS (request to send, clear to send)

Before a m2m node sends data it requests airtime by sending a short RTS packet, in order to reserve the channel. It waits until it receives a CTS signal. So A will send a short RTS packet, and B sends a short CTS packet. This way C will detect that there will be a transmission of a hidden node that it shouldn't interfere with.

However the current RTS/CTS mechanism of 802.11 works well only for 2 hop routes. On longer routes it can happen that multiple stations send RTS signals, which results in all the nodes stopping their transmissions and waiting for a CTS signal. This is called a RTS broadcast storm. For mesh networks of considerable size it is not recommended to use the RTS/CTS mechanism.

Timer synchronisation

The people designing the 802.11 ad-hoc protocol thought it would be smart if the WiFi devices would synchronise their MAC timer clocks by sending time stamps in beacons.

However bogus time stamps can occur due to bugs, and they often trigger race conditions in the hardware, drivers and the 802.11 ad-hoc protocol, which has not been designed to cope with such issues.

Failed attempts to synchronise time stamps often results in cell splitting (see below). There are some hacks which have been introduced to overcome the problem.

The best solution is to disable timer synchronisation entirely. However timer synchronisation is often done in the wireless interface hardware or firmware of a interface.

OpenWRT has a hack to disable timer synchronisation when using Atheros 802.11abg cards that work with the Madwifi driver. If you are using a recent Linux kernel, certain wireless drivers (namely ath9k) which are often used for mesh devices are quite robust against timer issues in ad-hoc mode.

However this doesn't help if the WiFi device comes with a closed source binary firmware which is not prepared to deal gracefully with timer issues. There is not much we can do about it, other than using drivers/firmwares/chipsets which are known to work reliably.

IBSS cell splitting

This is a typical problem of the way 802.11 suggests implementation of the m2m mode. Ad-hoc devices may fail to agree on a certain cell-id (IBSS-ID). If they don't manage to agree on using one certain cell-id, they are logically separated wireless cells.

This is a real show stopper, because wireless devices will not be able to communicate with each other. The problem is related to issues with timer synchronisation. Since Linux 2.6.31 it is possible to manually fix the IBSS-ID. This feature is also available in OpenWRT.

9. SECURITY FOR WIRELESS NETWORKS

Introduction

While unlicensed spectrum provides a huge cost savings to the user, it has the unfortunate side effect that Denial of Service (DoS) attacks are trivially simple. By simply turning on a high powered access point, cordless phone, video transmitter, or other 2.4GHz device, a malicious person could cause signifi cant problems on the network. Many network devices are vulnerable to other forms of denial of service attacks as well, such as disassociation fl ooding and ARP table overfl ows.

There are several categories of individuals who may cause problems on a wireless network:

Unintentional users.

Densely populated areas such as city centres and university campuses can lead to a density of wireless access points. In these populated areas, it is common for laptop users to accidentally associate to the wrong network. Most wireless clients will simply choose any available wireless network, often the one with the strongest signal, when their preferred network is unavailable.

The user may then make use of this network as usual, completely unaware that they may be transmitting sensitive data on someone elses network.

Malicious people may even take advantage of this by setting up access points in strategic locations, to try to attract unwitting users and capture their data.

The first step in avoiding this problem is educating your users, and stressing the importance of connecting only to known and trusted networks.

Many wireless clients can be configured to only connect to trusted networks, or to ask permission before joining a new network.

As we will see later in this chapter, users can safely connect to open public networks by using strong encryption.

War drivers.

The "war driving" phenomenon draws its name from the popular 1983 hacker film, "War Games".

War drivers are interested in finding the physical location of wireless networks.

They typically drive around with a laptop, GPS, and omnidirectional antenna, logging the name and location of any networks they find. These logs are then combined with logs from other war drivers, and are turned into graphical maps depicting the wireless "footprint" of a particular city.

The vast majority of war drivers likely pose no direct threat to networks, but the data they collect might be of interest to a network cracker. For example, it might be obvious that an unprotected access point detected by a war driver is located inside a sensitive building, such as a government or corporate office.

A malicious person could use this information to illegally access the network there. Arguably, such an AP should never have been set up in the first place, but war driving makes the problem all the more urgent. As we will see later in this chapter, war drivers who use the popular program NetStumbler can be detected with programs such as Kismet.

For more information about war driving, see sites such as:

http://wigle.net/,

http://www.nodedb.com/, or http://www.stumbler.net/.

Rogue access points.

There are two general classes of rogue access points:

Those incorrectly installed by legitimate users, and those installed by malicious people who intend to collect data or do harm to the network. In the simplest case, a legitimate network user may want better wireless coverage in their office, or they might find security restrictions on the corporate wireless network too difficult to comply with.

By installing an inexpensive consumer access point without permission, the user opens the entire network up to potential attacks from the inside. While it is possible to scan for unauthorised access points on your wired network, setting a clear policy that prohibits them is a very important first step.

The second class of rogue access point can be very difficult to deal with. By installing a high powered AP that uses the same ESSID as an existing network, a malicious person can trick people into using their equipment, and log or even manipulate all data that passes through it.

Again, if your users are trained to use strong encryption, this problem is significantly reduced.

Eavesdroppers.

As mentioned earlier, eavesdropping is a very difficult problem to deal with on wireless networks. By using a passive monitoring tool (such as Kismet), an eavesdropper can log all network data from a great distance away, without ever making their presence known.
Poorly encrypted data can simply be logged and cracked later, while unencrypted data can be easily read in real time.

If you have difficulty convincing others of this problem, you might want to demonstrate tools such as
Driftnet (http://www.ex-parrot.com/~chris/driftnet/).
Driftnet watches a wireless network for graphical data, such as GIF and JPEG files. While other users are browsing the Internet, these tools simply display all graphics found in a graphical collage. While you can tell a user that their email is vulnerable without encryption, nothing drives the message home like showing them the pictures they are looking at in their web browser. Again, while it cannot be completely prevented, proper application of strong encryption will discourage eavesdropping.

Protecting the wireless network

In a traditional wired network, access control is relatively straightforward: If a person has physical access to a computer or network hub, they can use (or abuse) the network resources. While software mechanisms are an important component of network security, limiting physical access to the network devices is the ultimate access control mechanism. Simply put, if all terminals and network components are physically only accessible to trusted individuals, the network can likely be trusted. The rules change significantly with wireless networks. While the apparent range of your access point may seem to be just a few hundred metres, a user with a high gain antenna may be able to make use of the network from several blocks away.
Should an unauthorised user be detected, is impossible to simply "trace the cable" back to the users location.
Without transmitting a single packet, a sufficiently talented nefarious user can capture and log traffic on a wireless network to disk.

This data can later be used to launch a more sophisticated attack against the network. Never assume that radio waves simply "stop" at the edge of your property line, or inside your building. Physical security in wireless networks is limited to preventing compromise of the active components, cables and power supply.

Where physical access to the network cannot be prevented we have to rely on electronic means for controlling access to the wireless infrastructure in order to only allow authorised persons and systems to use the wireless network. But remember, while a certain amount of access control and authentication is necessary in any network, you have failed in your job if legitimate users find it difficult to use the network to communicate. Lastly, it is usually unreasonable to completely trust all users of the network, also on wired networks. Disgruntled employees, uneducated network users, and simple mistakes on the part of honest users can cause significant harm to network operations. As the network architect, your goal is to facilitate private communication between legitimate users of the network and between legitimate users and services.

There's an old saying that the only way to completely secure a computer is to unplug it, lock it in a safe, destroy the key, and bury the whole thing in concrete. While such a system might be completely "secure", it is useless for communication. When you make security decisions for your network, remember that above all else, the network exists so that its users can communicate with each other. Security considerations are important, but should not get in the way of the network's users.

A simple rule-of-thumb as to whether or not the network is getting in the way of its users is to look at how much time you or other staff spend on helping people get on the network.

If regular users are repeatedly having problems simply gaining access to the network, even after they have been provided instruction and training on how to do so, it's possible that the access procedures are too cumbersome and a review of them is in order.

Taking all of this into account, our goal is to provide adequate physical security, control access protect the communication without sacrificing ease of use.

Physical security for wireless networks

When installing a network, you are building an infrastructure that people depend on. Security measures exist to ensure that the network is reliable. Wireless networks have physical components, such as wires and boxes, which are easily disturbed. In many installations, people will not understand the purpose of the installed equipment, or curiosity may lead them to experiment.

They may not realise the importance of a cable connected to a port. Someone may unplug an Ethernet cable so that they can connect their laptop for 5 minutes, or move a switch because it is in their way. A plug might be removed from a power bar because someone needs that receptacle. Assuring the physical security of an installation is paramount. Signs and labels will only be useful to those who can read your language. Putting things out of the way and limiting access is the best means to assure that accidents and tinkering do not occur. In your locality, proper fasteners, ties, or boxes may not be easy to find.

You should be able to find electrical supplies that will work just as well. Custom enclosures are also easy to manufacture and should be considered essential to any installation. It is often economical to pay a mason to make holes and install conduit. PVC can be embedded in cement walls for passing cable from room to room. This avoids the need to smash new holes every time a cable needs to be passed. Plastic bags can be stuffed into the conduit around the cables for insulation. Small equipment should be mounted on the wall and larger equipment should be put in a closet or in a cabinet.

Switches

Switches, hubs or interior access points can be screwed directly onto a wall with a wall plug. It is best to put this equipment as high as possible to reduce the chance that someone will be able to touch the device or its cables without significant effort.

Cables

At the very least, cables should be hidden and fastened. It is possible to find plastic cable conduit that can be used in buildings. If you cannot find it, simple cable attachments can be nailed into the wall to secure the cable. This will make sure that the cable doesn't hang where it can be snagged, pinched or cut.

When fastening cable to the wall, it is important to not nail or screw into the cable itself. The cable contains many tiny wires that the network data travels over. Nailing through the cable will damage it and make it useless for transmitting data. Take care also to not overly bend or twist the cable as this will damage it as well.

It is preferable to bury cables, rather than to leave them hanging across a yard. Hanging wires might be used for drying clothes, or be snagged by a ladder, etc. To avoid vermin and insects, use plastic electrical conduit. The marginal expense will be well worth the trouble. The conduit should be buried about 30 cm deep, or below the frost level in cold climates. It is worth the extra investment of buying larger conduit than is presently required, so that future cables can be run through the same tubing. Consider labeling buried cable with a "call before you dig" sign to avoid future accidental outages.

Power

It is best to have power bars locked in a cabinet. If that is not possible, mount the power bar under a desk, or on the wall and use duct tape (or gaffer tape, a strong adhesive tape) to secure the plug into the receptacle. On the UPS and power bar, do not leave any empty receptacles. Tape them if necessary. People will have the tendency to use the easiest receptacle, so make these critical ones difficult to use. If you do not, you might find a fan or light plugged into your UPS; though it is nice to have light, it is nicer to keep your server running!

Water

Protect your equipment from water and moisture. In all cases make sure that your equipment, including your UPS is at least 30 cm from the ground, to avoid damage from fl ooding. Also try to have a roof over your equipment, so that water and moisture will not fall onto it. In moist climates, it is important that the equipment has proper ventilation to assure that moisture can be exhausted. Small closets need to have ventilation, or moisture and heat can degrade or destroy your gear.

Masts

Equipment installed on a mast is often safe from thieves. Nevertheless, to deter thieves and to keep your equipment safe from winds, it is good to over-engineer mounts.

Painting equipment a dull white or grey colour reflects the sun and makes it look plain and uninteresting. Panel antennas are often preferred because they are much more subtle and less interesting than dishes. Any installation on walls should be high enough to require a ladder to reach. Try choosing well-lit but not prominent places to put equipment. Also avoid antennae that resemble television antennae, as those are items that will attract interest by thieves, whereas a wifi antenna will be useless to the average thief.

Authentication and access control

Talking about authentication a number of related terms like (digital) identity, authorisation, privacy etc. pop up. So before we get into authentication proper we need to introduce some terminology, without trying to be exhaustive.

Digital identity is the electronic entity that is a representation of a physical entity, like a person or a device. Authentication is the process of verifying the claim that an (electronic) entity is allowed to act on behalf of a given known (physical) entity. In other words, authentication is the process of proving that the physical entity corresponds to a certain electronic one. Authorisation in turn is the process of establishing the rights of the identity to access certain resources or to perform certain tasks. Privacy, finally, is a complex issue but has to do with the right of a person not to have their personal data and behaviour be known to parties that do not strictly need it to provide the service the users request. So for example, it is reasonable for a liquor shop to know that a customer is above a certain age before selling alocholic drinks, but not to know the name of the customer, and third parties should not have any knowledge of the transaction at all. Privacy is of particular concern in a world in which users increasingly use networks and services outside their home environment. Without proper attention to privacy aspects it is too easy to trace a users behaviour and movement.
It is worth mentioning that there is a trade-off between authentication and privacy. Verifying the identity of a user in itself invades the user's privacy, the authenticating party knows who is using what resource at a particular time and location, but the challenge is to minimise the amount of information about a user and the number of parties that are privvy to that information.

In the context of this book we are mainly interested in techniques for controlling access to the network. In other words, we want to be able to decide who (authenticated identity) gets to access what (authorisation) without sacrificing privacy.

Authentication is typically done by proof of knowledge of a secret (a password, a signature), possession of a token or characteristic (a certificate, a fingerprint) or both. Access control is often needed to make sure that only authorised users can use the network, to prevent exhaustion of scarce resources and/or compliance with rules and regulations. In addition to access controlled networks there may also be open networks with limited access or for a limited time, but due to the need for organisations to control access to their scarce resources and also anti-terrorist laws they become less ubiquitous. Over the years a number of techniques have been employed to control access to wireless networks. Subsequently they have been mostly abandoned due to security or scalability issues as WiFi became increasingly popular.

Mac filtering

Access to a WiFi network can be based on the MAC address. This is the 48-bit number assigned by the manufacturer to every wireless and Ethernet device, and that is supposed to be unique and persistent. By employing mac filtering on our access points, we can authenticate users based on their MAC address.

With this feature, the access point keeps an internal table of approved MAC addresses.

When a wireless user tries to associate to the access point, the MAC address of the client must be on the approved list, or the association will be denied. Alternatively, the AP may keep a table of known "bad" MAC addresses, and permit all devices that are not on the list. Unfortunately, this is not an ideal security mechanism. Maintaining MAC tables on every device can be cumbersome, requiring all client devices to have their MAC addresses recorded and uploaded to the APs. Even worse, MAC addresses can often be changed in software. By observing MAC addresses in use on a wireless network, a determined attacker can spoof (impersonate) an approved MAC address and successfully associate to the AP.

While MAC filtering will prevent unintentional users and even most curious individuals from accessing the network, MAC filtering alone cannot prevent attacks from determined attackers.

MAC filters are useful for temporarily limiting access from misbehaving clients. For example, if a laptop has a virus that sends large amounts of spam or other traffic, its MAC address can be added to the filter table to stop the traffic immediately. This will buy you time to track down the user and fix the problem.

Closed networks

Another at one time popular "authentication feature" of WiFi networks is the so-called closed network mode. In a typical network, APs will broadcast their ESSID many times per second, allowing wireless clients (as well as tools such as NetStumbler) to find the network and display its presence to the user. In a closed network, the AP does not beacon the ESSID ("hidden ESSID"), and users must know the full name of the network before the AP will allow association. This prevents casual users from discovering the network and selecting it in their wireless client. There are a number of drawbacks to this feature. Forcing users to type in the full ESSID before connecting to the network is error prone and often leads to support calls and complaints. Since the network isn't obviously present in site survey tools like NetStumbler, this can prevent your networks from showing up on war driving maps. But it also means that other network builders cannot easily find your network either, and specifically won't know that you are already using a given channel.

A conscientious neighbour may perform a site survey, see no nearby networks, and install their own network on the same channel you are using. This will cause interference problems for both you and your neighbour.

Finally, using closed networks ultimately adds little to your overall network security. By using passive monitoring tools (such as Kismet), a malicious user can detect frames sent from your legitimate clients to the AP. These frames necessarily contain the network name. The malicious user can then use this name to associate to the access point, just like a normal user would. Encryption is probably the best tool we have for authenticating wireless users. Through strong encryption, we can uniquely identify a user in a manner that is very difficult to spoof, and use that identity to determine further network access. Encryption also has the benefit of adding a layer of privacy by preventing eavesdroppers from easily watching network traffic. Encryption is the technique that is used for authenticating users in most current deployments.

WEP

The first widely employed encryption method on WiFi networks was WEP encryption. WEP stands for Wired Equivalent Privacy, and is supported by virtually all 802.11a/b/g equipment. Incidentally, this is a complete misnomer as the privacy that WEP provides is in no way equivalent to that of wired connections. WEP uses a shared 40-bit key to encrypt data between the access point and client. The key must be entered on the APs as well as on each of the clients.

With WEP enabled, wireless clients cannot associate with the AP until they use the correct key. An eavesdropper listening to a WEP-enabled network will still see traffic and MAC addresses, but the data payload of each packet is encrypted.

This provided an authentication mechanism while also adding a bit of privacy to the network. WEP is definitely not the strongest encryption solution available.

For one thing, the WEP key is shared between all users. If the key is compromised (say, if one user tells a friend what the password is, or if an employee is let go) then changing the password can be prohibitively difficult, since all APs and client devices need to be changed.

This also means that legitimate users of the network can still eavesdrop on each others' traffic, since they all know the shared key.

The key itself is often poorly chosen, making off ine cracking attempts feasible. But most importantly, WEP itself is broken, making it very easy to gain illegal access the network.

So WEP should not be used anymore.

For more details about the state of WEP encryption, see these papers:
http://www.isaac.cs.berkeley.edu/isaac/wep-faq.html
http://www.cs.umd.edu/~waa/wireless.pdf

"Switched" wireless networks

One critical difference between modern switched Ethernet networks and wireless is that wireless networks are built on a shared medium.

They more closely resemble the old network hubs than modern switches, in that every computer connected to the network can "see" the traffic of every other user.

To monitor all network traffic on an access point, one can simply tune to the channel being used, put the network card into monitor mode, and log every frame.

This data might be directly valuable to an eavesdropper (including data such as email, voice data, or online chat logs). It may also provide passwords and other sensitive data, making it possible to compromise the network even further. WPA and 802.1X are designed to make the wireless network behave like a switched rather than a shared network.

WPA

Another data-link layer authentication protocol is Wi-Fi Protected Access, or WPA. WPA was created specifically to deal with the known problems with WEP mentioned earlier. WPA was intended to be a backwards compatible interim solution while the full standard 802.11i (WPA2) was developed.

WPA and WPA2 can operate in combination with the 802.1X umbrella standard for wireless authentication (see below) but also much in the same mode as WEP, with a shared secret between all clients and the AP, the so-called Pre Shared Key (PSK) mode (the WiFi Alliance calls WPA-PSK "WPA Personal", as opposed to "WPA Enterprise" that is used in combination with 802.1X).

Overall, WPA provides significantly better authentication and privacy than standard WEP, mainly by leveraging the Temporary Key Integrity Protocol (TKIP) that continuously and automatically changes the keying material between clients and access points.

Unfortunately precisely the backwards compatibility of TKIP has given rise to some attack vectors against TKIP that allow for decrypting certain encrypted packets, that in turn can be manipulated for further attacks.

More information can be found in the following articles:
http://dl.aircrack-ng.org/breakingwepandwpa.pdf
http://download.aircrack-ng.org/wiki-files/doc/enhanced_tkip_michael.pdf

The consequence of these discoveries is that it is wise to move to the next generation of secure WiFi access protocols: WPA2.

WPA2-PSK

WPA2 is the full IEEE 802.11i standard.

The main difference with WPA is the use of the Advanced Encryption System (AES) instead of TKIP a (so far) not broken encryption standard.

So the use of WPA2 with AES can be considered secure for now!

Summary

The major downside of any of these last three authentication methods is that, regardless of the strength of the encryption, they are still built upon the notion of a common shared secret between all clients and the access point.

They don't allow for identifying individual users and frankly, a secret shared by potentially tens of thousands of users can hardly be called a secret.

Another serious problem with wireless networks to which access is controlled by any of the methods mentioned, is that its users are relatively anonymous.

While it is true that every wireless device includes a unique MAC address that is supplied by the manufacturer, as mentioned these addresses can often be changed with software.
And even when the MAC address is known, it can be very difficult to judge where a wireless user is physically located.

Multi-path effects, high-gain antennas, and widely varying radio transmitter characteristics can make it impossible to determine if a wireless user is sitting in the next room or is in an apartment building a mile away.

The concerns about security, accountability and scalability have led to the rise of what is commonly called identity-based networking.

Identity-based networking

In identity based networking individual users are being authenticated rather than secrets shared between many users.

Typically the authentication system verifies user credentials against some sort of enterprise database or directory.
Commonly by using the RADIUS protocol, a protocol originally designed for controlling access to dial-in modem pools but sufficiently versatile to serve as generic access control protocol for network access.

Captive portals

One common authentication tool used on wireless networks is the captive portal. A captive portal uses a standard web browser to give a wireless user the opportunity to present login credentials.
It can also be used to present information (such as an Acceptable Use Policy) to the user before granting further access.
By using a web browser instead of a custom program for authentication, captive portals work with virtually all laptops and operating systems. Captive portals are typically used on open networks with no other authentication methods (such as WEP or MAC filters).

To begin, a wireless user opens their laptop browser and is directed to the portal.
They will then be asked to accept the use policy or answer other questions such as entering a user name and password, and click on a "login" button, or perhaps type in numbers from a pre-paid ticket.

The user enters their credentials, which are checked by the access point or another server on the network.

All other network access is blocked until these credentials are verified.
After verification their computer will receive a DHCP lease.
They can then use their web browser to go to any site on the Internet.

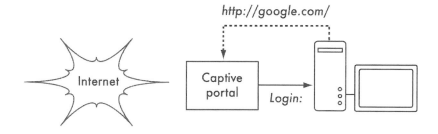

Figure SWN 1: The user requests a web page and is redirected.

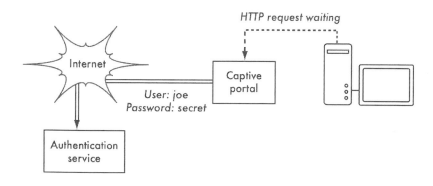

Figure SWN 2: The user's credentials are verified before further network access is granted. The authentication server can be the access point itself, another machine on the local network, or a server anywhere on the Internet.

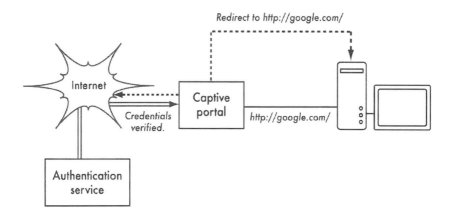

Figure SWN 3: After authenticating, the user is permitted to access the rest of the network and is typically redirected to the original site requested - in this case Google.

Captive portals provide no encryption for the wireless users, instead relying on the MAC and IP address of the client as a unique identifier, which can be spoofed easily, many implementations will therefore require the user to keep a connection window open. Since they, just like MAC- or WEP-based, do not provide protection against eavesdropping (they use a shared medium) and are vulnerable to session hijacking, captive portals are not a very good choice for networks that need to be locked down to only allow access from trusted users.

They are much more suited to cafes, hotels, and other public access locations where casual network users are expected. Another downside of captive portals is that they rely on the use of a browser for authentication, which can be very counterintuitive for users that just try to check their e-mail or send an instant message, not to mention the fact that many wireless devices like sensors, printers and cameras don't have a builtin browser.

In public or semi-public network settings, encryption techniques such as WEP and WPA are effectively useless. There is simply no way to distribute public or shared keys to members of the general public without compromising the security of those keys.

In these settings, a simple application such as a captive portal provides a level of service somewhere between completely open and completely closed. Many vendors and open source projects exist that provide captive portal capability, to name a few:

- CoovaChilli, CoovaAP (http://coova.org/CoovaChilli/), Coova is the successor of the no longer actively maintained Chillispot project. Coova allows for the use of a RADIUS authentication backend.

- WiFidog (http://www.wifidog.org/), WiFi Dog provides a very complete captive portal authentication package in very little space (typically under 30kb). From a user's perspective, it requires no pop-up or javascript support, allowing it to work on a wider variety of wireless devices.

- M0n0wall, pfSense (http://m0n0.ch/wall/), m0n0wall is a complete embedded operating system based on FreeBSD. It includes a captive portal with RADIUS support, as well as a PHP web server.

Many general networking vendors offer some form of integrated captive portals, e.g. Microtik, Cisco, Aruba, Aptilo.

802.1X

In enterprise and campus deployments, the most common wireless network authentication framework is that based on IEEE 802.1X. 802.1X is a layer 2 protocol that can be used both for wired and wireless network authentication and in fact comprises a number of technologies. 802.1X describes the interaction between the client device (supplicant in 802.1X) and the Access Point or Switch (Authenticator) as well as that between the Access Point or Switch and a backend RADIUS server (Authentication Server) that in turn verifies user credentials against an enterprise directory (or flat file for that matter). Finally, 802.1X describes how to transport user credentials from the supplicant to the authentication server transparently to the authenticator or any other device in the path by leveraging the Extensible Authentication Protocol (EAP).

The encryption between the supplicant and the authenticator can be done using rotating WEP-keys, WPA with TKIP or WPA2 with AES. For the reasons mentioned in the paragraph on WEP, WPA-PSK and WPA2-PSK, WPA2 with AES is highly recommended.

Probably the most interesting feature of the 802.1X is the use of EAP. EAP defines a generic way of encapsulating credentials and transporting them from a supplicant (client software) to an authentication server (RADIUS server). So-called EAP-methods define how specific authentication methods can be encapsulated into EAP. There are EAP-methods for all common types of authentication methods like certificates, SIM-cards, username/password, one-time passwords and hardware tokens. Due to key or token distribution problems with handing out tokens or certificates and revocation the vast majority of large scale deployments use what are called tunnelled EAP- methods: username/password based authentication using a TLS tunnel to the authentication server through which the actual username and password are transmitted. The user identity used for the TLS-envelope is typically of the form anonymous@domain (this is called the outer identity) whereas the inner identity (inside the TLS tunnel) is of the form username@domain. This distinction is particularly interesting for roaming to other organisations networks. It is possible to transport the authentication credentials of a user over the Internet while only revealing the home organisation of the user (the domain part), but that is the topic of the next paragraph. So what happens in a typical 802.1X authentication with a tunnelled EAP-method is the following:

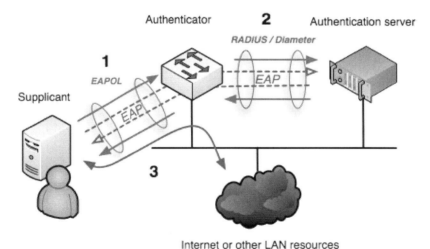

Figure SWN 4: The use of 802.1X with tunnelled EAP for network access
(courtesy SURFnet)

The client associates with the Access Point (authenticator).

The Access Point requests the client (supplicant) to authenticate.

The client sends an EAP message containing a TLS packet with an outer identity anonymous@domain and inside the TLS packet username@domain and the password to the access point over the 802.11 link (EAP over LAN or EAPoL).

The Access point receives the EAP-message, encapsulates it in RADIUS and sends it to the organisational RADIUS server (authentication server).

The RADIUS server decapsulates the EAP-message and verifies the user credntials against some sort of backend like a flat file, an LDAP directory, Active Directory or something else.

If the credentials are valid the RADIUS server sends back a RADIUS Access Accept message to the Access Point. The Access Point gives the client access to the wireless LAN.

The client performs a DHCP request, gets an IP-address and is connected to the network.

There are a number of tunnelled EAP-methods that essentially work the same, the differences are in the support in common operating systems, the vulnarability to dictionary and man-in-the-middle attacks and whether they require storage of clear-text passwords in the backend.

The most widely deployed tunnelled EAP-methods nowadays are EAP-TTLS (EAP Tunnelled Transport Layer Security) and PEAP (Protected EAP).

There have been incompatible implementations of PEAP due to disagreements between the proponents of PEAP (Apple, Cisco and Microsoft) resulting in a large uptake of TTLS. The fact that these incompatibilities are largely solved and the lack of native support for TTLS in a number of common OSes (Apple iOS and MS Windows variants) have resulted in an increase in uptake of PEAP. A newer EAP-method that is gaining traction is EAP-FAST due to its security properties. EAP-FAST has also been chosen as the basis for the new to be developed tunnelled EAP-method (TEAP) that the IETF expects to be the single endorsed one.

Inter-organisational roaming

RADIUS has the interesting property that RADIUS messages can be proxied to other RADIUS servers.

That means it is possible for an organisation to allow for each others users to gain access to the network by authenticating to their home organisation's RADIUS server.

When the RADIUS server of organisation A receives an authentication request from anonymous@organisationB.org it can forward the request to organisation B's RADIUS-server instead of verifying the credentials locally. Organisation B's RADIUS server in turn can verify the credentials and send the access accept back to organisation A's RADIUS server, that then tells the Access Point to allow the user access. This so-called federated access allows for the creation of very scalable and large deployments while at the same time allowing the individual organisations to apply their own authentication policies to their users.

While RADIUS proxying is certainly possible in captive portal deployments it definitely shines in an 802.1X/EAP environment.

By using EAP the user credentials can be protected so that only the home organisation of the user is able to see them.

This way large deployments can be created without the leakage of credentials and without teaching users to enter their secret credentials in every website that is thrown up in front of them. As an example, eduroam, the roaming wireless access federation in education, that extends the above concept slightly by instead of having direct RADIUS connections between organisations, it builds a hierarchical system of national and international RADIUS servers, allowing millions of students to gain access to over 5000 campus networks in many countries in all continents with the exception of Antarctica.

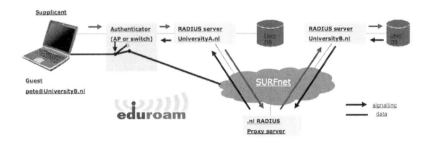

Figure SWN 5: The eduroam infrastructure for world-wide roaming in academia

End to end encryption

It should be noted that whereas WEP, WPA-PSK and WPA2-PSK are using encryption techniques to provide access control and protect against eavesdropping, they only protect the wireless traffic between client and access point, not on the wired part of the communication path. In order to protect the communication against unauthorised tampering or eavesdropping end to end encryption is needed.

Most users are blissfully unaware that their private email, chat conversations, and even passwords are often sent "in the clear" over dozens of untrusted networks before arriving at their ultimate destination on the Internet. However mistaken they may be, users still typically have some expectation of privacy when using computer networks. Privacy can be achieved, even on untrusted networks such as public access points and the Internet. The only proven effective method for protecting privacy is the use of strong end-to-end encryption. These techniques work well even on untrusted public networks, where eavesdroppers are listening and possibly even manipulating data coming from an access point.

To ensure data privacy, good end-to-end encryption should provide the following features:

Verified authentication of the remote end.

The user should be able to know without a doubt that the remote end is who it claims to be. Without authentication, a user could give sensitive data to anyone claiming to be the legitimate service.

Strong encryption methods.

The encryption algorithm should stand up to public scrutiny, and not be easily decrypted by a third party. There is no security in obscurity, and strong encryption is even stronger when the algorithm is widely known and subject to peer review.

A good algorithm with a suitably large and protected key can provide encryption that is unlikely to be broken by any effort in our lifetimes using current technology. Beware of product vendors who assure you that their proprietary encryption using trade-secret algorithms are better than open, peer-reviewed ones.

Public key cryptography.

While not an absolute requirement for end-to-end encryption, the use of public key cryptography instead of a shared key can ensure that an individual's data remains private, even if the key of another user of the service is compromised. It also solves certain problems with distributing keys to users over untrusted networks.

Data encapsulation.

A good end-to-end encryption mechanism protects as much data as possible. This can range from encrypting a single email transaction to encapsulation of all IP traffic, including DNS lookups and other supporting protocols. Some encryption tools simply provide a secure channel that other applications can use. This allows users to run any program they like and still have the protection of strong encryption, even if the programs themselves dont support it.

Be aware that laws regarding the use of encryption vary widely from place to place.
Some countries treat encryption as munitions, and may require a permit, escrow of private keys, or even prohibit its use altogether.
Before implementing any solution that involves encryption, be sure to verify that use of this technology is permitted in your local area.
In the following sections, we'll take a look at some specific tools that can provide good protection for your user's data.

TLS

The most widely available end-to-end encryption technology is Transport Layer Security, known simply as TLS (or its predecessor SSL: Secure Sockets Layer).
Built into virtually all web browsers and many other applications, TLS uses public key cryptography and a trusted public key infrastructure (PKI) to secure data communications on the web.
Whenever you visit a web URL that starts with https, you are using TLS.

The TLS implementation built into web browsers includes a collection of certificates from organisations called Certificate Authorities (CA).
A CA validates the identity of network users and/or providers and ensures that they are who they say they are and issues a certificate saying so.

Rather than doing this by a signed fancy document suitable for framing, this is done through the exchange of cryptographic keys.

As an example, someone wanting a certificate for their website submits a Certificate Request (CR), encoded ("signed") with a cryptographic key created specifically for signing the certificate request.

They submit this request to the CA, who then "signs" the request with their own key. These are encoded in the certificate along with the exact name of the website that the requester wants the certificate to be valid for.

For example, WWW.AIPOTU.GOV, from a certificate perspective, is not the same as AIPOTU.GOV. Each site would require its own certificate to present to a browser to do authenticated HTTPS transactions.
If the owner of AIPOTU.GOV domain only has their certificate issued for AIPOTU.GOV and not also WWW.AIPOTU.GOV, a user accessing the "WWW" address will see a warning for an invalid certificate for that site. This can be confusing to some users and, over time, lead them to just expect TLS Certificate warnings from their browser as normal when the truth is completely the opposite case.

When you browse to a website that uses TLS, the browser and the server first exchange certificates.

The browser then verifies that the hostname in the certificate provided by the server matches the DNS hostname that the browser knows the server by, that the certificate has not expired or been revoked and that it has been signed by a trusted certificate authority.

The server optionally verifies the validity of the browser's certificate. If the certificates are approved, both sides then negotiate a master session key using the previously exchanged certificates to protect the session that is being established.

That key is then used to encrypt all communications until the browser disconnects.
This kind of data encapsulation is known as a tunnel.

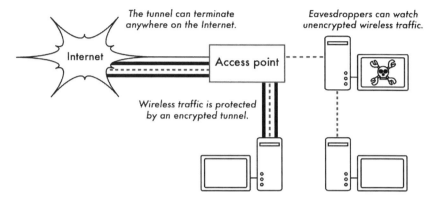

Figure SWN 6: Eavesdroppers must break strong encryption to monitor traffic over an encrypted tunnel. The conversation inside the tunnel is identical to any other unencrypted conversation.

The use of certificates with a PKI not only protects communication against eavesdroppers but is also used to prevent the so-called Man-in-the-Middle (MitM) attack. In a MitM attack, a malicious user intercepts all communication between a client and a server. By presenting counterfeit certificates to both the client and the server, a malicious user could carry on two simultaneous encrypted sessions. Since this user knows the secret on both connections, it is trivial to observe and manipulate the data being passed between the client and the server.

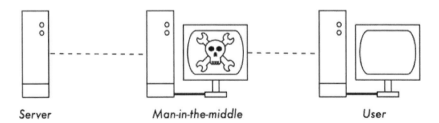

Figure SWN 7: The man-in-the-middle effectively controls everything the user sees, and can record and manipulate all traffic. Without a public key infrastructure to verify the authenticity of keys, strong encryption alone cannot protect against this kind of attack.

Use of a good PKI can significantly reduce the chances of this kind of attack. In order to be successful, the malicious user would have to be in possession of a certificate signed by a trusted CA that it can present to the client to accept as authentic.

This is only possible if they can trick the end-user into accepting it or if the CA is compromised.

Certificate Authorities bear a special burden to protect their systems and network from unauthorised access and malicious users.

If a CA were to be compromised, the attacker who performed the compromise could conduct MiTM attacks on any of the users trying to connect to systems with a certificate issued by that CA.

They could also issue counterfeit certificates in response to legitimate certificate requests, ensuring their ability to intercept or interfere with encrypted communications between browser users and servers.

While CA compromise was once held to be very unlikely to happen, there have been a number of incidents at the time of this writing proving this is no longer the case.

Companies whose primary activity was in acting as a commercial CA have gone out of business as a result of having their systems compromised and counterfeit certificates issued under their name.

In September 2011, the certificate authority DigiNotar was found to have been subverted by crackers, forcing all its certificates to be revoked and sending it into bankruptcy.

These compromises were not the result of sophisticated computer criminals employing exotic sophisticated attacks but simply lapses in the security of their overall infrastructure and security policies and procedures. Finally, it is good to point out that TLS is not only used for web browsing. Insecure email protocols such as IMAP, POP, and SMTP can be secured by wrapping them in an TLS tunnel.

Most modern email clients support IMAPS and POPS (secure IMAP and POP) as well as TLS protected SMTP. If your email server does not provide TLS support, you can still secure it with TLS using a package like Stunnel (http://www.stunnel.org/). TLS can be used to effectively secure just about any service that runs over TCP.

SSH

Most people think of SSH as a secure replacement for telnet, just as SCP and SFTP are the secure counterparts of RCP and FTP.

But SSH is much more than an encrypted remote shell. For example, it can also act as a general purpose encrypting tunnel, or even an encrypting web proxy.

By first establishing an SSH connection to a trusted location near (or even on) a remote server, insecure protocols can be protected from eavesdropping and attack.

Like TLS, it uses strong public key cryptography to verify the remote server and encrypt data. Instead of a PKI, it uses a key fingerprint cache that is checked before a connection is permitted.
It can use passwords and public keys for user authentication, and, through its support for the Pluggable Authentication Modules (PAM) system, it can also support other methods of authentication.

While this technique may be a bit advanced for many users, network architects can use SSH to encrypt traffic across untrusted links, such as wireless point-to-point links.

Since the tools are freely available and run over standard TCP, any educated user can implement SSH connections for themselves, providing their own end-to-end encryption without administrator intervention. OpenSSH (http://openssh.org/) is probably the most popular implementation on Unix-like platforms.

Free implementations such as Putty (http://www.putty.nl/)and WinSCP (http://winscp.net/) are available for Windows.
OpenSSH will also run on Windows under the Cygwin package (http://www.cygwin.com/) These examples will assume that you are using a recent version of OpenSSH.

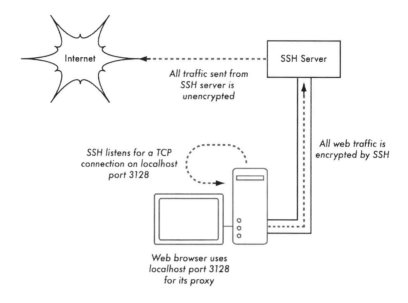

Figure SWN 8: The SSH tunnel protects web traffic up to the SSH server itself.

To establish an encrypted tunnel from a port on the local machine to a port on the remote side, use the -L switch. For example, suppose you want to forward web proxy traffic over an encrypted link to the squid server at squid.example.net. Forward port 3128 (the default proxy port) using this command:

ssh -fN -g -L3128:squid.example.net:3128 squid.example.net

The -fN switches instruct ssh to fork into the background after connecting.

The -g switch allows other users on your local segment to connect to the local machine and use it for encryption over the untrusted link.

OpenSSH will use a public key for authentication if you have set one up, or it will prompt you for your password on the remote side. You can then configure your web browser to connect to localhost port 3128 as its web proxy service.

All web traffic will then be encrypted before transmission to the remote side. SSH can also act as a dynamic SOCKS4 or SOCKS5 proxy. This allows you to create an encrypting web proxy, without the need to set up squid. Note that this is not a caching proxy; it simply encrypts all traffic:

ssh -fN -D 8080 remote.example.net

Configure your web browser to use SOCKS4 or SOCKS5 on local port 8080, and away you go.

SSH can encrypt data on any TCP port, including ports used for email. It can even compress the data along the way, which can decrease latency on low capacity links:

ssh -fNCg -L110:localhost:110 -L25:localhost:25 mailhost.example.net

The -C switch turns on compression.

You can add as many port forwarding rules as you like by specifying the -L switch multiple times.
Note that in order to bind to a local port less than 1024, you must have root privileges on the local machine. These are just a few examples of the flexibility of SSH. By implementing public keys and using the ssh forwarding agent, you can automate the creation of encrypted tunnels throughout your wireless network, and protect your communications with strong encryption and authentication.

OpenVPN

OpenVPN is a VPN implementation built on SSL encryption with both a commercially-licensed and an Open Source "community" edition.
There are OpenVPN client implementations for a wide range of operating systems including Linux, Window 2000/XP and higher, OpenBSD, FreeBSD, NetBSD, and Mac OS X. Many users find it easier to understand and configure than IPSEC VPNs.
OpenVPN has some disadvantages though, such as fairly high latency of traffic over the VPN tunnel.
Some amount of latency is unavoidable since all encryption/decryption is done in user space, but using relatively new computers on either end of the tunnel can minimise this.

While it can use traditional shared keys for authentication, OpenVPN really shines when used with SSL certificates and a certificate authority. OpenVPN has many advantages that make it a good option for providing end-to-end security. Some of these reasons include:

- it is based on proven, robust, encryption protocols (SSL and RSA)
- it is relatively easy to configure. It functions across many different platforms
- it is well-documented. An Open-Source "Community" version is maintained in addition to a for-pay commercial version.

OpenVPN needs to connect to a single TCP or UDP port on the remote side. Once established, it can encapsulate all data down to the Networking layer, or even down to the Data-Link layer, if your solution requires it. You can use it to create robust VPN connections between individual machines, or simply use it to connect network routers over untrusted wireless networks.

VPN technology is a complex field, and is a bit beyond the scope of this section to go into more detail. It is important to understand how VPNs fit into the structure of your network in order to provide the best possible protection without opening up your organisation to unintentional problems.

There are many good online resources that deal with installing OpenVPN on a server and client, we recommend this article from Linux Journal:
http://www.linuxjournal.com/article/7949
as well as the official HOWTO:
http://openvpn.net/howto.html

Tor & Anonymizers

The Internet is basically an open network based on trust.

When you connect to a web server across the Internet, your traffic passes through many different routers, owned by a great variety of institutions, corporations and individuals.

In principle, any one of these routers has the ability to look closely at your data, seeing the source and destination addresses, and quite often also the actual content of the data.

Even if your data is encrypted using a secure protocol, it is possible for your Internet provider to monitor the amount of data transferred, as well as the source and destination of that data.

Often this is enough to piece together a fairly complete picture of your activities on-line. Privacy and anonymity are important, and closely linked to each other. There are many valid reasons to consider protecting your privacy by anonymizing your network traffic.

Suppose you want to offer Internet connectivity to your local community by setting up a number of access points for people to connect to.

Whether you charge them for their access or not, there is always the risk that people use the network for something that is not legal in your country or region.

You could plead with the legal system that this particular illegal action was not performed by yourself, but could have been performed by anyone connecting to your network. The problem is neatly sidestepped if it were technically infeasible to determine where your traffic was actually headed.

And what about on-line censorship?

Publishing web pages anonymously may also be necessary to avoid government censorship. There are tools that allow you to anonymize your traffic in relatively easy ways.

The combination of Tor (http://www.torproject.org) and Privoxy (http://www.privoxy.org) is a powerful way to run a local proxy server that will pass your Internet traffic through a number of servers all across the net, making it very difficult to follow the trail of information.

Tor can be run on a local PC, under Microsoft Windows, Mac OSX, Linux and a variety of BSD's, where it anonymizes traffic from the browser on that particular machine.

Tor and Privoxy can also be installed on a gateway server, or even a small embedded access point (such as a Linksys WRT54G) where it provides anonymity to all network users automatically.

Tor works by repeatedly bouncing your TCP connections across a number of servers spread throughout the Internet, and by wrapping routing information in a number of encrypted layers (hence the term onion routing), that get peeled off as the packet moves across the network.

This means that, at any given point in the network, the source and destination addresses cannot be linked together.

This makes traffic analysis extremely difficult.

The need for the Privoxy privacy proxy in connection with Tor is due to the fact that name server queries (DNS queries) in most cases are not passed through the proxy server, and someone analysing your traffic would easily be able to see that you were trying to reach a specific site (say google.com) by the fact that you sent a DNS query to translate google.com to the appropriate IP address.

Privoxy connects to Tor as a SOCKS4a proxy, which uses hostnames (not IP addresses) to get your packets to the intended destination. In other words, using Privoxy with Tor is a simple and effective way to prevent traffic analysis from linking your IP address with the services you use online.

Combined with secure, encrypted protocols (such as those we have seen in this chapter), Tor and Privoxy provide a high level of anonymity on the Internet.

PLANNING AND DEPLOYMENT

10. DEPLOYMENT PLANNING

Estimating capacity

In order to estimate capacity, it is important to understand that a wireless device's listed speed (the so called data rate) refers to the rate at which the radios can exchange symbols, not the useable throughput you will observe. Throughput is also referred to as channel capacity, or simply bandwidth (although this term is different from radio bandwidth!

The bandwidth-throughput is measured in Mbps, while the radio bandwidth is measures in MHz.). For example, a single 802.11g link may use 54 Mbps radios, but it will only provide up to 22 Mbps of actual throughput. The rest is overhead that the radios need in order to coordinate their signals using the 802.11g protocol.

Note that throughput is a measurement of bits over time. 22 Mbps means that in any given second, up to 22 megabits can be sent from one end of the link to the other. If users attempt to push more than 22 megabits through the link, it will take longer than one second.
Since the data can't be sent immediately, it is put in a queue, and transmitted as quickly as possible.

This backlog of data increases the time needed for the most recently queued bits to the traverse the link.
The time that it takes for data to traverse a link is called latency, and high latency is commonly referred to as lag.

Your link will eventually send all of the queued traffic, but your users will likely complain as the lag increases.
How much throughput will your users really need?
It depends on how many users you have, and how they use the wireless link.
Various Internet applications require different amounts of throughput.

Application	Requirement/ User	Notes
Text messaging / IM	< 1 kbps	As traffic is infrequent and asynchronous, IM will tolerate high latency.
Email	1 to 100 kbps	As with IM, email is asynchronous and intermittent, so it will tolerate latency. Large attachments, viruses, and spam significantly add to bandwidth usage. Note that web email services (such as Yahoo or Hotmail) should be considered as web browsing, not as email.
Web browsing	50 - 100+ kbps	Web browsers only use the network when data is requested. Communication Is asynchronous, so a fair amount of lag can be tolerated. As web browsers request more data (large images, long downloads, etc.) bandwidth usage will go up significantly.
Streaming audio	96 - 160 kbps	Each user of a streaming audio service will use a constant amount of relatively large bandwidth for as long as it plays. It can tolerate some transient latency by using large buffers on the client. But extended periods of lag will cause audio "skips" or outright session failures.
Voice over IP (VoIP)	24 - 100+ kbps	As with streaming audio, VoIP commits a constant amount of bandwidth to each user for the duration of the call. But with VoIP, the bandwidth is used roughly equally in both directions. Latency on a VoIP connection is immediate and annoying to users. Lag greater than a few milliseconds is unacceptable for VoIP.
Streaming video	64 - 200+ kbps	As with streaming audio, some intermittent latency is avoided by using buffers on the client. Streaming video requires high throughput and low latency to work properly.
Peer-to-peer file sharing applications	0 - infinite Mbps	While peer to peer applications will tolerate any amount of latency, they tend to use up all available throughput by transmitting data to as many clients as possible, as quickly as possible. Use of these applications will cause latency and throughput problems for all other network users unless you use careful bandwidth shaping.

To estimate the necessary throughput you will need for your network, multiply the expected number of users by the requirements of applications they will probably use.

For example, 50 users who are chiefly browsing the web will likely consume 2.5 to 5 Mbps or more of throughput at peak times, and will tolerate some latency.

On the other hand, 50 simultaneous VoIP users would require 5 Mbps or more of throughput in both directions with low latency. Since 802.11g wireless equipment is half duplex (that is, it only transmits or receives, never both at once) you should accordingly double the required throughput, for a total of 10 Mbps.

Your wireless links must provide that capacity every second, or conversations will lag.
Since all of your users are unlikely to use the connection at precisely the same moment, it is common practice to oversubscribe available throughput by some factor (that is, allow more users than the maximum available bandwidth can support).

Oversubscribing by a factor of 5 to 10 is quite common. In all likelihood, you will oversubscribe by some amount when building your network infrastructure.

By carefully monitoring throughput throughout your network, you will be able to plan when to upgrade various parts of the network, and how much additional resources will be needed.

Calculating the link budget

Determining if a link is feasible or not is a process called *link budget calculation* and can be either performed manually or using specialised tools.
A basic communication system consists of two radios, each with its associated antenna, the two being separated by the path to be covered as shown in following Figure DP 1.

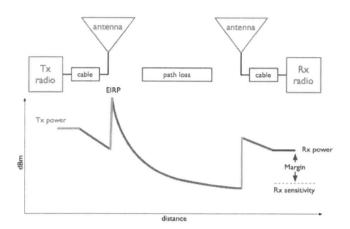

Figure DP 1: Components of a Basic Communication System.

In order to have a communication between the two, the radios require a certain minimum signal to be collected by the antennas and presented at their input ports.

Whether or not signals can be passed between the radios depends on the characteristics of the equipment and on the diminishment of the signal due to distance, called *path loss.* In such a system, some of the parameters can be modified (the equipment used) while others are fixed (the distance between the radios). Let's start by examining the parameters that can we can modify.

1. The characteristics of the equipment to be considered when calculating the link budget are:

Transmit (TX) Power. It is expressed in milliwatts or in dBm. TX power is often dependent on the transmission rate.

The TX power of a given device should be specified in the literature provided by the manufacturer.

An example is shown here following, where you can see that using 802.11g there is a difference of 5 dB in output power when using 6 Mbs or 54 Mbs.

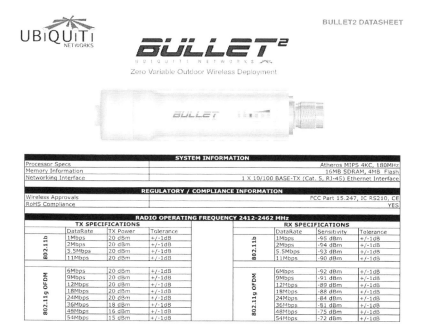

Figure DP 2: Ubiquiti Bullet2 Datasheet

Antenna Gain.

Antennas are passive devices that create the effect of amplification by virtue of their physical shape.

Antennas have the same characteristics when receiving and transmitting. So a 12 dBi antenna is simply a 12 dBi antenna, without specifying if it is in transmission or reception mode.

Typical values are as follows: parabolic antennas have a gain of 19-24 dBi; omnidirectional antennas have 5- 12 dBi; sectorial antennas have roughly a 12-15 dBi gain.

Minimum Received Signal Level (RSL), or simply, the sensitivity of the receiver.

The minimum RSL is always expressed as a negative dBm (-dBm) and is the lowest signal level the radio can distinguish.

The minimum RSL is dependent upon rate, and as a general rule the lowest rate (1 Mbps) has the greatest sensitivity.

The minimum will be typically in the range of -75 to -95 dBm.

Like TX power, the RSL specifications should be provided by the manufacturer of the equipment.
In the datasheet presented above you can see that there is a 20 dB difference in receiver sensitivity, with -92 dBm at 6 Mbs and -72 dBm at 54 Mbs. Don't forget that a difference of 20 dB means a ratio of 100 in terms of power!

Cable Losses. Some of the signal's energy is lost in the cables, the connectors and other devices, going from the radios to the antennas. T
he loss depends on the type of cable used and on its length.

Signal loss for short coaxial cables including connectors is quite low, in the range of 2-3 dB. It is better to have cables as short as possible. Equipment now tends to have embedded antennas and thus the cable length is very short.

2. When calculating the path loss, several effects must be considered. One has to take into account the *free space loss, attenuation* and *scattering.*

Free Space Loss.
Geometric spreading of the wavefront, commonly known as free space loss, diminishes signal power. Ignoring everything else, the further away the two radios, the smaller the received signal is due to free space loss. This is independent of the environment, it depends only on the distance. This loss happens because the radiated signal energy expands as a function of the distance from the transmitter.

Using decibels to express the loss and using a generic frequency f, the equation for the Free Space Loss is:

$$L_{fsl} = 32.4 + 20*log_{10}(D) + 20*log_{10}(f)$$

where L_{fsl} is expressed in dB and D is in kilometres and f is in MHz.
When plotting the free space loss vs the distance, one gets a figure like the one below. What should be noted is that the difference between using 2400 MHz and 5300 MHz is 6 dB in terms of free space loss.
So the higher frequency gives a higher free space loss, which is usually balanced by a higher gain in the parabolic antennas.

A parabolic antenna operating at 5 GHz is 6 dB more powerful than one operating at 2.4 GHz, for the same antenna dimensions.

Having two antennas of 6 dB higher gain on each side, provides a net 6 dB advantage when migrating from 2.4 to 5 GHz.

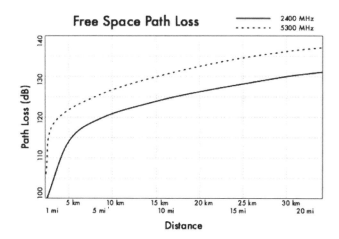

Figure DP 3: Free Space Path Loss calculation graph.

Attenuation

The second contribution to the path loss is attenuation.

This takes place as some of the power is absorbed when the wave passes through solid objects such as trees, walls, windows and floors of buildings. Attenuation can vary greatly depending upon the structure of the object the signal is passing through, and it is very difficult to quantify.

Scattering

Along the link path, the RF energy leaves the transmitting antenna and energy spreads out. Some of the RF energy reaches the receiving antenna directly, while some bounces off the ground. Part of the RF energy which bounces off the ground reaches the receiving antenna. Since the reflected signal has a longer way to travel, it arrives at the receiving antenna later than the direct signal. This effect is called *multipath,* or signal dispersion. In some cases reflected signals add together and cause no problem.

When they add together out of phase, the received signal is almost worthless. In some cases, the signal at the receiving antenna can be zeroed by the reflected signals.

This is known as extreme fading, or **_nulling_**. There is a simple technique that is used to deal with multipath, called **_antenna diversity._**
It consists of adding a second antenna to the radio. Multipath is in fact a very location-specific phenomenon.

If two signals add out of phase at one location, they will not add destructively at a second, nearby location.
If there are two antennas, at least one of them should be able to receive a useable signal, even if the other is receiving a faded one. In commercial devices, antenna switching diversity is used: there are multiple antennas on multiple inputs, with a single receiver.

The signal is received through only one antenna at a time.
When transmitting, the radio uses the antenna last used for reception. The distortion given by multipath degrades the ability of the receiver to recover the signal in a manner much like signal loss.
Putting all these parameters together leads to the **link budget calculation**. If you are using different radios on the two sides of the link, you should calculate the path loss twice, once for each direction (using the appropriate TX power, RX power, TX antenna gain, RX antenna gain for each calculation).

Adding up all the gains and subtracting all the losses gives:

TX Power	*Radio 1*
+Antenna Gain	*Radio 1*
-Cable Losses	*Radio 1*
+Antenna Gain	*Radio 2*
-Cable Losses	*Radio 2*
=Total Gain	

Subtracting the Path Loss from the Total Gain:

Total Gain - Path Loss = Signal Level at receiving side of the link

If the resulting signal level is greater than the minimum received signal level of the receiving radio, then the link is feasible!
The received signal is powerful enough for the radio to use it.

Remember that the minimum RSL is always expressed as a negative dBm, so -56 dBm is greater than -70 dBm.

On a given path, the variation in path loss over a period of time can be large, so a certain margin should be considered.
This margin is the amount of signal above the sensitivity of radio that should be received in order to ensure a stable, high quality radio link during bad weather and other atmospheric disturbances.

A margin of 10 to 15 dB is fine. To give some space for attenuation and multipath in the received radio signal, a margin of 20dB should be safe enough.

Once you have calculated the link budget in one direction, repeat the calculation for the other direction.
Substitute the transmit power for that of the second radio, and compare the result against the minimum received signal level of the first radio.

Example link budget calculation
As an example, we want to estimate the feasibility of a 5 km link, with one access point and one client radio.

- The access point is connected to an omni directional antenna with a 10dBi gain, while the client is connected to a directional antenna with 14 dBi gain.
- The transmitting power of the AP is 100 mW (or 20 dBm) and its sensitivity is -89 dBm.
- The transmitting power of the client is 30 mW (or 15 dBm) and its sensitivity is -82 dBm.
- The cables are short, so we can estimate a loss of 2 dB at each side.

Let's start by calculating the link budget from the AP to the client, as shown in Figure DP 4.

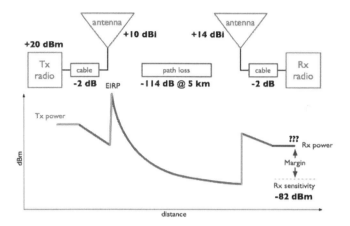

Figure DP 4: Link Budget calculation from AP to client.

Adding up all the gains and subtracting all the losses for the AP to client link gives:

20 dBm	*(TX Power Radio 1)*
+10 dBi	*(Antenna Gain Radio 1)*
-2 dB	*(Cable Losses Radio 1)*
+14 dBi	*(Antenna Gain Radio 2)*
-2 dB	*(Cable Losses Radio 2)*
40 dB=	*Total Gain*

The path loss for a 5 km link, considering the free space loss is -114dB. Subtracting the path loss from the total gain:

$$40\ dB - 114\ dB = -74\ dB$$

Since -74 dB is greater than the minimum receive sensitivity of the client radio (-82 dBm), the signal level is just enough for the client radio to be able to hear the access point. There is only 8 dB of margin (82 dB - 74 dB) which will likely work fine in fair weather, but may not be enough to protect against extreme weather conditions.

Next we calculate the link from the client back to the access point, as shown on the next page.

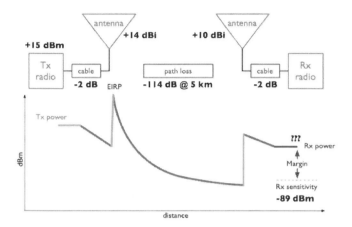

Figure DP 5: Link Budget calculation from client to AP.

15 dBm	*(TX Power Radio 2)*
+14 dBi	*(Antenna Gain Radio 2)*
-2 dB	*(Cable Losses Radio 2)*
+10 dBi	*(Antenna Gain Radio 1)*
-2 dB	*(Cable Losses Radio 1)*
35 dB =	*Total Gain*

Obviously, the path loss is the same on the return trip. So our received signal level on the access point side is:

$$35 \ dB - 114 \ dB = -79 \ dB$$

Since the receive sensitivity of the AP is -89dBm, this leaves us 10dB of margin (89dB - 79dB). Overall, this link will probably work fine.

By using a 24dBi dish on the client side rather than a 14dBi antenna, we will get an additional 10dBi of gain on both directions of the link (remember, antenna gain is reciprocal).

A more expensive option would be to use higher power radios on both ends of the link, but note that adding an amplifier or higher-powered card to one end generally does not help the overall quality of the link.

Tables for calculating link budget

To calculate the link budget, simply approximate your link distance, and then fill in the following tables:

Free Space Path Loss at 2.4 GHz

Distance (m)	100	500	1000	3000	5000	10000
Loss (dB)	80	94	100	110	114	120

Antenna Gain:

Radio 1 Antenna	+ Radio 2 Antenna	= Total Antenna Gain

Losses:

Radio 1 + Cable Loss (dB)	Radio 2 + Cable Loss (dB)	Free Space Path Loss (dB)	= Total Loss (dB)

Link Budget for Radio 1 Radio 2:

Radio 1 TX Power	+ Antenna Gain	- Total Loss	= Signal	> Radio 2 Sensitivity

Link Budget for Radio 2 Radio 1:

Radio 2 TX Power	+ Antenna Gain	- Total Loss	= Signal	> Radio 1 Sensitivity

If the received signal is greater than the minimum received signal strength in both directions of the link, as well as any noise received along the path, then the link is possible.

Link planning software

While calculating a link budget by hand is straightforward, there are a number of tools available that will help automate the process. In addition to calculating free space loss, these tools will take many other relevant factors into account as well (such as tree absorption, terrain effects, climate, and even estimating path loss in urban areas). Most commercial tools are very expensive and are often designed to be used with specific hardware. In this section we will discuss a free tool called RadioMobile.

RadioMobile

Radio Mobile is a tool for the design and simulation of wireless systems. It predicts the performance of a radio link by using information about the equipment and a digital map of the area. It is public domain software but it is not open source. Radio Mobile uses a digital terrain elevation model for the calculation of coverage, indicating received signal strength at various points along the path. It automatically builds a profile between two points in the digital map showing the coverage area and first Fresnel zone. An example is shown in Figure DP 6 below.

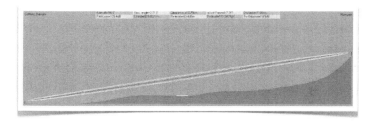

Figure DP 6: Radio Mobile simulation showing a digital terrain elevation and first Fresnel zone.

During the simulation, it checks for line of sight and calculates the Path Loss, including losses due to obstacles.

It is possible to create networks of different topologies, including net master/slave, point-to-point, and point-to-multipoint. The software calculates the coverage area from the base station in a point-to-multipoint system. It works for systems having frequencies from 100 kHz to 200 GHz. Digital elevation maps (DEM) are available for free from several sources, and are available for most of the world.

DEMs do not show coastlines or other readily identifiable landmarks, but they can easily be combined with other kinds of data (such as aerial photos or topographical charts) in several layers to obtain a more useful and readily recognisable representation. You can digitise your own maps and combine them with DEMs. The digital elevation maps can be merged with scanned maps, satellite photos and Internet map services (such as Google Maps) to produce accurate prediction plots.

There are two versions of RadioMobile: a version running on Windows and a version running online via a web interface. Here are the main differences between the two.

Web version:

- it can run on any machine (Linux, Mac, Windows, tablet, phone, etc)
- it does not require big downloads. As it runs online, data is stored on the server and only the necessary data is downloaded.
- it saves sessions. If you run a simulation and login after some time, you will still find your simulation and the results.
- it is easier to use, especially for beginners.
- it requires connectivity. It is not possible to run a simulation if you are offline.
- as it has been developed for the community of radio amateurs, it can work only for certain frequency bands. As an example, it is not possible to simulate links at 5.8GHz but only at 5.3GHz. This is good enough from a practical point of view, but one has to keep it in mind.

Windows version:

- it can run offline. Once the maps are downloaded, there is no need to be online to run the simulation.
- one can use an external GPS to get the exact position of the station. While this is not often used, it could be useful sometimes.
- it runs on Windows (it does run in Linux but not directly).
- it requires big downloads. If your bandwidth is limited, downloading many maps might be a problem. The online version requires smaller downloads.
- It is not user friendly, especially for beginners.

The main RadioMobile webpage, with examples and tutorials, is available at: http://www.cplus.org/rmw/english1.html
Follow the instructions to install the software on Windows.

RadioMobile online

To use the online version of RadioMobile you must first create an account. Go to: http://www.cplus.org/rmw/rmonline.html and create a new account. You will receive a confirmation email in few minutes and will be ready to go.
Simulating a link requires some steps that require following the menu on the left, from top to bottom as shown in Figure DP 7 below.

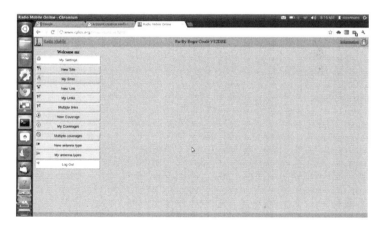

Figure DP 7: Preparing to simulate a link using Radio Mobile.

The first step is to click on New Site. You will be presented a Map, similar to Google Maps. You can zoom into the map to find the location of your first site. Drag the Orange Placemark and place it in the desired position. Once you are done, click on Submit.

Give the location a meaningful name, and click on Add to My Sites. In this way, you will be able to use this location for the simulation.
Repeat the same process for the second site.
Once you have at least two sites, you can proceed with the next step. The interface will not allow you to enter directly the coordinates of the site, so you might want to enter an approximate position for the cursor and then afterward correct the value of the coordinates in the table.

The second step involves entering information about the link: the equipment characteristics, the antennas, etc.

Click on New Link in the menu on the left. Select the two sites from the drop-down menus. Enter a meaningful name for the link and enter the information about the equipment use. The sensitivity of the receiver is expressed in microVolts, while we usually use dBm.

To translate from microVolts to dBm, here are some examples:

-90dBm is 7.07 microvolts
-85dBm is 12.6 microvolts
-80dBm is 22.4 microvolts
-75dBm is 39.8 microvolts
-70dBm is 70.7 microvolts

It is very important to choose a frequency that RadioMobile online can handle.
Here are the most important frequencies for WiFi links:

Use 2300 MHz for 2.4 GHz links and 5825 MHz for 5.8 GHz links.

Once you have entered all the information, click on Submit. In a short time, you will be presented a figure similar to the one below.

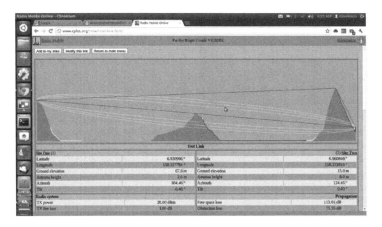

Figure DP 8: Output of the simulation.

This page has all the necessary information to understand if the link is feasible or not.

It gives information on: the link length, the azimuth, elevation and tilt you should give to antennas, the free space loss, the total space loss including the statistical loss and, most importantly, the received signal level.

With the receiver sensitivity you entered, you will be presented with the fade margin and you will be able to determine if the link is feasible or not. If you are happy with the result, at the top of the page you should click on Add to my sites and the link will be stored. If you are not satisfied and want to simulate a different equipment setup, click on Modify this link.

Radio Mobile for Windows in simple steps

This is an abridged guide to start using Radio Mobile after installation. Every parameter not specified here can be left at the default value and later modified if needed.

Step 1: download the digital elevation maps (DEMs) of your area of interest. choose SRTM format.

Step 2: create a map. In "File" ➜" Map properties", choose the midpoint of your area of interest as the coordinates for your map and a size in km large enough to encompass all your points. Use 514X514 pixels for now. You can add another type of map (like one with roads) to the raw DEM if you like.

Step 3: Create systems. In "File" ➜" Network properties" ➜"Systems". Each one is a combination of TX power, RX sensitivity and antenna gain. Select omni antenna even if your antenna is directional, but insert the real gain.

Step 4: create units. Each unit has a name and a geographical position. You can use degrees, minutes, seconds or degrees and fractions, but make sure to select the proper hemisphere (N or S, E or W).

Step 5: Assign roles: select item "Networks properties" from the menu "File". Then go to the tab "Membership" where you will be allowed to edit the system and role for each unit. Enable each unit in the list with a checkmark. Assign a name to your network and in the "parameters" tab set the minimum and maximum frequency of operation in MHz.

Step 6: View your network on the map. Select "View" ➜"Show networks"➜"All"

Step 7: Obtain the profile and point to point link budget. "Tools" ↦ "Radio link". You can switch to the detailed view that gives you a textual description of the output of the simulation.

Step 8: View coverage: go to "Tools" ↦ "Radio coverage" ↦ "Single polar" to obtain the coverage of each station. Here the type of antenna becomes relevant. If it is not an omni you should edit the antenna pattern and the orientation of the boresight (direction where the beam is pointing).

Using Google Earth to obtain an elevation profile

Google Earth is a very popular mapping application. It can be used to obtain the elevation profile between two points and therefore determine the existence (or not) of the optical line of sight. The radioelectric line of sight can be derived from the optical one by adding the effect of the earth curvature (by using the modified radius of the earth) and the requirements for clearing the first Fresnel zone.

The procedure is the following:

Install Google Earth in your device, launch the application and zoom into the map so you can see the two points that you want to link.

1. In the upper menu, click on "Add path"
2. Click to set the first point and then for the second point
3. Give the connection a name ("Link" for example) and click OK in the pop-up window
4. The connection appears in the menu on the left
5. Right click on the name of the connection ("Link" in our example)
6. Select "Show elevation profile"
7. The elevation profile appears at the bottom on the screen
8. If you move along the profile, you will see a red arrow showing where the point is in the map

Avoiding noise

The unlicensed ISM and U-NII bands represent a very tiny piece of the known electromagnetic spectrum. Since this region can be utilised without paying license fees, many consumer devices use it for a wide range of applications.

Cordless phones, analog video senders, Bluetooth, baby monitors, and even microwave ovens compete with wireless data networks for use of the very limited 2.4 GHz band.

These signals, as well as other local wireless networks, can cause significant problems for long-range wireless links. Here are some steps you can use to reduce reception of unwanted signals.

Increase antenna gain on both sides of a point-to-point link. Antennas not only add gain to a link, but their increased directionality tends to reject noise from areas around the link.

Two high gain dishes that are pointed at each other will reject noise from directions that are outside the path of the link. Using omnidirectional antennas will receive noise from all directions.

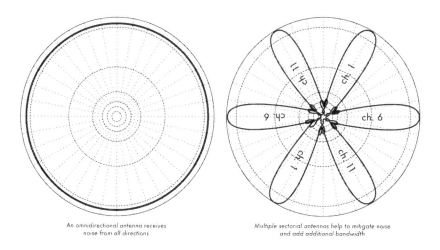

An omnidirectional antenna receives
noise from all directions

Multiple sectorial antennas help to mitigate noise
and add additional bandwidth

Figure DP 9: A single omnidirectional antenna vs. multiple sectorials.

Use sectorials instead of using an omnidirectional.

By making use of several sectorial antennas, you can reduce the overall noise received at a distribution point.

By staggering the channels used on each sectorial, you can also increase the available bandwidth to your clients.

Don't use an amplifier. Amplifiers can make interference issues worse by indiscriminately amplifying all received signals, including sources of interference. Amplifiers also cause interference problems for other nearby users of the band.

Use the best available channel. Remember that 802.11b/g channels are 22 MHz wide, but are only separated by 5 MHz.

Perform a site survey, and select a channel that is as far as possible from existing sources of interference. Remember that the wireless landscape can change at any time as people add new devices (cordless phones, other networks, etc.) If your link suddenly has trouble sending packets, you may need to perform another site survey and pick a different channel.

If possible, use 5.8 GHz. While this is only a short-term solution, there is currently far more consumer equipment installed in the field that uses 2.4 GHz. Using 802.11a you will avoid this congestion altogether.

If all else fails, use licensed spectrum. There are places where all available unlicensed spectrum is actually used. In these cases, it may make sense to spend the additional money for obtaining the respective license and deploy equipment that uses a less congested band. For long distance point-to-point links that require very high throughput and maximum uptime, this is certainly an option. Of course, these features come at a much higher price tag compared to unlicensed equipment.

Recently equipment in the 17 GHz and 24 GHz bands have become available.

Although it is considerably more expensive, it also offers much greater bandwidth, and in many countries these frequencies are unlicensed.

To identify sources of noise, you need tools that will show you what is happening in the air at 2.4 GHz. We will see some examples of these tools in the chapters called **Network Monitoring** and **Maintenance and Troubleshooting.**

Repeaters

The most critical component to building long distance network links is line of sight (often abbreviated as LOS).

Terrestrial microwave systems simply cannot tolerate large hills, trees, or other obstacles in the path of a long distance link. You must have a clear idea of the lay of the land between two points before you can determine if a link is even possible. But even if there is a mountain between two points, remember that obstacles can sometimes be turned into assets.

Mountains may block your signal, but assuming power can be provided they also make very good repeater sites.

Repeaters are nodes that are configured to rebroadcast traffic that is not destined for the node itself. In a mesh network, every node is a repeater. In a traditional infrastructure network, repeater nodes must be configured to pass along traffic to other nodes. A repeater can use one or more wireless devices.

When using a single radio (called a one-arm repeater), overall efficiency is slightly less than half of the available bandwidth, since the radio can either send or receive data, but never both at once. These devices are cheaper, simpler, and have lower power requirements.

A repeater with two (or more) radio cards can operate all radios at full capacity, as long as they are each configured to use non-overlapping channels.

Of course, repeaters can also supply an Ethernet connection to provide local connectivity.

Repeaters can be purchased as a complete hardware solution, or easily assembled by connecting two or more wireless nodes together with Ethernet cable.

When planning to use a repeater built with 802.11 technology, remember that nodes must be configured for master, managed, or ad-hoc mode.

Typically, both radios in a repeater are configured for master mode, to allow multiple clients to connect to either side of the repeater. But depending on your network layout, one or more devices may need to use ad-hoc or even client mode.

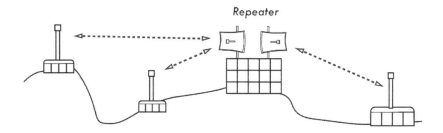

Figure DP 10: The repeater forwards packets over the air between nodes that have no direct line of sight.

Typically, repeaters are used to overcome obstacles in the path of a long distance link. For example, there may be buildings in your path, but those buildings contain people. Arrangements can often be worked out with building owners to provide bandwidth in exchange for roof rights and electricity. If the building owner isn't interested, tenants on high floors may be able to be persuaded to install equipment in a window. If you can't go over or through an obstacle, you can often go around it. Rather than using a direct link, try a multi-hop approach to avoid the obstacle.

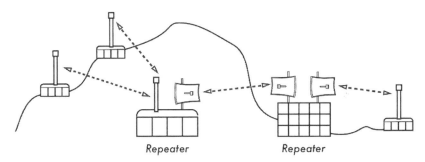

Figure DP 11: No power was available at the top of the hill, but it was circumvented by using multiple repeater sites around the base.

Finally, you may need to consider going backwards in order to go forwards. If there is a high site available in a different direction, and that site can see beyond the obstacle, a stable link can be made via an indirect route.

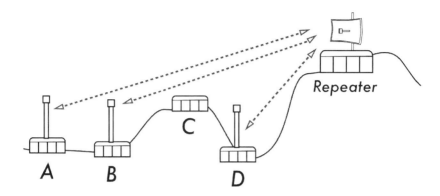

Figure DP 12: Site D could not make a clean link to site A or B, since site C is in the way and is not hosting a node. By installing a high repeater, nodes A, B, and D can communicate with each other. Note that traffic from node D actually travels further away from the rest of the network before the repeater forwards it along.

Deployment planning for IPv6

As we mentioned in the chapter called **Networking**, most regions of the world have either exhausted or nearly exhausted their IPv4 addresses. Therefore it is important that you build into your planning the deployment of IPv6 based networks.

It is understood that at the time of writing there are still many sites and services that are still only available by IPv4.

So in order to become a leader in deploying IPv6, you will need to be able to interconnect with legacy IPv4 networks as well as teach and guide your users and developers how to handle IPv6 alongside IPv4.

By leading the way in deploying IPv6 in your network you will be at the forefront of the Internet and recognised as someone who is prepared to be aware of and support the next generation of networking.

In preparing for IPv6 here are some steps you can take to help move you in the right direction -

1. Dont buy routers, firewalls and other IP equipment that process IPv4 packets in hardware at full speed and only process IPv6 packets more slowly in software or worse still dont handle IPv6 at all. The vast majority of available network devices support IPv6. RIPE has prepared some requirements to be added into any call for tender to ensure that IPv6 is included:
   http://www.ripe.net/ripe/docs/current -ripe-documents/ripe-554
   You may also look for the IPv6-Ready logo on the data sheets of the devices.
2. When deploying new software, make sure it works over IPv6.
3. When discussing your backhaul link with a local ISP, check that they have deployed or have plans to deploy and offer IPv6 services. If not discuss with them how you might co-operate and interconnect your IPv6 network with them. The cost of IPv6 should be included in the overall cost; this means that you will not pay anything more to get IPv6. The Service Level Agreement (SLA) for IPv6 should be identical (throughput, latency, incident response time, ...) as for IPv4. There are several IPv4/IPv6 transition techniques that can be deployed.

Here are some urls that you can read that will give you up to date information about these -
http://www.petri.co.il/ipv6-transition.htm
http://en.wikipedia.org/wiki/IPv6_transition_mechanisms
http://www.6diss.org/tutorials/transitioning.pdf

There is more information about the growth of IPv6 and the lack of available IPv4 adresses in this article published in late 2012.
http://arstechnica.com/business/2013/01/ipv6-takes-one-step-forward-ipv4-two-steps-back-in-2012/

There is also an EC funded project called 6Deploy who offer training and helpdesk services for network engineers starting their IPv6 deployments. It is highly recommended that you contact them to discuss your project.
http://www.6deploy.eu/index.php?page=home

11. HARDWARE SELECTION AND CONFIGURATION

In the last few years, an unprecedented surge of interest in wireless networking hardware has brought a huge variety of inexpensive equipment to the market.

So much variety, in fact, that it would be impossible to catalog every available component. In this chapter, we'll look at the sort of features and attributes that are desirable in a wireless component.

Wired wireless

With a name like "wireless", you may be surprised at how many wires are involved in making a simple point-to-point link.

A wireless node consists of many components, which must all be connected to each other with appropriate cabling.

You obviously need at least one computer connected to an Ethernet network, and a wireless router or bridge attached to the same network. Radio components need to be connected to antennas, but along the way they may need to interface with a lightning arrestor or other device.

Many components require power, either via an AC mains line or using a DC transformer.

All of these components use various sorts of connectors, not to mention a wide variety of cable types and thicknesses.

Now multiply those cables and connectors by the number of nodes you will bring online, and you may well be wondering why this stuff is referred to as "wireless".

The diagram on the next page will give you some idea of the cabling required for a typical point-to-point link.

Note that this diagram is not to scale, nor is it necessarily the best choice of network design. But it will introduce you to many common interconnects and components that you will likely encounter in the real world.

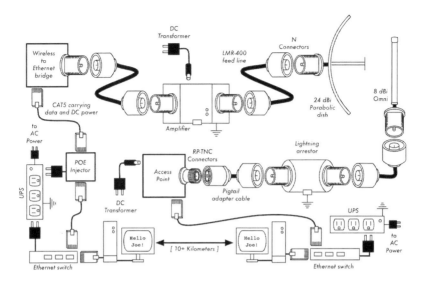

Figure HW 1: Common Interconnects and Components for a Wireless PtP link.

While the actual components used will vary from node to node, every installation will incorporate these parts:

1. An existing computer or network connected to an Ethernet switch.
2. A device that connects that network to a wireless device (a wireless router, bridge, or repeater).
3. An antenna that is connected via feed line, or is integrated into the wireless device itself.
4. Electrical components consisting of power supplies, conditioners, and lightning arrestors.

The actual selection of hardware should be determined by establishing the requirements for the project, determining the available budget, and verifying that the project is feasible using the available resources (including providing for spares and ongoing maintenance costs).

As discussed in other chapters of this book, establishing the scope of your project is critical before any purchasing decisions are made.

Choosing wireless components

Unfortunately, in a world of competitive hardware manufacturers and limited budgets, the price tag is the single factor that usually receives the most attention.

The old saying that "you get what you pay for" often holds true when buying high tech equipment, but should not be considered an absolute truth. While the price tag is an important part of any purchasing decision, it is vital to understand precisely what you get for your money so you can make a choice that fits your needs.

When comparing wireless equipment for use in your network, be sure to consider these variables:

Interoperability. Will the equipment you are considering work with equipment from other manufacturers? If not, is this an important factor for this segment of your network? If the gear in question is standard (such as 802.11b/g), then it will likely interoperate with equipment from other sources.

Range. This is not something inherent in a particular piece of equipment. A device's range depends on the antenna connected to it, the surrounding terrain, the characteristics of the device at the other end of the link, and other factors.

Rather than relying on a semi- fictional "range" rating supplied by the manufacturer, it is more useful to know the transmission power of the radio as well as the antenna gain (if an antenna is included).

With this information, you can calculate the theoretical range as done when looking at link budget calculations as we did in the chapter called **Deployment Planning**.

Radio sensitivity. How sensitive is the radio device at a given bit rate? The manufacturer should supply this information, at least at the fastest and slowest speeds.

This can be used as a measure of the quality of the hardware, as well as allow you to complete a link budget calculation.

When considering radio sensitivity, a lower number is better.

Throughput. Manufacturers consistently list the highest possible bit rate as the "speed" of their equipment. Keep in mind that the radio symbol rate (eg. 54 Mbps) is never the actual throughput rating of the device (eg. about 22 Mbps for 802.11g).

If throughput rate information is not available for the device you are evaluating, a good rule of thumb is to divide the device "speed" by two, and subtract 20% or so. When in doubt, perform throughput testing on an evaluation unit before committing to purchasing a large amount of equipment that has no official throughput rating.

Required accessories. To keep the initial price tag low, vendors often leave out accessories that are required for normal use. Does the price tag include all power adapters? For example DC supplies are typically included; Power over Ethernet injectors typically are not. Double-check input voltages as well, as equipment is sometimes provided with a US power supply.

What about pigtails, adapters, cables, antennas, and radio cards? If you intend to use it outdoors, does the device include a weatherproof case?

Availability. Will you be able to easily replace failed components? Can you order the part in large quantity, should your project require it? What is the projected life span of this particular product, both in terms of useful running time in-the-field and likely availability from the vendor?

Power consumption. For remote installations, power consumption is the most critical issue. If the devices are to be powered with solar panels, it is very important to select the ones that require the lowest possible power.

The cost of solar panels and batteries can be much higher than the cost of wireless devices; so lower power consumption can result in a much lower overall budget.

Other factors. Be sure that other features are provided to meet your particular needs. For example, does the device include an external antenna connector? If so, what type is it? Are there user or throughput limits imposed by software, and if so, what is the cost to increase these limits?

What is the physical form factor of the device?

Does it support POE as a power source? Does the device provide encryption, NAT, bandwidth monitoring tools, or other features critical to the intended network design?

By answering these questions first, you will be able to make intelligent buying decisions when the time comes to choosing networking hardware. It is unlikely that you will be able to answer every possible question before buying gear, but if you prioritise the questions and press the vendor to answer them before committing to a purchase, you will make the best use of your budget and build a network of components that are well suited to your needs.

Commercial vs. DIY solutions

Your network project will almost certainly consist of components purchased from vendors as well as parts that are sourced or even fabricated locally. This is a basic economic truth in most areas of the world. At this stage of human technology, global distribution of information is quite trivial compared to global distribution of goods. In many regions, importing every component needed to build a network is prohibitively expensive for all but the largest budgets.

You can save considerable money in the short term by finding local sources for parts and labour, and only importing components that must be purchased.

Of course, there is a limit to how much work can be done by any individual or group in a given amount of time. To put it another way, by importing technology, you can exchange money for equipment that can solve a particular problem in a comparatively short amount of time. The art of building local telecommunications infrastructure lies in finding the right balance of money to effort needed to be expended to solve the problem at hand.

Some components, such as radios and antenna feed line, are likely far too complex to consider having them fabricated locally. Other components, such as antennas and towers, are relatively simple and can be made locally for a fraction of the cost of importing. Between these extremes lie the communication devices themselves.

By using off-the-shelf radio cards, motherboards, and other components, you can build devices that provide features comparable (or even superior) to most commercial implementations. Combining open hardware platforms with open source software can yield significant "bang for the buck" by providing custom, robust solutions for very low cost.

This is not to say that commercial equipment is inferior to a do-it-yourself solution.

By providing so-called "turn-key solutions", manufacturers not only save development time, but they can also allow relatively unskilled people to install and maintain equipment.

The chief strengths of commercial solutions are that they provide support and a (usually limited) equipment warranty. They also provide a consistent platform that tends to lead to very stable, often interchangeable network installations.

If a piece of equipment simply doesn't work or is difficult to configure or troubleshoot, a good manufacturer will assist you.

Should the equipment fail in normal use (barring extreme damage, such as a lightning strike) then the manufacturer will typically replace it.

Most will provide these services for a limited time as part of the purchase price, and many offer support and warranty for an extended period for a monthly fee. By providing a consistent platform, it is simple to keep spares on hand and simply "swap out" equipment that fails in the field, without the need for a technician to configure equipment on-site.

Of course, all of this comes at comparatively higher initial cost for the equipment compared to off-the-shelf components.

From a network architect's point of view, the three greatest hidden risks when choosing commercial solutions are vendor lock-in, discontinued product lines, and ongoing licensing costs.

It can be costly to allow the lure of ill-defined new "features" drive the development of your network.

Manufacturers will frequently provide features that are incompatible with their competition by design, and then issue marketing materials to convince you that you simply cannot live without them (regardless of whether the feature contributes to the solution of your communications problem).

As you begin to rely on these features, you will likely decide to continue purchasing equipment from the same manufacturer in the future.

This is the essence of vendor lock-in.

If a large institution uses a significant amount of proprietary equipment, it is unlikely that they will simply abandon it to use a different vendor.

Sales teams know this (and indeed, some rely on it) and use vendor lock-in as a strategy for price negotiations.

When combined with vendor lock-in, a manufacturer may eventually decide to discontinue a product line, regardless of its popularity. This ensures that customers, already reliant on the manufacturer's proprietary features, will purchase the newest (and nearly always more expensive) model. The long term effects of vendor lock-in and discontinued products are difficult to estimate when planning a networking project, but should be kept in mind.

Finally, if a particular piece of equipment uses proprietary computer code, you may need to license use of that code on an ongoing basis. The cost of these licenses may vary depending on features provided, number of users, connection speed, or other factors. If the license fee is unpaid, some equipment is designed to simply stop working until a valid, paid-up license is provided! Be sure that you understand the terms of use for any equipment you purchase, including ongoing licensing fees.

By using equipment that supports standards and open source software, you can avoid some of these pitfalls.

For example, it is very difficult to become locked-in to a vendor that uses open protocols (such as TCP/IP over 802.11a/b/g). If you encounter a problem with the equipment or the vendor, you can always purchase equipment from a different vendor that will interoperate with what you have already purchased. It is for these reasons that we recommend using proprietary protocols and licensed spectrum only in cases where the open equivalent (such as 802.11a/b/g) is not technically feasible.

Likewise, while individual products can always be discontinued at any time, you can limit the impact this will have on your network by using generic components.

For example, a particular motherboard may become unavailable on the market, but you may have a number of PC motherboards on hand that will perform effectively the same task.

Obviously, there should be no ongoing licensing costs involved with open source software (with the exception of a vendor providing extended support or some other service, without charging for the use of the software itself).

There have occasionally been vendors who capitalise on the gift that open source programmers have given to the world by offering the code for sale on an ongoing licensed basis, thereby violating the terms of distribution set forth by the original authors.

It would be wise to avoid such vendors, and to be suspicious of claims of "free software" that come with an ongoing license fee.

The disadvantage of using open source software and generic hardware is clearly the question of support.

As problems with the network arise, you will need to solve those problems for yourself. This is often accomplished by consulting free online resources and search engines, and applying code patches directly.

If you do not have team members who are competent and dedicated to designing a solution to your communications problem, then it can take a considerable amount of time to get a network project off the ground.

Of course, there is never a guarantee that simply "throwing money at the problem" will solve it either.

While we provide many examples of how to do much of the work yourself, you may find this work very challenging. You will need to find the balance of commercial solutions and the do-it-yourself approach that works for your project.

In short, always define the scope of your network first, identify the resources you can bring to bear on the problem, and allow the selection of equipment to naturally emerge from the results.

Consider commercial solutions as well as open components, while keeping in mind the long- term costs of both.

When considering which equipment to use, always remember to compare the expected useful distance, reliability, and throughput, in addition to the price.

And finally, make sure that the radios you purchase operate in an unlicensed band where you are installing them, or if you must use licensed spectrum, that you have budget and permission to pay for the appropriate licenses.

Professional lightning protection

Lightning is a natural predator of wireless equipment. A map showing the global distribution of lightning from 1995 to 2003 is shown here following.

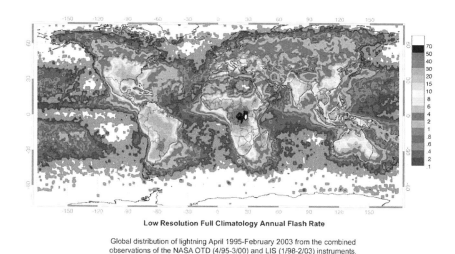

Low Resolution Full Climatology Annual Flash Rate

Global distribution of lightning April 1995-February 2003 from the combined observations of the NASA OTD (4/95-3/00) and LIS (1/98-2/03) instruments.

Figure HW 2: Global distribution of lightning from 1995 to 2003

There are two different ways lightning can strike or damage equipment: direct hits or induction hits.

Direct hits happen when lightning actually hits the tower or antenna. Induction hits are caused when lightning strikes near the tower.

Imagine a negatively charged lightning bolt.
Since like charges repel each other, that bolt will cause the electrons in the cables to move away from the strike, creating current on the lines.
This can be much more current than the sensitive radio equipment can handle.
Either type of strike will usually destroy unprotected equipment.

Figure HW 3: A tower with a heavy copper grounding wire.

Protecting wireless networks from lightning is not an exact science, and there is no guarantee that a lightning strike will not happen, even if every single precaution is taken. Many of the methods used will help prevent both direct and induction strikes.

While it is not necessary to use every single lightning protection method, using more methods will help further protect the equipment.

The amount of lightning historically observed within a service area will be the biggest guide to how much needs to be done.

Start at the very bottom of the tower. Remember, the bottom of the tower is below the ground. After the tower foundation is laid, but before the hole is backfilled, a ring of heavy braided ground wire should have been installed with the lead extending above ground surfacing near a tower leg.

The wire should be American Wire Gauge (AWG) #4 or thicker.

In addition, a backup ground or earthing rod should be driven into the ground, and a ground wire run from the rod to the lead from the buried ring.

It is important to note that not all steel conducts electricity the same way.

Some types of steel act as better electrical conductors than others, and different surface coatings can also affect how tower steel handles electrical current. Stainless steel is one of the worst conductors, and rust proof coatings like galvanizing or paint lessen the conductivity of the steel.

For this reason, a braided ground wire is run from the bottom of the tower all the way to the top. The bottom needs to be properly attached to the leads from both the ring and the backup ground rod.

The top of the tower should have a lightning rod attached, and the top of that needs to be pointed. The finer and sharper the point, the more effective the rod will be.

The braided ground wire from the bottom needs to be terminated at the grounding of the lightning rod. It is very important to be sure that the ground wire is connected to the actual metal. Any sort of coating, such as paint, must be removed before the wire is attached. Once the connection is made, the exposed area can be repainted, covering the wire and connectors if necessary to save the tower from rust and other corrosion.

The above solution details the installation of the basic grounding system. It provides protection for the tower itself from direct hits, and installs the base system to which everything else will connect.

The ideal protection for indirect induction lightning strikes are gas tube arrestors at both ends of the cable.

These arrestors need to be grounded directly to the ground wire installed on the tower if it is at the high end.

The bottom end needs to be grounded to something electrically safe, like a ground plate or a copper pipe that is consistently full of water. It is important to make sure that the outdoor lightning arrestor is weatherproofed.

Many arresters for coax cables are weatherproofed, while many arresters for CAT5 cable are not. In the event that gas arrestors are not being used, and the cabling is coax based, then attaching one end of a wire to the shield of the cable and the other to the ground wire installed on the towers will provide some protection.

This can provide a path for induction currents, and if the charge is weak enough, it will not affect the conductor wire of the cable. While this method is by no means as good a protection as using the gas arrestors, it is better than doing nothing at all.

Access Point Configuration

This section will provide a simple procedure for the basic configuration of WiFi Access Points and Clients by reviewing the main settings and analysing their effects on the behaviour of the network.
It will also suggest some practical tips and tricks and troubleshooting advice.

Before you start

When you receive some new wireless equipment take some time to get acquainted with its main features and make sure you:

- Download or otherwise obtain all **user's manuals** and **specification sheets** available for the devices you are going to deploy.
- If you have second-hand devices, be sure to receive full information on their current –or last-known– configuration (e.g. password, IP addresses, etc.)
- You should already have a plan on hand for the network you are going to deploy (including **link budget**, **network topology**, **channels** and **IP settings**).
- Be ready to take written notes of all settings that you are going to apply (especially **passwords**!)
- Make backups of **last known good** configuration files.

Getting in touch with the device

As a first step, it is important that you learn the meaning of all LEDs on the device.
The figure below shows the typical front of an Access Point, with several LEDs lit.

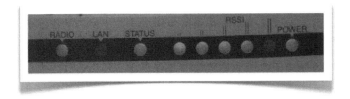

Figure HW 4: The front of a typical Access Point.

LEDs typically indicate:

- Presence of power
- Active ports / traffic (yellow/green colour)
- Error status (red colour)
- Received signal strength (LED bars, sometimes multicolour; some devices can even be set to light each LED at specific thresholds, e.g. *Ubiquiti*)

Sometimes, different meanings are associated with the same LED, using different colours and dynamics (e.g. LED is on/off/blinking at different speeds).

You should identify the different ports and interfaces on the device:

- Radio interfaces, sometimes called WLANs. They should have one or more antenna connectors (or non-detachable antennas).
- One or more Ethernet interfaces:
- One or more ports for local network (LAN)
- One port for uplink (also called WAN)
- Power input (5, 6, 7.5, 12V or other, usually DC). It is really (really) important that the power supply matches the voltage! Sometimes the power is provided to the device through the same UTP cable that carries the Ethernet data: this is called Power-over-Ethernet (PoE).
- Power button (not always present).
- Reset button (often "hidden" in a small hole, can be pressed using a straightened paperclip).

The reset button may have different effects (from a simple restart to a factory full reset) if pressed briefly vs. for a longer time. It can take 30 seconds or more to trigger a full reset.
NOTE: to fully reset (i.e. reset to factory settings) a device that is in an *unknown* state may be a painful task!
Be sure to keep written notes of critical parameters like the device IP address and network mask, and the administrator username and password. The following figure shows a common Linksys Access Point, with the power input, the network ports, the reset button and two antennas.

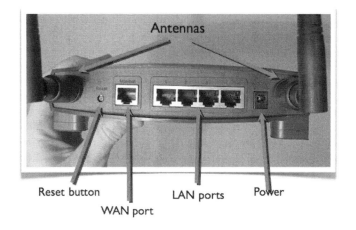

Figure HW 5: Linksys Access Point

User interfaces

You can interact (i.e. issue commands and change settings) with the Access Point in several ways, depending on the hardware you are using. The possible ways are the following:

- Graphical User Interface (web page)
- Graphical User Interface (proprietary software application)
- Command Line Interface (telnet, ssh)
- Software interface embedded in the system (when the AP/client is a computer or smartphone with a display and its own OS)

User interfaces: GUI (web page)

This system is used in Linksys, Ubiquiti and most modern Aps.
Once you are connected to the AP, you interact with the device using a normal browser.

Advantages: it works with most browsers and operating systems
Disadvantages: the static interface does not reflect changes immediately; poor feedback; can be incompatible with some web browsers; requires a working TCP/IP configuration. Some recent implementations (e.g. Ubiquiti, shown below) are very good and use modern dynamic web features to provide feedback and advanced tools.

Figure HW 6: Ubiquiti user interface.

User interfaces: GUI (software application)

In this case you need a special piece of software to interact with the device. This system is used in Mikrotik (called Winbox), Apple (called Airport Utility), Motorola (called Canopy), many old APs.

Advantages: usually powerful and appealing interfaces; allow batch configuration of multiple devices.

Disadvantages: proprietary solutions; usually available for only one operating system; software must be installed before starting configuration. Mikrotik Winbox, shown below, is a very powerful solution and can manage large networks.

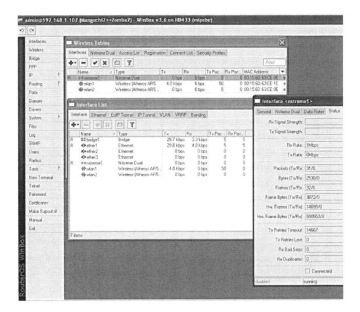

Figure HW 7: Mikrotik WinBox.

User interfaces: Command Line Interface (sometimes called text shell)

In this case you connect to the device using a serial or Ethernet connection, via telnet or ssh. ssh is much safer than telnet from a security perspective (the latter should be avoided if possible).

Configuration is performed with commands executed in the host operating system, usually a flavour of Linux or a proprietary OS, as shown on the next page.

This is system is used by Mikrotik (called RouterOS), Ubiquiti (called AirOS), high-end APs (Cisco) and embedded PC-based Aps.

Advantages: very powerful as it can be scripted.

Disadvantages: difficult to learn.

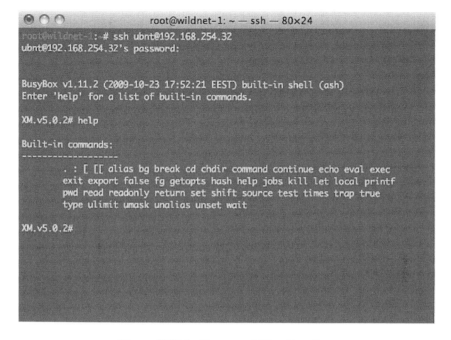

Figure HW 8: Command Line Interface.

Configure the AP

Before configuring the AP, remember to:

- Start from a known state, or reset the device to factory settings (always a good idea!).
- Connecting to the device via Ethernet is usually easier than via wireless. Most devices with a web GUI have a default IP configuration on the network 192.168.0.0, but this is not a rule! Read the manual.
- If convenient, upgrade the firmware to the latest stable version (but be careful!)
- Change the default admin username and password first!
- Change the device name with something that clearly identifies it (e.g. something like "AP_conference_room_3" or "hotspot_public_area"). This will help you to recognise the AP in future when you connect to it over the network.

- Firmware upgrade is often a risky procedure, be sure to adopt all precautions before attempting it - such as - connecting the device and the computer to a UPS, then do not perform the firmware update while doing other tasks on the computer, and check that you have a valid firmware binary image, READ THE INSTRUCTIONS CAREFULLY! If the procedure fails, you can end up with an unusable device that cannot be recovered (the so-called "brick").
- Remember to write down (and store in a safe place) all these settings, especially the admin username/password.

Configure the AP - IP layer

If you are confident with what you are doing, you can perform the IP configuration after the wireless setup to avoid the reconfiguration and reconnection of your PC or laptop. But in this way, if you make any mistake in both the wired and wireless setup, you may end up with an unreachable AP. We'd strongly advise that you do the critical configuration steps one at a time, checking the status of the device after each step. Remember to write down (and store in a safe place) all IP settings!

Configure the Ethernet interface of the AP according to your wired network setup:

- IP address/netmask/gateway *or* DHCP
- DNS address(es)
- Double check the new settings and apply them (sometimes you may have to reboot the AP)
- Now you may need to reconfigure your PC/laptop to match the new Ethernet setup, and reconnect to the AP.

Configure the AP - physical layer

- Configure the mode: *master* (or *access point* or *base station*). The device's mode can usually be configured as: "master" (also called "access point" or "base station" or "BS"), "client" (also called "managed" or "station" or "client station" or "CPE"), "monitor", "WDS" (Wireless Distribution System), and rarely some other variants.

- Configure the SSID (Service Set Identifier, the name of the wireless network created by the AP, up to 32 characters long): *it is best to choose a meaningful name.* Remember that "security through obfuscation is not real security", so a hidden or fake/random SSID does not add much security to your network.
- Configure the wireless channel, according to the local regulations and the result of the site survey.
- Do not use a channel that is already occupied by another AP or other sources of RF power. *You should already have planned the channel in advance, during the design phase.* The choice of the best channel is sometimes a hard task, and you may need to perform a site survey with software tools (wireless sniffers) or hardware spectrum analysers (like the WiSpy from Metageek and the AirView from Ubiquiti).
- Configure the transmit power and network speed (these values may also be set to "*automatic*" in some devices). The value of transmit power is also subject to local regulation, check in advance the maximum power allowed by law, and try always to use the minimum value that fit your needs, in order to avoid interference with other networks (yours or others).

The choice of network speed is limited to the values that are part of 802.11a/b/g/n standards (up to 54 Mbps), but some vendors created extensions to the standards (often called "turbo" modes) of 100 Mbps or higher. These are non-standard and may not be able to interoperate with equipment from other vendors.

Configuring "backwards compatibility" modes (such as supporting 802.11b on 802.11g networks) will reduce the overall throughput available to your fastest clients. The Access Point must send the preamble at a slower rate for 802.11b clients, and actual communications between the client and the AP happen at 802.11b speeds. This takes more time, and slows down the otherwise faster 802.11g clients.

Configure the AP – security

Security settings are often a hard choice, and it may be difficult to balance good protection from unintentional users with easy access for authorised users.

More complex security needs will require a more complex configuration and additional software.

Configure the security features of the network:

- No encryption (all traffic is in clear)
- WEP (*Wired Equivalent Privacy*), 40 or 104 bits keys, it is flawed and therefore deprecated
- WPA / WPA2 (*WiFi Protected Access*): PSK, TKIP and EAP
- Enable or disable (hide) the SSID broadcast ("beacon"). Hidden SSID and MAC filtering do not add much security, and are hard to maintain and a source of troubles for inexperienced users of the network.
- Enable or disable an Access Control List (based on MAC addresses of clients). MAC filtering is a weak security measure:– A malicious client can capture packets and find out which MAC addresses have the right to associate– It can then change its own MAC address to one of the accepted ones and "fool" the access point.

For more information about how to design the security of your wireless network please read the chapter called **Security in Wireless Networks.**

Configure the AP - routing/NAT

Advanced IP layer and routing configuration features are often included in modern access points. This can include functionality for routing and Network Address Translation (NAT), in addition to basic bridging.

Advanced IP configuration includes:

- Static routing
- Dynamic routing
- NAT (masquerading, port forwarding)
- Firewalling
- Some APs can also act as file servers and print servers (external HD and printers can be connected via USB)

Configure the AP – advanced

A few more advanced settings may be available for your AP, depending on the model/vendor/firmware/etc.:

- Beacon interval
- RTS/CTS. RTS/CTS (ready to send, clear to send) mechanism can help alleviate the problem of hidden nodes (clients that can "hear" the AP but not the other clients, therefore creating interferences).
- Fragmentation. Configuration of the fragmentation can be used to increase performance in the case of low signal areas, those with marginal coverage, or long links.
- Interference robustness
- Vendor extensions to the WiFi standards
- Other settings for long distance links (10 to 100 kilometres) and better security.

Configure the client

The client side configuration is much simpler:

- Configure the mode: **client** (or **managed, station, client station, CPE**)
- Configure the SSID of the network to be joined
- The channel, speed, and other parameters will be set automatically to match the AP
- If WEP or WPA is enabled on the AP, you will have to enter the matching password (key)
- Clients may also have additional (often vendor-specific) settings. *For example, some clients can be configured to associate only with an AP with a specified MAC address.*

Hints - working outdoors

- You should try to configure the devices (both APs and clients) well in advance and in a comfortable place such as a laboratory. Working outdoors is more difficult and may lead to mistakes (**"on-site" configuration = trouble**).

- If you **must** configure devices outdoors, be sure to have enough battery charge on your laptop, to carry all the required information with you (on paper, not only in electronic format!) and to carry a notepad to take notes. Good documentation is paramount for future maintenance in the field!

Troubleshooting

- Organise your work in logical steps and follow them.
- Read the manual, study the meaning of parameters and settings, run tests and experiments (don't be scared!).
- In the case of problems, do a factory reset and try again.
- If the problem persists, try again **changing one parameter/setting at a time.**
- It still doesn't work? Try to search on the web for relevant keywords (name of the device, etc.), search in forums and on manufacturer / vendors websites.
- Upgrade the firmware to the latest version.
- If you still have problems try with a different client/AP, to check if you have a hardware issue with the original one.

12. INDOOR INSTALLATION

Introduction

Previous editions of this book have focused on wide-area outdoor wireless as a means of connecting communities with each other and to the Internet. However, with the availability of WiFi access points for a low price and the proliferation of portable devices that are WiFi capable, WiFi has become the de facto standard for indoor network access in Enterprises and schools. This chapter introduces the main focus points in choosing and installing WiFi networks indoors.

Preparations

Before installing a wireless LAN it is a good idea to first think things through a bit:

- What is it that you are planning to do with the wireless network? Is it an addition to the wired network or a replacement for it? Are you going to run applications over the network that are not delay tolerant or sensitive to bandwidth variations (like voice and video conferencing).
- The main difference between indoor and outdoor wireless is the absorption and reflection of radio waves by the building itself. What building features do you need to take into account? Do the walls have metal, water or heavy concrete in them? Do windows have metal in them (e.g. metal coating or metal grids)? Is the building long and stretched or compact?
- Do you expect users to be mainly static, or will they move a lot? And when they move, is it important to have uninterrupted handover (this means, a handover so quick that you will not notice the interruption of a voice call)?
- Are there good places to hang the Access Points? Are wired sockets and electricity for the APs readily available? Is electricity stable? If not, even indoor APs might need stable solar/battery power supplies and/or UPS.
- Are there sources of interference, like ad-hoc APs brought in by users, bluetooth devices, microwaves?

Bandwidth requirements

Step one in designing an indoor wireless network is to determine the need in terms of the number of concurrent users to support, the number and type of devices and the type of applications they are running. It is also important to understand the distribution of the users. Lecture halls or meeting rooms have different usage patterns than corridors. A wireless network that is hardly used and that needs to support a low number of users is easy to deploy and will not run into trouble easily. The trouble starts when the number of users and their use of the network increases. This chapter therefore focuses on high-density wireless networks.

To give you an idea for the bandwidth requirement for some typical applications:

web surfing:	500 - 1000 kb/s
audio:	100 - 1000 kb/s
streaming video:	1 - 4 Mb/s
file sharing:	1 - 8 Mb/s
device backup:	10 - 50 Mb/s

Typical installations in an office environment are dimensioned to support 20-30 users per cell and have about 1 access point per 250-500 square metres, but as mentioned before, depending on characteristics of the environment this may not be sufficient. In a dense environment there may be up to 1 device per 20 square metres. Bottom line, you need to calculate the throughput needed per coverage area.

So if you have let's say 10 users in a 100 square metre area, of which 8 are surfing the web and 2 are watching online video you will need: 8 * 1000 kb/s + 2 * 4000 kb/s = 16000 kb/s for the 100 square metre area or 160 kb/s per square metre.

Frequencies and data rates

The 2.4 GHz and the 5 GHz solutions differ in a few key aspects. The 2.4 GHz band has a better range and less attenuation and is supported in most devices.

The main downside of the 2.4 GHz band is that there are only 3 non-overlapping channels, which severely limits the the number of access points that can be placed in a certain area.

This is unfortunate since making smaller cells (formed by having the APs broadcast with less power) is the easiest way of creating more throughput per area.

Note: sometimes 4 slightly overlapping channels are advised, but research shows that in fact this decreases performance. Performance in general, degrades fast with overlapping channels (co- channel interference). The 5 GHz band on the other hand has a worse range but has in most geographies around 20 channels which makes it much easier to deploy without interference from adjacent channels.

The other important element is the choice of WiFi standard, considering that average throughput in Mb/s for the most common technologies are:

11b:	7.2 Mb/s
11g:	25 Mb/s
11a:	25 Mb/s
11n:	25 - 160 Mb/s

It should be noted that performance drops when, for example, both 802.11b and 11g devices are served by the same AP. In a network where client devices are using a mix of 802.11g and 11b, the AP will shift down to lower speeds.

5 GHz is a preferred choice for high performance and high density networks. As you would like to limit the coverage of each AP to one small well defined area anyway, signal attenuation by walls etc is an advantage rather than a problem.

The deployment of 2.4 GHz for the majority of devices, combined with 5 GHz for the "important devices" is worth considering too.

Access Points choice and placement

When it comes to choosing Access Points (APs) for indoor wireless there are essentially 2 architectural choices: controller based and "fat clients". Fat clients are stand-alone APs that have all the intelligence on board to manage a WiFi network (choosing SSIDs, encryption method, routing/switching etc.). The controller-based solution on the other hand has APs that implement the minimal functionality to offer a wireless service along with a central controller that is common for all APs at a location.

The central controller also has all the intelligence and all traffic from the APs directed to it.

The choice between one of the two architectures is a trade-off between cost, ease of management and scalability. In general one can say that the more complex the environment and the larger the size the more attractive it becomes to run a controller-based solution.

Access Points should in general be placed in the areas with a high density of users, the signal will probably "leak" sufficiently to also serve the less dense areas. Unfortunately the overall performance of the system will mostly be determined by the clients, not the APs, so the placement of the APs, while important, can only do so much for the performance of the overall system. Other radio sources in the WiFi bands have a very strong influence on the performance of the WiFi network, so in general, if possible the APs should, as much as possible, be isolated from other radio sources, by using walls, ceilings and people as "shields".

In order to improve performance it is possible to use external antennas. Omni directional antennas are the most commonly used and they provide a coverage area roughly circular around the AP. In most indoor cases, however, APs will be installed on walls, ceilings, or columns, and omnidirectional antennas are a bad choice, if you look at where the radio waves are going, and where the users are.

So for these cases in which the AP is not at the centre of the area to be covered, directional antennas are an alternative. Some hotels and conference halls, for example, place small directional antennas at the corners of large, open areas or overhead to provide a "canopy" or "umbrella" of signal coverage in large spaces. Keep in mind that the many reflections typically encountered in an indoor environment make it difficult to try to fully control a specific coverage.

Access Points can be mounted on the ceiling, on the walls or in furniture, each with different characteristics. Ceiling mounting gives a good blanket coverage, wall mounting often gets the APs closer to the users and APs under tables or chairs or within furniture can use the natural isolation to create small cells with little interference from neighbouring APs, but there might be concerns about the possibility of harmful effects from the radiation emitted.

Lastly, for really demanding high performance networks, APs with smart adaptive antenna technology might be an option. They come at a price, but offer the advantage of adapting the radio signal to the locations of users dynamically - they will direct the radio waves to where they are needed, at each point in time.

SSID and Network Architecture

Indoor networks are likely to serve many concurrent users. Larger complexes like a university campus typically consist of many buildings, each with their own indoor network, and outdoor networks in between.

It is therefore important to make a good plan for your SSIDs.
Note that the SSID defines the broadcast domain on Layer 2 of the network. SSID planning needs to play together with your Layer 3 network architecture. If you would like users to roam seamlessly across your whole wireless network area, within or even beyond one building, then all APs should offer the same SSID, for example "UniversityWireless", or "eduroam" for a university that wants to participate in the global roaming service that eduroam offers.
However, users who stay within one SSID will not require or request a new DHCP lease, so you will have to accomodate all users within ONE Layer 3 subnet.
For a large campus, this might require a large flat subnet for all wireless users.
This is a trade-off situation - you can either have huge subnets with seamless roaming, or a more manageable subnet architecture with separate SSIDs such as "Library", "LectureHall", "Cafeteria", ...

Post Installation

Now that the infrastructure is in place it is important to make sure that everything works as expected and remains like that. This can be done in the form of a site survey, measuring signal strengths and throughput. But in the end the main reason to install a wireless network is to serve the users of it, so listening to users complaints or issues is equally important. Demand constantly changes and so does the state-of-the- art. It is important to keep up to date with user requirements and match that to planned upgrades of the technology you are deploying.

13. OUTDOOR INSTALL

Although WiFi technology was designed for local area networks, its impact in developing countries is more dramatic in long-distance applications.

In developed countries, fibre optic cables offering large bandwidths have been installed satisfying the communication needs of most cities. The penetration of optical fibre in the developing world is not nearly as great and nowhere near enough to cover the needs. And the cost of its expansion often does not meet the ROI (Return on Investment) goals of telcos within a reasonable period of time. Wireless technologies, on the other hand, have been much more successful in developing countries and the potential for increasing the penetration using wireless networks is enormous.

Telcos have installed traditional microwave radio links in most countries. This is a mature technology that offers high reliability and availability reaching 99.999%. However, these systems cost many thousands of dollars and require specially trained personnel for installation.

Satellite systems have proved well suited for broadcast traffic like TV and certain other applications. However, satellite solutions are still expensive for bidirectional traffic, while WiFi is quite cost effective in outdoor point to point networks as well as in typical access networks where a Base Station (BS) is serving many Clients/CPEs (point to multipoint). In this chapter we will be concentrating on the outdoor long- distance point to point links.

Two significant hurdles had to be overcome before applying WiFi to long distance: Power budget limitations and timing limitations. The remaining limitations for using WiFi over long distances are the requirement for the existence of radio line of sight between the endpoints and the vulnerability to interference in the unlicensed band. The first limitation can often be addressed by taking advantage of the terrain elevations, or by using towers to overcome obstacles such as the curvature of the earth and to provide Fresnel zone clearance.

For indoor applications, line of sight is not required since the stations are very close together and most obstacles can be cleared by reflections on walls, ceiling, etc. But for long distance applications, line of sight is absolutely critical. The second limitation is less pronounced in rural areas and can be alleviated by migrating to the less crowded 5 GHz band.

The power budget issue can be handled by using high gain antennas and powerful and highly sensitive radios attached directly to the antenna to avoid RF cable loss. The timing limitation has to do with the media access techniques. WiFi uses a random access method to share the communications medium. This makes it subject to collisions, which cannot be detected over the air, therefore the transmitter relies on receiving an acknowledgment for every successfully received frame.

If, after a specified amount of time, called the "*ACKtimeout*", the acknowledge frame is not received, the transmitter will resend the frame.

Since the transmitter will not send a new frame until the ACK for the previous one has been received, the *ACKtimeout* must be kept short.

This works well in the original scenario intended for WiFi (indoor networks), in which the propagation time of 33.3 microseconds per kilometre is negligible, but breaks down for links over a few kilometres.

Although many WiFi devices do not have provisions for modifying the *ACKtimeout*, newer equipment meant for outdoor applications (or third party firmware like Open WRT) will give you this possibility, often by means of a *distance* field in the GUI (Graphical User Interface).

Changing this parameter will allow for a reasonable throughput, which will anyway decrease proportionally to the distance. The contention window slot-time also needs to be increased to adapt to longer distances.

Other manufacturers have chosen to move from random access to *Time Division Multiple Access* (*TDMA*) instead. TDMA divides access to a given channel into multiple time slots, and assigns these slots to each node on the network. Each node transmits only in its assigned slot, thereby avoiding collisions. In a point to point link this provides a great advantage since ACKs are not needed because each station takes turns at transmitting and receiving.

While this method is much more efficient, it is not compliant with the WiFi standard, so several manufacturers offer it as an optional proprietary protocol, alongside the standard WiFi. WiMAX and proprietary protocols (such as Mikrotik Nstreme, or Ubiquiti Networks AirMAX) use TDMA to avoid these ACK timing issues.

The 802.11 standard defines the receiver sensitivity as the received signal level required to guarantee a BER (Bit Error Rate) below 10^{B5}.

This determines the amount of energy per bit required to overcome the ambient noise plus the noise generated by the receiver itself. As the number of bits/second transmitted increases, more receiver power will be needed to provide the same energy per bit. Therefore the receiver sensitivity decreases as the transmitter rate increases, so to maintain the same signal/noise ratio as the distance increases the throughput diminishes, or, alternatively, for longer distances one should choose lower data rates to compensate for the reduction of the signal strength with distance.

What is needed for a long distance link?

There are four aspects that need to be considered to adapt WiFi devices to long distance: increase the radio dynamic range; increase the antenna gain; decrease the antenna cable loss; and make provisions for the the signal propagation time.

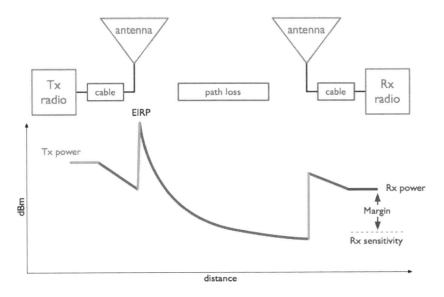

Figure OI 1: Power in dBm vs distance in a radio link (Power budget).

The graph above shows the power level at each point in a wireless link. The transmitter provides some amount of power.

A small amount is lost in attenuation between the transmitter and the antenna in the RF cable or waveguide. The antenna then focuses the power, providing a gain. At this point, the power is at the maximum possible value for the link. This value is called the EIRP (Equivalent Isotropic Radiated Power) since it corresponds to the power that a transmitter would have to emit if the antenna had no gain.

Between the transmitting and receiving antennas there are free space and environmental losses, which increase with the distance between the link endpoints. The receiving antenna provides some additional gain. Then there is a small amount of loss between the receiving antenna and the receiving radio.

If the received amount of energy at the far end is greater than the receive sensitivity of the radio, the link is possible. Increasing the transmission power can lead to violations of the regulatory framework of the country.
Increasing the antenna gain is by far the most effective way to improve range. Make sure that the radio to be employed has connectors for an external antenna (some devices have an embedded or otherwise non removable antenna).

Decreasing loss in antenna cables is still an important issue, and the most radical way to attain it is to place the radio outside, directly attached to the antenna, employing a weatherproof box. Often this lends to powering the radio using PoE (Power over Ethernet).

Improving the receiver sensitivity implies choosing a model with better performance, or settle for lower transmission speeds where sensitivity is higher.

Although high gain antennas can be expensive, in many countries one can find satellite antennas that are no longer being used and can be modified for the WiFi bands.

In a perfect world, we would use the highest gain antennas with the loudest and most sensitive radios possible. But a number of practical considerations make this impossible. Amplifiers introduce an additional point of failure, in addition they might violate maximum power permitted by local regulations and add noise in reception, so they should be avoided.

High power transmitters are available from many manufacturers that offer up to 1 W of output power which could be used instead of amplifiers in those countries where this is legal.

In general, it is better to use high gain antennas than high transmitter power. Greater antenna gain will help both in transmission and reception making a double impact in the link budget. It will also cause less interference to other users and receive less interference from other users and limit multipath effects But a high antenna gain implies a very narrow beamwidth, which means that special alignment techniques are required.

Antenna alignment

For short distances, when the corresponding antenna is visible, the antenna alignment procedure reduces to pointing the antenna in the direction of the correspondent, both in the horizontal plane (azimuth) and in the vertical plane (elevation). This should suffice to establish the connection. Once the connection is attained a fine adjustment can be made by reading the RSSL (Receiver Signal Strength Level) in the local radio. This value is provided by the user interface, and can also be obtained from programs like *netstumbler*. The procedure consists of moving the antenna in the horizontal plane in small steps while reading the RSSL. Do not touch the antenna when reading, since your body will affect the measurement. Once satisfied that a maximum value is obtained, the procedure is repeated in the vertical plane, moving the antenna first up and then down until a maximum value of received power is obtained, at which point the bolts that secure the antenna are tightened. This is all that is needed to aim a client device at an Access Point or Base Station. If you have a point to point link, the same procedure should be repeated at the other end of the link.

For long distances and when the other end of the link is not visible, some extra steps are required. First, the horizontal direction (bearing) to aim the antenna must be obtained from the coordinates of the end points. Then a compass is used to determine the direction in which the antenna should be aimed.
Keep in mind that in general there is a difference between the *magnetic* bearing measured by the compass and the *geographical* bearing obtained from the coordinates of the end points or from a map.

This difference is called the *magnetic declination*, it can be very significant in some places and must be accounted for to properly aim the antenna. Fig. OI 2 shows the 10° difference between the magnetic north shown by the compass and the geographical or true north indicated by the brass plate.

Figure OI 2: Difference between Magnetic and Geographical North at El Baul, Venezuela in 2006.

Keep in mind that iron and other magnetic metals will affect the reading of the compass, so stay away from those when making the measurement. If the antenna is to be mounted in an steel tower, it might be impossible to get an accurate reading near it. Instead, one must walk away a certain distance, use the compass to determine the direction the antenna must be aimed and then try to locate some easily recognisable object that can be used as a reference for pointing the antenna at later. Since the beamwidth of a highly directive antenna might be just a few degrees, after pointing with the compass we need to do some fine adjustment for the proper aiming of the antenna by measuring the strength of the received signal.

Unfortunately the RSSL indicated by the radio software will only work after a proper packet has been satisfactorily received and decoded, and this will only happen when the antenna is well aimed.

So we need an instrument that can reveal the strength of the received signal independently from the modulation that it might have.

The instrument needed for this task is the Spectrum Analyser.

There are a great variety of spectrum analysers on the market, some of them costing thousands of dollars, but if we are only interested in the WiFi bands we can make do with some inexpensive solutions like the following:

"RF Explorer" offers inexpensive devices for several frequency bands.
The "RF Explorer model 2.4G" costs 120$ from
http://www.seeedstudio.com/depot/-p-924.html?cPath=174
and is a stand alone unit that can measure signals from 2.4 to 2.485 GHz, with a sensitivity of -105 dBm. It has an SMA connector for the antenna and therefore is well suited for antenna alignment.

"WiSpy" is a spectrum analyser in a USB dongle that attaches to a laptop. You will need the models with SMA RP connector, there is one for 2.4 GHz moderately priced and another one that covers both the 2.4 and the 5 GHz bands sold for 600$ at www.metageek.net.

"Ubiquiti Networks" , www.ubnt.com, used to sell USB dongle spectrum analysers for 2.4 GHz at 70$.

Unfortunately they seem to have discontinued this product after incorporating the spectrum analyser capability in their M series radios.
So when using these radios you can take advantage of their "airView" alignment tool. In principle one of these inexpensive radios like the "Bullet M" which comes with a N Male connector can also be used to align antennas for other radios in both the 2.4 and 5 GHz bands.
Unfortunately the digitally modulated signal transmitted by WiFi radios is not well suited for antenna alignment, since its power is spread over the 20 MHz bandwidth. For antenna alignment a single frequency with a stable output power is required.
This type of signal is produced by a microwave signal generator, but they are quite expensive.
The "RF Explorer model 2.4G" incorporates a 2.4 GHz signal generator, but the maximum output power of 1 dBm is not well suited for long distance antenna pointing.Instead, we have repurposed devices called "video senders", meant for transmitting video signals, which act as powerful microwave single frequency signal sources when no modulation is applied.

They are available for both the 2.4 GHz and the 5 GHz bands with output power up to 33 dBm. For our purposes it is necessary to buy a model with an antenna connector, so that we can attach our own antenna. There are many vendors to choose from, see for instance: http://www.lightinthebox.com/Popular/Wifi_Video_Transmitter.html

As an example of long distance link using modified WiFi devices, we can mention an experiment performed in April 2005 in Venezuela between Pico del Aguila (8.83274638° N, 70.83074570°W,4100 m elevation) and El Baul (8.957667° N, 68.297528° W, 155 m elevation).

Using the Radio Mobile software, we find that the distance to El Baul is 280 km, the azimuth is 97°, the antenna elevation angle is -2.0°, and the place at which the beam is closest to the ground happens at 246 km, where it clears 1.7 times the first Fresnel zone at the 2.412 GHz frequency.

Fig. OI 3 shows the output of the program:

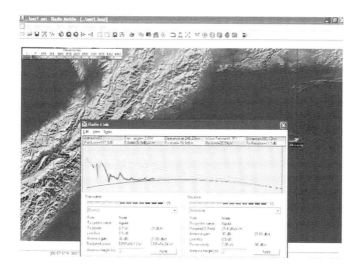

Figure OI 3: Profile of a 280 km path over which standard WiFi gear with OpenWRT firmware which allows for the ACKtimeout increase was used to transfer files at about 65 kb/s in April 2006 between Pico del Aguila and El Baul in Venezuela.

Notice that the earth curvature is quite apparent, and was overcome because one of the stations was at 4100 m altitude and the other at 155 m. Frequency was 2412 MHz, output power 100 mW, antenna gain around 30 dBi. Streaming video was successfully transmitted despite the limited bandwidth.

A year later the experiment was repeated with the same WiFi gear but with commercial 32 dBi antennas at both ends and similar results were obtained. Then, another type of firmware developed by the TIER group at UC Berkeley University that implements TDD (Time Division Duplexing) was tried which showed a remarkable bidirectional throughput of 6 Mbit/s with standard 802.11b hardware.

Moving the remote site to a 1400 m high hill called Platillon (9.88905350°N, 67.50552400°W), provided a 380 km testbed over which the experiment was again successfully repeated as described in the **Case Studies** section in this book.

This can be illustrated by using an on-line version of Radio Mobile, available at http://www.cplus.org/rmw/rmonline.html, which is simpler to use, although it has some limitation as compared with the downloadable version. One must register in the site, enter the coordinates of the points over which the radio link has to be established, the power values for the radios and the antennas gains and height, and the software will fetch the relevant elevation data required to perform the simulation of the link.

Keep in mind that only radio amateurs frequency are supported in the web version, so 2.3 GHz should be used instead of 2.4 GHZ, but the results are close enough and where validated by the experiment on the field.

In Figure OI 4 we show the output of the Radio Mobile on- line for this experiment that can be replicated by the reader as an exercise.

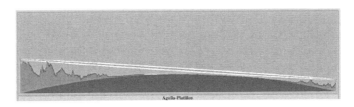

Figure OI 4: Profile of a 380 km test at 2.4 GHz performed in April and August 2007, Venezuela.

Notice that the earth curvature is even more noticeable over the 380 km path, but the height of the end points combined with flat land in between allows for ample clearance of the first Fresnel zone.

Figure OI 5 shows the numerical values of the Radio Mobile on-line simulation:

Figure OI 5: Results of the Radio Mobile on-line simulation for the 380 km link between Aguila and Platillon.

14. OFF-GRID POWER

Solar Power

This chapter provides an introduction to standalone photovoltaic systems. Standalone systems work without connection to an established power grid. This chapter presents the basic concepts of the generation and storage of photovoltaic solar energy. We will also provide a method for designing a solar system with limited access to information and resources. This chapter only discusses solar systems for the direct production of electricity (photovoltaic solar energy). Thermal solar energy systems are beyond the scope of this chapter.

Solar Energy

A photovoltaic system is based on the ability of photovoltaic panels to convert sun radiation directly into electrical energy. The total amount of solar energy that lights a given area is known as irradiance (G) and it is measured in watts per square metre (W/m^2). The instantaneous values are normally averaged over a period of time, so it is common to talk about total irradiance per hour, day or month. The amount of irradiance that arrives at the surface of the Earth varies due to natural weather variations and depends on the location. Therefore it is necessary to work with statistical data based on the "solar history" of a particular place. For many areas it can be difficult to acquire detailed information, so we will need to work with approximate values in this case.

A few organisations produce maps that include average values of daily global irradiation for different regions. These values are known as peak sun hours or PSHs. You can use the PSH value for your region to simplify your calculations. One unit of "peak sun hours" corresponds to a radiation of 1000 Watts per square metre for the duration of an hour. If an area has 4 PSH per day in the worst of the months, it means we can expect a daily irradiation of 4000 Wh/m^2. Low resolution PSH maps/calculation tools are available from a number of online sources such as: http://www.wunderground.com/calculators/solar.html

For more detailed information, consult a local solar energy vendor or weather station.

Airports normally record meteorological data including insolation.

What about wind power?

It is possible to use a wind generator in place of solar panels when an autonomous system is being designed for installation on a hill or mountain. To be effective, the average wind speed over the year should be at least 3 to 4 metres per second, and the wind generator should be 6 metres higher than other objects within a distance of 100 metres.

A location far away from the coast usually lacks sufficient wind energy to support a wind powered system.

Generally speaking, photovoltaic systems are more reliable than wind generators, as sunlight is more available than consistent wind in most places. On the other hand, wind generators are able to charge batteries even at night, as long as there is sufficient wind. It is of course possible to use wind in conjunction with solar power to help cover times when there is extended cloud cover, or when there is insufficient wind. In the Highlands and Islands of Scotland there is a project that uses both solar and wind power generation. See the following link for more information: http://www.wirelesswhitespace.org/projects/wind-firenewable-energy-basestation.aspx

For most locations however, the cost of a good wind generator is not justified by the meagre amount of power it will add to the overall system. This chapter will therefore focus on the use of solar panels for generating electricity.

Photovoltaic system components

A basic photovoltaic system consists of four main components: solar panel, battery, regulator, and load.

The panel is generating electricity. The battery stores electrical energy. The regulator protects the battery against excessive charge and discharge. The load refers to any device that requires electrical power. It is important to remember that solar panels and batteries produce direct current (DC). If the range of operational voltage of your equipment does not fit the voltage supplied by your battery, it will also be necessary to include some type of converter. If the equipment that you want to power uses a different DC voltage than the one supplied by the battery, you will need to use a DC/DC converter. If some of your equipment requires AC power, you will need to use a DC/AC converter, also known as an inverter. Every electrical system should also incorporate various safety devices in the event that something goes wrong.

These devices include proper wiring, circuit breakers, surge protectors, fuses, ground rods, lighting arrestors, etc.

The solar panel

The solar panel is composed of solar cells that collect solar radiation and transform it into electrical energy. This part of the system is sometimes referred to as a solar module or photovoltaic generator.

Solar panel arrays can be made by connecting a set of panels in series and/or parallel in order to provide the necessary energy for a given load. The electrical current supplied by a solar panel varies proportionally to the solar radiation. This will vary according to climatic conditions, the hour of the day, and the time of the year.

Figure OGP 1: A solar panel

Several technologies are used in the manufacture of solar cells.

The most common is crystalline silicon, and can be either monocrystalline or polycrystalline. Amorphous silicon can be cheaper but is less efficient at converting solar energy to electricity. With a reduced life expectancy and a 6 to 8% transformation efficiency, amorphous silicon is typically used for low power equipment, such as portable calculators.

New solar technologies, such as silicon ribbon and thin film photovoltaics, are currently under development. These technologies promise higher efficiencies but are not yet widely available.

The battery

The battery stores the energy produced by the panels that is not immediately consumed by the load. This stored energy can then be used during periods of low solar irradiation. The battery component is also sometimes called the accumulator. Batteries store electricity in the form of chemical energy. The most common type of batteries used in solar applications are maintenance-free lead-acid batteries, also called recombinant or VRLA (valve regulated lead acid) batteries.

Figure OGP 2: A 200 Ah lead-acid battery. The negative terminal was broken due to weight on the terminals during transportation.

Aside from storing energy, sealed lead-acid batteries also serve two important functions:

- They are able to provide an instantaneous power superior to that of the array of panels. This instantaneous power is needed to start some appliances, such as the motor of a refrigerator or a pump.
- They determine the operating voltage of your installation.

For a small power installation and where space constraints are important, other types of batteries (such as NiCd, NiMh, or Li-ion) can be used. These types of batteries need a specialised charger/regulator and cannot directly replace lead-acid batteries.

The regulator

The regulator (or more formally, the solar power charge regulator) assures that the battery is working in appropriate conditions. It avoids overcharging or overdischarging the battery, both of which are very detrimental to the life of the battery. To ensure proper charging and discharging of the battery, the regulator maintains knowledge of the state of charge (SoC) of the battery. The SoC is estimated based on the actual voltage of the battery.

By measuring the battery voltage and being programmed with the type of storage technology used by the battery, the regulator will know the precise points where the battery would be overcharged or excessively discharged.

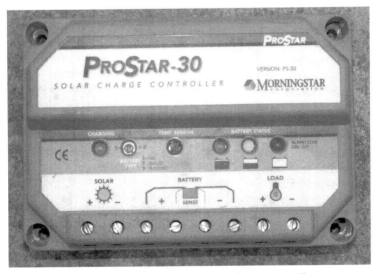

Figure OGP 3: A 30 Amp solar charge controller

The regulator can include other features that add valuable information and security control to the equipment. These features include ammeters, voltmeters, measurement of ampere-hour, timers, alarms, etc. While convenient, none of these features are required for a working photovoltaic system.

The converter

The electricity provided by the panel array and battery is DC at a fixed voltage. The voltage provided might not match what is required by your load. A direct/alternating (DC/AC) converter, also known as inverter, converts the DC current from your batteries into AC. This comes at the price of losing some energy during the conversion. If necessary, you can also use converters to obtain DC at voltage level other than what is supplied by the batteries. DC/DC converters also lose some energy during the conversion. For optimal operation, you should design your solar-powered system so that the generated DC voltage matches the load.

Figure OGP 4: An 800 Watt DC/AC converter (power inverter)

The load

The load is the equipment that consumes the power generated by your energy system.

The load may include wireless communications equipment, routers, workstations, lamps, TV sets, VSAT modems, etc.

Although it is not possible to precisely calculate the exact total consumption of your equipment, it is vital to be able to make a good estimate.

In this type of system it is absolutely necessary to use efficient and low power equipment to avoid wasting energy.

Putting it all together

The complete photovoltaic system incorporates all of these components. The solar panels generate power when solar energy is available.
The regulator ensures the most efficient operation of the panels and prevents damage to the batteries. The battery bank stores collected energy for later use.
Converters and inverters adapt the stored energy to match the requirements of your load. Finally, the load consumes the stored energy to do work.

When all of the components are in balance and are properly maintained, the system will support itself for years.

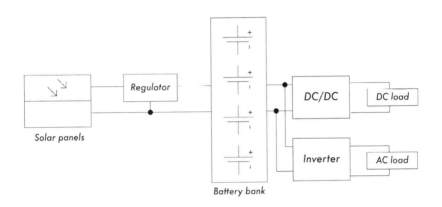

Figure OGP 5: A solar installation with DC and AC loads

We will now examine each of the individual components of the photovoltaic system in greater detail.

The solar panel

An individual solar panel is made of many solar cells.
The cells are electrically connected to provide a particular value of current and voltage.
The individual cells are properly encapsulated to provide isolation and protection from humidity and corrosion.

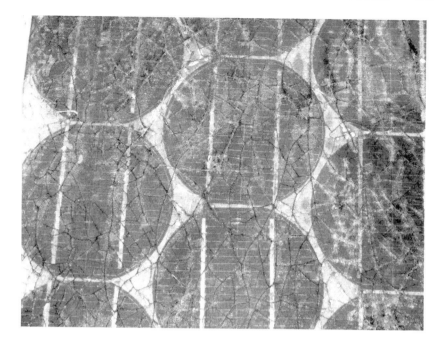

Figure OGP 6: The effect of water and corrosion in a solar panel

There are different types of modules available on the market, depending on the power demands of your application. The most common modules are composed of 32 or 36 solar cells of crystalline silicon. These cells are all of equal size, wired in series, and encapsulated between glass and plastic material, using a polymer resin (EVA) as a thermal insulator.

The surface area of the module is typically between 0.1 and 0.5 m^2. Solar panels usually have two electrical contacts, one positive and one negative.

Some panels also include extra contacts to allow the installation of bypass diodes across individual cells.

Bypass diodes protect the panel against a phenomenon known as "hot-spots". A hot-spot occurs when some of the cells are in shadow while the rest of the panel is in full sun. Rather than producing energy, shaded cells behave as a load that dissipates energy. In this situation, shaded cells can see a significant increase in temperature (about 85 to 100°C.)

Bypass diodes will prevent hot-spots on shaded cells, but reduce the maximum voltage of the panel. They should only be used when shading is unavoidable. It is a much better solution to expose the entire panel to full sun whenever possible.

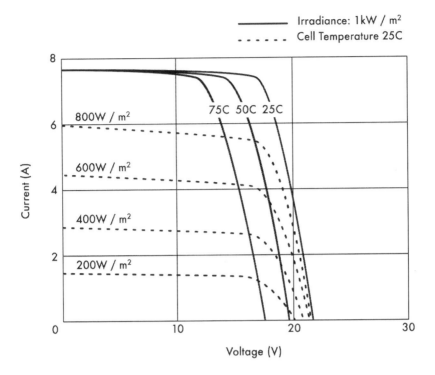

Figure OGP 7: Different IV Curves. The current (A) changes with the irradiance, and the voltage (V) changes with the temperature.

The electrical performance of a solar module is represented by the IV characteristic curve, which represents the current that is provided based on the voltage generated for a certain solar radiation.

The curve represents all the possible values of voltage-current.
The curves depend on two main factors: the temperature and the solar radiation received by the cells.
For a given solar cell area, the current generated is directly proportional to solar irradiance (G), while the voltage reduces slightly with an increase of temperature.

A good regulator will try to maximise the amount of energy that a panel provides by tracking the point that provides maximum power (V x I). The maximum power corresponds to the knee of the IV curve.

Solar Panel Parameters

The main parameters that characterise a photovoltaic panel are:

1. **Short Circuit Current** (I_{SC}): the maximum current provided by the panel when the connectors are short circuited.
2. **Open Circuit Voltage** (V_{OC}): the maximum voltage that the panel provides when the terminals are not connected to any load (an open circuit). This value is normally 22 V for panels that are going to work in 12 V systems, and is directly proportional to the number of cells connected in series.
3. **Maximum Power Point** (P_{max}): the point where the power supplied by the panel is at maximum, where $P_{max} = I_{max} \times V_{max}$.

The maximum power point of a panel is measured in Watts (W) or peak Watts (W_p). It is important not to forget that in normal conditions the panel will not work at peak conditions, as the voltage of operation is fixed by the load or the regulator. Typical values of V_{max} and I_{max} should be a bit smaller than the I_{SC} and V_{OC}.

1. **Fill Factor** (FF): the relation between the maximum power that the panel can actually provide and the product $I_{SC} \cdot V_{OC}$. This gives you an idea of the quality of the panel because it is an indication of the type of IV characteristic curve. The closer FF is to 1, the more power a panel can provide. Common values usually are between 0.7 and 0.8.
2. **Efficiency** (η): the ratio between the maximum electrical power that the panel can give to the load and the power of the solar radiation (p_l) incident on the panel. This is normally around 10-12%, depending on the type of cells (monocrystalline, polycrystalline, amorphous or thin film).

Considering the definitions of point of maximum power and the fill factor we see that:

$$\eta = P_{max} / P_L = FF \cdot I_{SC} \cdot V_{OC} / P_L$$

The values of I_{SC}, V_{OC}, I_{Pmax} and V_{Pmax} are provided by the manufacturer and refer to standard conditions of measurement with irradiance G = 1000 W/m², at sea-level, for a temperature of cells of T_c = 25ºC.

The panel parameter values change for other conditions of irradiance and temperature. Manufacturers will sometimes include graphs or tables with values for conditions different from the standard. You should check the performance values at the panel temperatures that are likely to match your particular installation.

Be aware that two panels can have the same W_p but very different behaviour in different operating conditions. When acquiring a panel, it is important to verify, if possible, that their parameters (at least, I_{SC} and V_{OC}) match the values promised by the manufacturer.

Panel parameters for system sizing

To calculate the number of panels required to cover a given load, you just need to know the current and voltage at the point of maximum power: I_{Pmax} and V_{Pmax}.

You should always be aware that the panel is not going to perform under perfect conditions as the load or regulation system is not always going to work at the point of maximum power of the panel. You should assume a loss of efficiency of 5% in your calculations to compensate for this.

Interconnection of panels

A solar panel array is a collection of solar panels that are electrically interconnected and installed on some type of support structure. Using a solar panel array allows you to generate greater voltage and current than is possible with a single solar panel. The panels are interconnected in such a way that the voltage generated is close to (but greater than) the level of voltage of the batteries, and that the current generated is sufficient to feed the equipment and to charge the batteries.

Connecting solar panels in series increases the generated voltage. Connecting panels in parallel increases the current.

The number of panels used should be increased until the amount of power generated slightly exceeds the demands of your load.

It is very important that all of the panels in your array are as identical as possible. In an array, you should use panels of the same brand and characteristics because any difference in their operating conditions will have a big impact on the health and performance of your system.

Even panels that have identical performance ratings will usually display some variance in their characteristics due to manufacturing processes.

The actual operating characteristics of two panels from the same manufacturer can vary by as much as ±10%.
Whenever possible, it is a good idea to test the real-world performance of individual panels to verify their operating characteristics before assembling them into an array.

Figure OGP 8: Interconnection of panels in parallel. The voltage remains constant while the current duplicates. (Photo: Fantsuam Foundation, Nigeria)

How to choose a good panel

One obvious metric to use when shopping for solar panels is to compare the ratio of the nominal peak power (W_p) to the price. This will give you a rough idea of the cost per Watt for different panels. But there are a number of other considerations to keep in mind as well.

If you are going to install solar panels in geographical areas where soiling (from dust, sand, or grit) will likely be a problem, consider purchasing panels with a low affinity for soil retention.

These panels are made of materials that increase the likelihood that the panel will be automatically cleaned by wind and rain.

Always check the mechanical construction of each panel. Verify that the glass is hardened and the aluminum frame is robust and well built. The solar cells inside the panel can last for more than 20 years, but they are very fragile and the panel must protect them from mechanical hazards. Look for the manufacturer's quality guarantee in terms of expected power output and mechanical construction.

Finally, be sure that the manufacturer provides not only the nominal peak power of the panel (W_p) but also the variation of the power with irradiation and temperature. This is particularly important when panels are used in arrays, as variations in the operating parameters can have a big impact on the quality of power generated and the useful lifetime of the panels.

The battery

The battery "hosts" a certain reversible chemical reaction that stores electrical energy that can later be retrieved when needed. Electrical energy is transformed into chemical energy when the battery is being charged, and the reverse happens when the battery is discharged. A battery is formed by a set of elements or cells arranged in series. Lead-acid batteries consist of two submerged lead electrodes in an electrolytic solution of water and sulphuric acid. A potential difference of about 2 volts takes place between the electrodes, depending on the instantaneous value of the charge state of the battery.

The most common batteries in photovoltaic solar applications have a nominal voltage of 12 or 24 volts. A 12 V battery therefore contains 6 cells in series.

The battery serves two important purposes in a photovoltaic system: to provide electrical energy to the system when energy is not supplied by the array of solar panels, and to store excess energy generated by the panels whenever that energy exceeds the load. The battery experiences a cyclical process of charging and discharging, depending on the presence or absence of sunlight.

During the hours that there is sun, the array of panels produces electrical energy. The energy that is not consumed immediately is used to charge the battery.

During the hours of absence of sun, any demand of electrical energy is supplied by the battery, thereby discharging it.

These cycles of charge and discharge occur whenever the energy produced by the panels does not match the energy required to support the load. When there is sufficient sun and the load is light, the batteries will charge.

Obviously, the batteries will discharge at night whenever any amount of power is required. The batteries will also discharge when the irradiance is insufficient to cover the requirements of the load (due to the natural variation of climatological conditions, clouds, dust, etc.)

If the battery does not store enough energy to meet the demand during periods without sun, the system will be exhausted and will be unavailable for consumption. On the other hand, oversizing the system (by adding far too many panels and batteries) is expensive and inefficient. When designing a stand-alone system we need to reach a compromise between the cost of components and the availability of power from the system.

One way to do this is to estimate the required number of days of autonomy. In the case of a telecommunications system, the number of days of autonomy depends on its critical function within your network design. If the equipment is going to serve as a repeater and is part of the backbone of your network, you will likely want to design your photovoltaic system with an autonomy of up to 5-7 days.

On the other hand, if the solar system is responsible for a providing energy to client equipment you can probably reduce number of days of autonomy to two or three. In areas with low irradiance, this value may need to be increased even more. In any case, you will always have to find the proper balance between cost and reliability.

Types of batteries

Many different battery technologies exist, and are intended for use in a variety of different applications. The most suitable type for photovoltaic applications is the stationary battery, designed to have a fixed location and for scenarios where the power consumption is more or less irregular. "Stationary" batteries can accommodate deep discharge cycles, but they are not designed to produce high currents in brief periods of time.

Stationary batteries can use an electrolyte that is alkaline (such as Nickel-Cadmium) or acidic (such as Lead-Acid).

Stationary batteries based on Nickel-Cadmium are recommended for their high reliability and resistance whenever possible. Unfortunately, they tend to be much more expensive and difficult to obtain than sealed lead-acid batteries. In many cases when it is difficult to find local, good and cheap stationary batteries (importing batteries is not cheap), you will be forced to use batteries targeted to the automobile market.

Using car batteries

Automobile batteries are not well suited for photovoltaic applications as they are designed to provide a substantial current for just few seconds (when starting then engine) rather than sustaining a low current for long period of time. This design characteristic of car batteries (also called traction batteries) results in a shortened effective life when used in photovoltaic systems. Traction batteries can be used in small applications where low cost is the most important consideration, or when other batteries are not available.

Traction batteries are designed for vehicles and electric wheelbarrows. They are cheaper than stationary batteries and can serve in a photovoltaic installation, although they require very frequent maintenance. These batteries should never be deeply discharged, because doing so will greatly reduce their ability to hold a charge. A truck battery should not be discharged to more than 70% of its total capacity. This means that you can only use a maximum of 30% of a lead-acid battery's nominal capacity before it must be recharged.

You can extend the life of a lead-acid battery by using distilled water. By using a densimeter or hydrometer, you can measure the density of the battery's electrolyte. A typical battery has specific gravity of 1.28.

Adding distilled water and lowering the density to 1.2 can help reduce the anode's corrosion, at a cost of reducing the overall capacity of the battery. If you adjust the density of battery electrolyte, you must use distilled water, as tap water or well water will permanently damage the battery.

States of charge

There are two special states of charge that can take place during the cyclic charge and discharge of the battery. They should both be avoided in order to preserve the useful life of the battery.

1. Overcharge

Overcharge takes place when the battery arrives at the limit of its capacity. If energy is applied to a battery beyond its point of maximum charge, the electrolyte begins to break down.

This produces bubbles of oxygen and hydrogen, in a process is known as gasification. This results in a loss of water, oxidation on the positive electrode, and in extreme cases, a danger of explosion.

On the other hand, the presence of gas avoids the stratification of the acid. After several continuous cycles of charge and discharge, the acid tends to concentrate itself at the bottom of the battery thereby reducing the effective capacity. The process of gasification agitates the electrolyte and avoids stratification.

Again, it is necessary to find a compromise between the advantages (avoiding electrolyte stratification) and the disadvantages (losing water and production of hydrogen). One solution is to allow a slight overcharge condition every so often. One typical method is to allow a voltage of 2.35 to 2.4 Volts for each element of the battery every few days, at 25ºC. The regulator should ensure a periodic and controlled overcharges.

2. Overdischarge

In the same way that there is a upper limit, there is also a lower limit to a battery's state of charge. Discharging beyond that limit will result in deterioration of the battery. When the effective battery supply is exhausted, the regulator prevents any more energy from being extracted from the battery. When the voltage of the battery reaches the minimum limit of 1.85 Volts per cell at 25°C, the regulator disconnects the load from the battery.

If the discharge of the battery is very deep and the battery remains discharged for a long time, three effects take place: the formation of crystallised sulphate on the battery plates, the loosening of the active material on the battery plate, and plate buckling. The process of forming stable sulphate crystals is called hard sulphation. This is particularly negative as it generates big crystals that do not take part in any chemical reaction and can make your battery unusable.

Battery Parameters

The main parameters that characterise a battery are:

Nominal Voltage, V_{NBat}: the most common value being 12 V.

Nominal Capacity, C_{NBat}: the maximum amount of energy that can be extracted from a fully charged battery. It is expressed in Ampere-hours (Ah) or Watt-hours (Wh).

The amount of energy that can be obtained from a battery depends on the time in which the extraction process takes place. Discharging a battery over a long period will yield more energy compared to discharging the same battery over a short period. The capacity of a battery is therefore specified at different discharging times. For photovoltaic applications, this time should be longer than 100 hours (C100).

Maximum Depth of Discharge, DoD_{max}: The depth of discharge is the amount of energy extracted from a battery in a single discharge cycle, expressed as a percentage. The life expectancy of a battery depends on how deeply it is discharged in each cycle. The manufacturer should provide graphs that relate the number of charge-discharge cycles to the life of the battery. As a general rule you should avoid discharging a deep cycle battery beyond 50%. Traction batteries should only be discharged by as little as 30%.

Useful Capacity, C_{UBat}: It is the real (as in usable) capacity of a battery. It is equal to the product of the nominal capacity and the maximum DoD. For example, a stationary battery of nominal capacity (C100) of 120 Ah and depth of discharge of 70% has a useful capacity of (120 x 0.7) = 84 Ah.

Measuring the state of charge of the battery

A sealed lead-acid battery of 12 V provides different voltages depending on its state of charge. When the battery is fully charged in an open circuit, the output voltage is about 12.8 V.

The output voltage lowers quickly to 12.6 V when loads are attached. As the battery is providing constant current during operation, the battery voltage reduces linearly from 12.6 to 11.6 V depending on the state of charge.

A sealed lead- acid batteries provides 95% of its energy within this voltage range. If we make the broad assumption that a fully loaded battery has a voltage of 12.6 V when "full" and 11.6 V when "empty", we can estimate that a battery has discharged 70% when it reaches a voltage of 11.9 V.

These values are only a rough approximation since they depend on the life and quality of the battery, the temperature, etc.

State of Charge	12V Battery Voltage	Volts per Cell
100%	12.7	2.12
90%	12.5	2.08
80%	12.42	2.07
70%	12.32	2.05
60%	12.2	2.03
50%	12.06	2.01
40%	11.9	1.98
30%	11.75	1.96
20%	11.58	1.93
10%	11.31	1.89

According to this table, and considering that a truck battery should not be discharged more than 20% to 30%, we can determine that the useful capacity of a 170 Ah truck battery is 34 Ah (20%) to 51 Ah (30%). Using the same table, we find that we should program the regulator to prevent the battery from discharging below 12.3 V.

Battery and regulator protection

Thermomagnetic circuit breakers or one time fuses must be used to protect the batteries and the installation from short circuit and malfunctions. There are two types of fuses: slow blow, and quick blow. Slow blow fuses should be used with inductive or capacitive loads where a high current can occur at power up. Slow blow fuses will allow a higher current than their rating to pass for a short time. Quick blow fuses will immediately blow if the current flowing through them is higher than their rating.

The regulator is connected to the battery and the loads, so two different kinds of protection need to be considered. One fuse should be placed between the battery and the regulator, to protect the battery from short-circuit in case of regulator failure. A second fuse is needed to protect the regulator from excessive load current.

This second fuse is normally integrated into the regulator itself.

Figure OGP 9: A battery bank of 3600 Ah, currents reach levels of 45 A during charging

Every fuse is rated with a maximum current and a maximum usable voltage. The maximum current of the fuse should be 20% bigger than the maximum current expected. Even if the batteries carry a low voltage, a short circuit can lead to a very high current which can easily reach several hundred amperes.

Large currents can cause fire, damage the equipment and batteries, and possibly cause electric shock to a human body. If a fuse breaks, never replace a fuse with a wire or a higher rated fuse. First determine the cause of the problem, then replace the fuse with another one which has the same characteristics.

Temperature effects

The ambient temperature has several important effects on the characteristics of a battery:

- The nominal capacity of a battery (that the manufacturer usually gives for 25°C) increases with temperature at the rate of about 1%/°C. But if the temperature is too high, the chemical reaction that takes place in the battery accelerates, which can cause the same type of oxidation that takes place during overcharging. This will obviously reduce the life expectancy of a battery. This problem can be compensated partially in car batteries by using a low density of dissolution (a specific gravity of 1.25 when the battery is totally charged).

- As the temperature is reduced, the useful life of the battery increases. But if the temperature is too low, you run the the risk of freezing the electrolyte. The freezing temperature depends on the density of the solution, which is also related to the state of charge of the battery. The lower the density, the greater the risk of freezing. In areas of low temperatures, you should avoid deeply discharging the batteries (that is, DoDmax is effectively reduced.)

- The temperature also changes the relation between voltage and charge. It is preferable to use a regulator which adjusts the low voltage disconnect and reconnect parameters according to temperature. The temperature sensor of the regulator should be fixed to the battery using tape or some other simple method.

- In hot areas it is important to keep the batteries as cool as possible. The batteries must be stored in a shaded area and never get direct sunlight. It's also desirable to place the batteries on a small support to allow air to flow under them, thus increase the cooling.

How to choose a good battery

Choosing a good battery can be very challenging.
High capacity batteries are heavy, bulky and expensive to import.

A 200 Ah battery weighs around 50 kg (120 pounds) and it cannot be transported as hand luggage.

If you want long-life (as in > 5 years) and maintenance free batteries be ready to pay the price.

A good battery should always come with its technical specifications, including the capacity at different discharge rates (C20, C100), operating temperature, cut-off voltage points, and requirements for chargers.

The batteries must be free of cracks, liquid spillage or any sign of damage, and battery terminals should be free of corrosion.

As laboratory tests are necessary to obtain complete data about real capacity and aging, expect lots of low quality batteries (including fakes) in the local markets. A typical price (not including transport and import tax) is $3-4 USD per Ah for 12 V lead-acid batteries.

Life expectancy versus number of cycles

Batteries are the only component of a solar system that should be amortised over a short period and regularly replaced.

You can increase the useful lifetime of a battery by reducing the depth of discharge per cycle. Even deep cycle batteries will have an increased battery life if the number of deep discharge (>30%) cycles is reduced.

If you completely discharge the battery every day, you will typically need to change it after less than one year. If you use only 1/3 of the capacity the battery, it can last more than 3 years. It can be cheaper to buy a battery with 3 times the capacity than to change the battery every year.

The charge regulator

The charge regulator is also known as a charge controller, a voltage regulator, a charge-discharge controller or a charge-discharge and load controller. The regulator sits between the array of panels, the batteries, and your equipment or loads.

Remember that the voltage of a battery, although always close to 2 V per cell, varies according to its state of charge. By monitoring the voltage of the battery, the regulator prevents overcharging or overdischarging.

Regulators used in solar applications should be connected in series: they disconnect the array of panels from the battery to avoid overcharging, and they disconnect the battery from the load to avoid overdischarging.

The connection and disconnection is done by means of switches which can be of two types: electromechanical (relays) or solid state (bipolar transistor, MOSFET). Regulators should never be connected in parallel.

In order to protect the battery from gasification, the switch opens the charging circuit when the voltage in the battery reaches its high voltage disconnect (HVD) or cut-off set point. The low voltage disconnect (LVD) prevents the battery from overdischarging by disconnecting or shedding the load. To prevent continuous connections and disconnections the regulator will not connect back the loads until the battery reaches a low reconnect voltage (LRV).

Typical values for a 12 V lead-acid battery are:

Voltage Point	Voltage
LVD	11.5
LRV	12.6
Constant Voltage Regulated	14.3
Equalisation	14.6
HVD	15.5

The most modern regulators are also able to automatically disconnect the panels during the night to avoid discharging of the battery.

They can also periodically overcharge the battery to improve its life, and they may use a mechanism known as pulse width modulation (PWM) to prevent excessive gassing.

As the peak power operating point of the array of panels will vary with temperature and solar illumination, new regulators are capable of constantly tracking the maximum point of power of the solar array. This feature is known as maximum power point tracking (MPPT).

Regulator Parameters

When selecting a regulator for your system, you should at least know the operating voltage and the maximum current that the regulator can handle. The operating voltage will be 12, 24, or 48 V. The maximum current must be 20% bigger than the current provided by the array of panels connected to the regulator.

Other features and data of interest include:

- Specific values for LVD, LRV and HVD.
- Support for temperature compensation. The voltage that indicates the state of charge of the battery varies with temperature. For that reason some regulators are able to measure the battery temperature and correct the different cut-off and reconnection values.
- Instrumentation and gauges. The most common instruments measure the voltage of the panels and batteries, the state of charge (SoC) or Depth of Discharge (DoD). Some regulators include special alarms to indicate that the panels or loads have been disconnected, LVD or HVD has been reached, etc.

Converters

The regulator provides DC power at a specific voltage.
Converters and inverters are used to adjust the voltage to match the requirements of your load.

DC/DC Converters

DC/DC converters transform a continuous voltage to another continuous voltage of a different value. There are two conversion methods which can be used to adapt the voltage from the batteries: linear conversion and switching conversion.Linear conversion lowers the voltage from the batteries by converting excess energy to heat. This method is very simple but is obviously inefficient. Switching conversion generally uses a magnetic component to temporarily store the energy and transform it to another voltage. The resulting voltage can be greater, less than, or the inverse (negative) of the input voltage. The efficiency of a linear regulator decreases as the difference between the input voltage and the output voltage increases. For example, if we want to convert from 12 V to 6 V, the linear regulator will have an efficiency of only 50%.
A standard switching regulator has an efficiency of at least 80%.

DC/AC Converter or Inverter

Inverters are used when your equipment requires AC power. Inverters chop and invert the DC current to generate a square wave that is later filtered to approximate a sine wave and eliminate undesired harmonics.

Very few inverters actually supply a pure sine wave as output. Most models available on the market produce what is known as "modified sine wave", as their voltage output is not a pure sinusoid. When it comes to efficiency, modified sine wave inverters perform better than pure sinusoidal inverters. Be aware that not all the equipment will accept a modified sine wave as voltage input. Most commonly, some laser printers will not work with a modified sine wave inverter. Motors will work, but they may consume more power than if they are fed with a pure sine wave. In addition, DC power supplies tend to warm up more, and audio amplifiers can emit a buzzing sound.

Aside from the type of waveform, some important features of inverters include:

Reliability in the presence of surges. Inverters have two power ratings: one for continuous power, and a higher rating for peak power. They are capable of providing the peak power for a very short amount of time, as when starting a motor. The inverter should also be able to safely interrupt itself (with a circuit breaker or fuse) in the event of a short circuit, or if the requested power is too high.

Conversion efficiency. Inverters are most efficient when providing 50% to 90% of their continuous power rating.

You should select an inverter that most closely matches your load requirements.

The manufacturer usually provides the performance of the inverter at 70% of its nominal power.

Battery charging. Many inverters also incorporate the inverse function: the possibility of charging batteries in the presence of an alternative source of current (grid, generator, etc). This type of inverter is known as a charger/inverter.

Automatic fall-over. Some inverters can switch automatically between different sources of power (grid, generator, solar) depending on what is available.

When using telecommunication equipment, it is best to avoid the use of DC/AC converters and feed them directly from a DC source. Most communications equipment can accept a wide range of input voltage.

Equipment or load

It should be obvious that as power requirements increase, the expense of the photovoltaic system also increases. It is therefore critical to match the size of the system as closely as possible to the expected load.

When designing the system you must first make a realistic estimate of the maximum consumption.

Once the installation is in place, the established maximum consumption must be respected in order to avoid frequent power failures.

Home Appliances

The use of photovoltaic solar energy is not recommended for heat-exchange applications (electrical heating, refrigerators, toasters, etc.). Whenever possible, energy should be used sparingly using low power appliances.

Here are some points to keep in mind when choosing appropriate equipment for use with a solar system:

- The photovoltaic solar energy is suitable for illumination. In this case, the use of halogen light bulbs or fluorescent lamps is mandatory. Although these lamps are more expensive, they have much better energy efficiency than incandescent light bulbs. LED lamps are also a good choice as they are very efficient and are fed with DC.

- It is possible to use photovoltaic power for appliances that require low and constant consumption (as in a typical case, the TV). Smaller televisions use less power than larger televisions. Also consider that a black-and-white TV consumes about half the power of a colour TV.

- Photovoltaic solar energy is not recommended for any application that transforms energy into heat (thermal energy). Use solar heating or butane as an alternative.

- Conventional automatic washing machines will work, but you should avoid the use of any washing programs that include centrifugal water heating.

- If you must use a refrigerators, it should consume as little power as possible. There are specialised refrigerators that work on DC, although their consumption can be quite high (around 1000 Wh/day).

The estimation of total consumption is a fundamental step in sizing your solar system. Here is a table that gives you a general idea of the power consumption that you can expect from different appliances.

Equipment	Consumption (Watts)
Portable Computer	30-50
Low Power Lamp	6-10
Router with one radio	4-10
VSAT modem	15-30
PC without LCD	20-30
PC with LCD	200-300
16 port Network switch	6-8

Wireless telecommunications equipment

Saving power by choosing the right gear saves a lot of money and trouble. For example, a long distance link doesn't necessarily need a strong amplifier that draws a lot of power.

A Wi-Fi card with good receiver sensitivity and a Fresnel zone that is at least 60% clear will work better than an amplifier, and save power consumption as well.

A well known saying of radio amateurs applies here, too: the best amplifier is a good antenna. Further measures to reduce power consumption include throttling the CPU speed, reducing transmit power to the minimum value that is necessary to provide a stable link, increasing the length of beacon intervals, and switching the system off during times it is not needed.

Most autonomous solar systems work at 12 or 24 volts. Preferably, a wireless device that runs on DC voltage should be used, operating at the 12 Volts that most lead acid batteries provide.

Transforming the voltage provided by the battery to AC or using a voltage at the input of the access point different from the voltage of the battery will cause unnecessary energy loss. A router or access point that accepts 8-20 Volts DC is perfect.

Most cheap Access Points have a switched mode voltage regulator inside and will work through such a voltage range without modification or becoming hot (even if the device was shipped with a 5 or 12 Volt power supply).

WARNING: Operating your Access Point with a power supply other than the one provided by your manufacturer will certainly void any warranty, and may cause damage to your equipment. While the following technique will typically work as described, remember that should you attempt it, you do so at your own risk.

Open your Access Point and look near the DC input for two relatively big capacitors and an inductor (a ferrite toroid with copper wire wrapped around it). If they are present then the device has a switched mode input, and the maximum input voltage should be somewhat below the voltage printed on the capacitors. Usually the rating of these capacitors is 16 or 25 volts.

Be aware that an unregulated power supply has a ripple and may feed a much higher voltage into your Access Point than the typical voltage printed on it may suggest. So, connecting an unregulated power supply with 24 Volts to a device with 25 Volt-capacitors is not a good idea.

Of course, opening your device will void any existing warranty. Do not try to operate an Access Point at higher voltage if it doesn't have a switched mode regulator. It will get hot, malfunction, or burn.

Equipment based on traditional Intel x86 CPUs are power hungry in comparison with RISC-based architectures as ARM or MIPS.

One of the boards with lowest power consumptions is the Soekris platform that uses an AMD ElanSC520 processor. Another alternative to AMD (ElanSC or Geode SC1100) is the use of equipment with MIPS processors. MIPS processors have a better performance than an AMD Geode at the price of consuming between 20-30% of more energy.

The amount of power required by wireless equipment depends not only on the architecture but on the number of network interfaces, radios, type of memory/storage and traffic.

As a general rule, a wireless board of low consumption consumes 2 to 3 W, and a 200 mW radio card consumes as much as 3 W. High power cards (such as the 400 mW Ubiquiti) consume around 6 W. A repeating station with two radios can range between 8 and 10 W.

Although the standard IEEE 802.11 incorporates a power saving mode (PS) mechanism, its benefit is not as good as you might hope. The main mechanism for energy saving is to allow stations to periodically put their wireless cards to "sleep" by means of a timing circuit.

When the wireless card wakes up it verifies if a beacon exists, indicating pending traffic. The energy saving therefore only takes place on the client side, as the access point always needs to remain awake to send beacons and store traffic for the clients. Power saving mode may be incompatible between implementations from different manufacturers, which can cause unstable wireless connections. It is nearly always best to leave power saving mode disabled on all equipment, as the difficulties created will likely outweigh the meagre amount of saved power.

Selecting the voltage

Most low power stand-alone systems use 12 V battery power as that is the most common operational voltage in sealed lead-acid batteries.
When designing a wireless communication system you need to take into consideration the most efficient voltage of operation of your equipment. While the input voltage can accept a wide range of values, you need to ensure that the overall power consumption of the system is minimal.

Wiring

An important component of the installation is the wiring, as proper wiring will ensure efficient energy transfer. Some good practices that you should consider include:

- Use a screw to fasten the cable to the battery terminal. Loose connections will waste power.
- Spread Vaseline or mineral jelly on the battery terminals. Corroded connections have an increased resistance, resulting in loss.

Wire size is normally given in American Wire Gauge (AWG).

During your calculations you will need to convert between AWG and mm^2 to estimate cable resistance. For example, an AWG #6 cable has a diameter of 4.11 mm and can handle up to 55 A.
A **conversion chart,** including an estimate of resistance and current carrying capacity, is available in **Appendix D: Cables Sizes**. Keep in mind that the current carrying capacity can also vary depending on the type of insulation and application. When in doubt, consult the manufacturer for more information.

Orientation of the panels

Most of the energy coming from the sun arrives in a straight line. The solar module will capture more energy if it is "facing" the sun, perpendicular to the straight line between the position of the installation and the sun. Of course, the sun's position is constantly changing relative to the earth, so we need to find an optimal position for our panels. The orientation of the panels is determined by two angles, the *azimuth a* and the *inclination* or *elevation β*. The azimuth is the angle that measures the deviation with respect to the south in the northern hemisphere, and with respect to the north in the southern hemisphere. The inclination is the angle formed by the surface of the module and the horizontal plane.

Azimuth

You should have the module turned towards the terrestrial equator (facing south in the northern hemisphere, and north in the southern) so that during the day the panel catches the greatest possible amount of radiation (a = 0°). It is very important that no part of the panels are ever in shade! Study the elements that surround the panel array (trees, buildings, walls, other panels, etc.) to be sure that they will not cast a shadow on the panels at any time of the day or year. It is acceptable to turn the panels ±20° towards the east or the west if needed (a = ±20°).

Inclination

Once you have fixed the azimuth, the parameter that is key in our calculations is the inclination of the panel, which we will express as the angle beta (β). The maximum height that the sun reaches every day will vary, with the maximum on the day of the summer solstice and the minimum on the day of the winter solstice.

Ideally, the panels should track this variation, but this is usually not possible for cost reasons. In installations with telecommunications equipment it is normal to install the panels at a fixed inclination. In most telecommunications scenarios the energy demands of the system are constant throughout the year. Providing for sufficient power during the "worst month" will work well for the rest of the year.

The value of β should maximise the ratio between the offer and the demand for energy.

For installations with consistent (or nearly consistent) consumption throughout the year, it is preferable to optimise the installation to capture the maximum radiation during "the winter" months.

You should use the absolute value of the latitude of the place (angle F) increased by 10° ($\boxtimes = |\,F\,| + 10\,°$).

For installations with less consumption during the winter, the value of the latitude of the place can be used as the solar panel inclination. This way the system is optimised for the months of spring and autumn ($\boxtimes = |\,F\,|$).

For installations that are only used during summer, you should use the absolute value of the latitude of the place (angle F) decreased by 10° ($\boxtimes = |\,F\,| - 10°$).

The inclination of the panel should never be less than 15° to avoid the accumulation of dust and/or humidity on the panel. In areas where snow and ice occur, it is very important to protect the panels and to incline them at an angle of 65° or greater.

If there is a considerable increase in consumption during the summer, you might consider arranging for two fixed inclinations, one position for the months of summer and another for the months of winter.

This would require special support structures and a regular schedule for changing the position of the panels.

How to size your photovoltaic system

When choosing equipment to meet your power needs, you will need to determine the following, at a minimum:

- The number and type of solar panels required to capture enough solar energy to support your load.
- The minimum capacity of the battery. The battery will need to store enough energy to provide power at night and through days with little sun, and will determine your number of days of autonomy.

- The characteristics of all other components (the regulator, wiring, etc.) needed to support the amount of power generated and stored.

System sizing calculations are important, because unless the system components are balanced, energy (and ultimately, money) is wasted. For example, if we install more solar panels to produce more energy, the batteries should have enough capacity to store the additional energy produced. If the bank of batteries is too small and the load is not using the energy as it is generated, then energy must be thrown away. A regulator of a smaller amperage than needed, or one single cable that is too small, can be a cause of failure (or even fire) and render the installation unusable.

Never forget that the ability of the photovoltaic energy to produce and store electrical energy is limited. Accidentally leaving on a light bulb during the day can easily drain your reserves before nighttime, at which point no additional power will be available.

The availability of "fuel" for photovoltaic systems (i.e. solar radiation) can be difficult to predict. In fact, it is never possible to be absolutely sure that a standalone system is going to be able to provide the necessary energy at any particular moment. Solar systems are designed for a certain consumption, and if the user exceeds the planned limits the provision of energy will fail. The design method that we propose consists of considering the energy requirements, and based on them to calculate a system that works for the maximum amount of time so it is as reliable as possible.

Of course, if more panels and batteries are installed, more energy will be able to be collected and stored.

This increase of reliability will also have an increase in cost.

In some photovoltaic installations (such as the provision of energy for telecommunications equipment on a network backbone) the reliability factor is more important that the cost.

In a client installation, low cost is likely to be the most important factor. Finding a balance between cost and reliability is not a easy task, but whatever your situation, you should be able to determine what it is expected from your design choices, and at what price.

The method we will use for sizing the system is known as the ***method of the worst month***. We simply calculate the dimensions of the standalone system so it will work in the month in which the demand for energy is greatest with respect to the available solar energy. It is the worst month of the year, as this month will have the largest ratio of demanded energy to available energy.

Using this method, *reliability* is taken into consideration by fixing the maximum number of days that the system can work without receiving solar radiation (that is, when all consumption is made solely at the expense of the energy stored in the battery.) This is known as the *maximum number of days of autonomy* (N), and can be thought of as the number of consecutive cloudy days when the panels do not collect any significant amount of energy. When choosing N, it is necessary to know the climatology of the place, as well as the economic and social relevance of the installation.

Will it be used to illuminate houses, a hospital, a factory, for a radio link, or for some other application? Remember that as N increases, so does the investment in equipment and maintenance.

It is also important to evaluate all possible logistics costs of equipment replacement.

It is not the same to change a discharged battery from an installation in the middle of a city versus one at the top of a telecommunication tower that is several hours or days of walking distance. Fixing the value of N is not an easy task as there are many factors involved, and many of them cannot be evaluated easily. Your experience will play an important role in this part of the system sizing. One commonly used value for critical telecommunications equipment is N = 5, whereas for low cost client equipment it is possible to reduce the autonomy to N = 3.

In **Appendix E: Solar Dimensioning**, we have included several tables that will facilitate the collection of required data for sizing the system. The rest of this chapter will explain in detail what information you need to collect and explain how to use the method of the "worst month".

Data to collect

Latitude of the installation. Remember to use a positive sign in the northern hemisphere and negative in the south.

Solar radiation data. For the method of the "worst month" it is enough to know just twelve values, one for every month. The twelve numbers are the monthly average values of daily global irradiation on the horizontal plane, $G_{dm}(0)$, in kWh/m^2 per day. The monthly value is the sum of the values of global irradiation for every day of the month, divided by the number of days of the month.

If you have the data in Joules (J), you can apply the following conversion:

$$1 J = 2.78 \times 10^{-7} \ kWh$$

The irradiation data $G_{dm}(0)$ of many places of the world is gathered in tables and databases.

You should check for this information from a weather station close to your implementation site, but do not be surprised if you cannot find the data in electronic format.

It is a good idea to ask companies that install photovoltaic systems in the region, as their experience can be of great value.

Do not confuse "sun hours" with the number of "peak sun hours".

The number of peak sun hours has nothing to do with the number of hours without clouds, but refers to the amount of daily irradiation.

A day of 5 hours of sun without clouds does not necessarily have all those hours with the sun at its zenith.

A peak sun hour is a normalised value of solar radiation of 1000 W/m² at 25degC. So when we refer to 5 peak sun hours, this implies a daily solar radiation of 5000 W/m².

Electrical characteristics of system components

The electrical characteristics of the components of your system should be provided by the manufacturer. It is advisable to make your our own measurements to check for any deviation from the nominal values. Unfortunately, deviation from promised values can be large and should be expected.

These are the minimum values that you need to gather before starting your system sizing:

Panels

You need to know the voltage V_{Pmax} and the current I_{Pmax} at the point of maximum power in standard conditions.

Batteries

Nominal capacity (for 100 hours discharge) C_{NBat}, operational voltage V_{NBat}, and either the maximum depth of discharge DoD_{max} or useful capacity C_{UBat}.

You also need to know the type of battery that you plan to use, whether sealed lead-acid, gel, AGM, modified traction etc.

The type of battery is important when deciding the cut-off points in the regulator.

Regulator

You need to know the nominal voltage V_{NReg}, and the maximum operational current I_{maxReg}.

DC/AC Converter/Inverter

If you are going to use a converter, you need to know the nominal voltage V_{NConv}, instantaneous power P_{IConv} and performance at 70% of maximum load H_{70}.

Equipment or load

It is necessary to know the nominal voltage V_{NC} and the nominal power of operation PC for every piece of equipment powered by the system.

In order to know the total energy that our installation is going to consume, it is also very important to consider the average time each load will be used.

Is it constant? Or will it be used daily, weekly, monthly or annually?

Consider any changes in the usage that might impact the amount of energy needed (seasonal usage, training or school periods, etc.).

Other variables

Aside from the electrical characteristics of the components and load, it is necessary to decide on two more pieces of information before being able to size a photovoltaic system.

These two decisions are the required number of days of autonomy and the operational voltage of the system.

N, number of days of autonomy

You need to decide on a value for N that will balance meteorological conditions with the type of installation and overall costs. It is impossible to give a concrete value of N that is valid for every installation, but the next table gives some recommended values.

Take these values as a rough approximation, and consult with an experienced designer to reach a final decision.

Available Sunlight	Domestic Installation	Critical Installation
Very Cloudy	5	10
Variable	4	8
Sunny	3	6

V_N, nominal voltage of the installation

The components of your system need to be chosen to operate at a nominal voltage V_N. This voltage is usually 12 or 24 Volts for small systems, and if the total power of consumption surpasses 3 kW, the voltage will be 48 V. The selection of V_N is not arbitrary, and depends on the availability of equipment. If the equipment allows it, try to fix the nominal voltage to 12 or 24 V. Many wireless communications boards accept a wide range of input voltage and can be used without a converter. If you need to power several types of equipment that work at different nominal voltages, calculate the voltage that minimises the overall power consumption including the losses for power conversion in DC/DC and DC/AC converters.

Procedure of calculation

There are three main steps that need to be followed to calculate the proper size of a system:

1. **Calculate the available solar energy (the offer)**. Based on statistical data of solar radiation, and the orientation and optimal inclination of the solar panels, we can calculate the solar energy available.
   The estimation of solar energy available is done in monthly intervals, reducing the statistical data to 12 values. This estimation is a good compromise between precision and simplicity.

2. **Estimate the required electrical energy (the demand)**. Record the power consumption characteristics of the equipment chosen as well as estimated usage.
   Then calculate the electrical energy required on a monthly basis. You should consider the expected fluctuations of usage due to the variations between winter and summer, the rainy period / dry season, school / vacation periods, etc. The result will be 12 values of energy demand, one for each month of the year.

3. **Calculate the ideal system size (the result).**

With the data from the "worst month", when the relation between the solar demanded energy and the energy available is greatest, we calculate:

- The current that the array of panels needs to provide, which will determine the minimum number of panels.
- The necessary energy storage capacity to cover the minimum number of days of autonomy, which will determine the required number of batteries.
- The required electrical characteristics of the regulator.
- The length and the necessary sections of cables for the electrical connections.

Required current in the worst month

For each month you need to calculate the value of I_m, which is the maximum daily current that an array of panels operating at nominal voltage of V_N needs to provide, in a day with a irradiation of G_{dm} for month "m", for panels with an inclination of ⊠ degrees.

The I_m(WORST MONTH) will be the largest value of I_m, and the system sizing is based on the data of that worst month.

The calculations of $G_{dm}(⊠)$ for a certain place can be made based on $G_{dm}(0)$ using computer software such as PVSYST (http://www.pvsyst.com/) or PVSOL (http://www.solardesign.co.uk/).

Due to losses in the regulator and batteries, and due to the fact that the panels do not always work at the point of maximum power, the required current I_{mMAX} is calculated as:

$$I_{mMAX} = 1.21\ I_m\ (WORST\ MONTH)$$

Once you have determined the worst month, the value of I_{mMAX}, and the total energy that you require E_{TOTAL}(WORST MONTH) you can proceed to the final calculations. E_{TOTAL} is the sum of all DC and AC loads, in Watts. To calculate E_{TOTAL} see **Appendix E: Solar Dimensioning.**

Number of panels

By combining solar panels in series and parallel, we can obtain the desired voltage and current.

When panels are connected in series, the total voltage is equal to the sum of the individual voltages of each module, while the current remains unchanged. When connecting panels in parallel, the currents are summed together while the voltage remains unchanged.

It is very important, to use panels of nearly identical characteristics when building an array.

You should try to acquire panels with V_{Pmax} a bit bigger than the nominal voltage of the system (12, 24 or 48 V).

Remember that you need to provide a few volts more than the nominal voltage of the battery in order to charge it. If it is not possible to find a single panel that satisfies your requirements, you need to connect several panels in series to reach your desired voltage.

The number of panels in series N_{ps} is equal to the nominal voltage of the system divided by the voltage of a single panel, rounded up to the nearest integer.

$$N_{ps} = V_N / V_{Pmax}$$

In order to calculate the number of panels in parallel (N_{pp}), you need to divide the I_{mMAX} by the current of a single panel at the point of maximum power I_{pmax}, rounded up to the nearest integer.

$$N_{pp} = I_{mMAX} / I_{Pmax}$$

The total number of panels is the result of multiplying the number of panels in series (to set the voltage) by the number of panels in parallel (to set the current).

$$N_{TOTAL} = N_{ps} * N_{pp}$$

Capacity of the battery or accumulator

The battery determines the overall voltage of the system and needs to have enough capacity to provide energy to the load when there is not enough solar radiation.

To estimate the capacity of our battery, we first calculate the required energy capacity of our system (necessary capacity, CNEC).

The necessary capacity depends on the energy available during the "worst month" and the desired number of days of autonomy (N).

$$C_{NEC} \ (Ah) = E_{TOTAL}(WORST \ MONTH)(Wh) \ / \ V_N(V) \ * \ N$$

The nominal capacity of the battery C_{NOM} needs to be bigger than the C_{NEC} as we cannot fully discharge a battery.

To calculate the size of the battery we need to consider the maximum depth of discharge (DoD) that the battery allows:

$$C_{NOM}(Ah) = C_{NEC}(Ah) \ / \ DoD_{MAX}$$

In order to calculate the number of batteries in series (N_{bs}), we divide the nominal voltage of our installation (V_N) by the nominal voltage of a single battery (V_{NBat}):

$$N_{bs} = V_N \ / \ V_{NBat}$$

Regulator

One important warning: always use regulators in series, never in parallel. If your regulator does not support the current required by your system, you will need to buy a new regulator with a larger working current.

For security reasons, a regulator needs to be able to operate with a current I_{maxReg} at least 20% greater than the maximum intensity that is provided by the array of panels:

$$I_{maxReg} = 1.2 \ N_{pp} \ I_{PMax}$$

DC/AC Inverter

The total energy needed for the AC equipment is calculated including all the losses that are introduced by the DC/AC converter or inverter.

When choosing an inverter, keep in mind that the performance of the inverter varies according to the amount of requested power.

An inverter has better performance characteristics when operating close to its rated power.

Using a 1500 Watt inverter to power a 25 Watt load is extremely inefficient.

In order to avoid this wasted energy, it is important to consider not the peak power of all your equipment, but the peak power of the equipment that is expected to operate simultaneously.

Cables

Once you know the numbers of panels and batteries, and type of regulators and inverters that you want to use, it is necessary to calculate the length and the thickness of the cables needed to connect the components together.

The **length** depends on the location of your the installation. You should try to minimise the length of the cables between the regulator, panels, and batteries. Using short cables will minimise lost power and cable costs.

The **thickness** chosen is based on the length of the cable and the maximum current it must carry.

The goal is to minimise voltage drops.

In order to calculate the thickness S of the cable it is necessary to know:

- the maximum current I_{MC} that is going to circulate in the cable. In the case of the panel-battery subsystem, it is I_{mMAX} calculated for every month. In the battery-load subsystem it depends on the way that the loads are connected;

- the voltage drop (V_a-V_b) that we consider acceptable in the cable. The voltage drop that results from adding all possible individual drops is expressed as a percent of the nominal voltage of the installation.

Typical maximum values are:

Component	Voltage Drop (%of V_N)
Panel Array -> Battery	1,00%
Battery -> Converter	1,00%
Main Line	3,00%
Main Line (Illumination)	3,00%
Main Line (Equipment)	5,00%

Typical acceptable voltage drops in cables

The section of the cable is determined by Ohm's Law:

$$S(mm^2) = r(\Omega mm^2/m)L(m) \, I_{mMAX}(A)/ \, (V_a - V_b)(V)$$

where S is the section, r is resistivity (intrinsic property of the material: for copper, 0.01286 $\Omega mm^2/m$), and L the length.

S is chosen taking into consideration the cables available in the market.

You should choose the immediately superior section to the one that is obtained from the formula. For security reasons there are some minimum values, for the cable that connects the panels and battery, this is a minimum of 6 mm^2. For the other sections, the minimum is 4 mm^2.

Cost of a solar installation

While solar energy itself is free, the equipment needed to turn it into useful electric energy is not.

You not only need to buy equipment to transform the solar energy in electricity and store it for use, but you must also replace and maintain various components of the system.
The problem of equipment replacement is often overlooked, and a solar system is sometimes implemented without a proper maintenance plan.
In order to calculate the real cost of your installation, we include an illustrative example.
The first thing to do it is to calculate the initial investment costs.

Description	Number	Unit Cost	Subtotal
60W Solar Panel (about $4/W)	4	$300	$1,200
30A Regulator	1	$100	$100
Wiring (metres)	25	$1/metre	$25z
50 Ah Deep Cycle Batteries	6	$150	$900
		Total	$2,225

The calculation of the investment cost is relatively easy once the system has been dimensioned.

You just need to add the price for each piece equipment and the labour cost to install and wire the equipment together.

For simplicity, we do not include the costs of transport and installation but you should not overlook them.

To figure out how much a system will really cost to operate we must estimate how long each part will last and how often you must replace it. In accounting terminology this is known as ***amortisation***.

Our new table will look like this:

Description	Number	Unit Cost	Subtotal	Lifetime(yrs)	Yearly Cost
60W Solar Panel	4	$300	$1,200	20	$60
30A Regulator	1	$100	$100	5	$20
Wiring (metres)	25	$1/ metre	$25	10	$2.50
50Ah Deep cycle batteries	6	$150	$900	5	$180
		Total	$2,225	Annual Cost	$262.50

As you see, once the first investment has been done, an annual cost of $262.50 is expected.

The annual cost is an estimation of the required capital per year to replace the system components once they reach the end of their useful life.

MAINTENANCE, MONITORING AND SUSTAINABILITY

15. MAINTENANCE AND TROUBLESHOOTING

Introduction

How you establish the support infrastructure for your network is as important as what type of equipment you use. Unlike wired connections, problems with a wireless network are often invisible, and can require more skill and more time to diagnose and remedy. Interference, wind, and new physical obstructions can cause a long-running network to fail. This chapter details a series of strategies to help you build a team that can support your network effectively.

We also describe some standard troubleshooting techniques that have proven successful in solving problems in networks in general.

Building your team

Every village, company or family has individuals who are intrigued by technology.

They are the ones found splicing the television cable, re- wiring a broken television or welding a new piece to a bicycle. These people will take interest in your network and want to learn as much about it as possible. Though these people are invaluable resources, you must avoid imparting all of the specialised knowledge of wireless networking to only one person. If your only specialist loses interest or finds better paying work somewhere else, they take the knowledge with them when they go.

There may also be many ambitious young adults who will be interested and have the time to listen, help, and learn about the network.

Again, they are very helpful and will learn quickly, but the project team must focus their attention on those who are best placed to support the network in the coming months and years.

Young adults and teenagers will go off to university or find employment taking their knowledge and skills about the network with them.

These youngsters may also have little influence in the community, where an older individual is likely to be more well-known and less impulsive regarding making decisions that affect the network as a whole.

Even though these individuals might have less time to learn and might appear to be less interested, their involvement and proper education about the system can be critical.

Therefore, a key strategy in building a support team is to balance and to distribute the knowledge among those that are best placed to support the network for the long term.

You should involve young people, but alongside them involve a mix of more mature contributors. Find people who are committed to the community, who have roots in the community, who are motivated to learn and teach. A complementary strategy is to divide up the duties and functions, and to document all methodology and procedures. In this way, people can be trained easily, and substituted with little effort.

On one project the training team selected a bright young university graduate who had returned to his village. He was very motivated and learned quickly. Soon he was an expert in IT and networking skills as well as the local network. He was able to deal with many different problems, from fixing a PC to rewiring Ethernet cable. Unfortunately, two months after the project launch he was offered a government job and left the community. Even a better salary could not keep him, since the prospect of a stable government job was too appealing. All of the knowledge about the network and how to support it left with him. The training team had to return and begin the training again, this time with local people who were known to be remaining in the village. Although this retraining took much longer, the community could be assured that the knowledge and skills would remain in the village for a longer time.

It is often best to find a local partner organisation or a local manager, and work with them to find the right technical team. Values, history, local politics, and many other factors will be important to them, while remaining completely unfathomable to people who are not from that community. The best approach is to coach your local partner, providing them with sound criteria, make sure that they understand that criteria, and set firm boundaries.

Such boundaries should include rules about nepotism and patronage, considering, of course, your own local situation. It may be impossible to say that you cannot hire kin, but it is best to provide a means of checks and balances.

Where a candidate is kin, there should be clear criteria and a second authority in deciding upon their candidacy.

It is also important that the local partner is given authority and is not undermined by the project organisers, thus compromising their ability to manage. They will be best able to judge who will work best with them.

If they are well educated in this process, then your requirements should be satisfied. Troubleshooting and support of technology is an abstract art.

The first time you look at an abstract painting, it may just look to you like a bunch of random paint splatters. After reflecting on the composition for a time, you may come to appreciate the work as a whole, and the "invisible" coherence becomes very real.

The neophyte looking at a wireless network may see the antennas and wires and computers, but it can take a while for them to appreciate the point of the "invisible" network. In rural areas, it can often take a huge leap of understanding before locals will appreciate an invisible network that is simply dropped into their village. Therefore, a phased approach is needed to ease people into supporting technology systems. Again, the best method is involvement.

Once the participants are chosen and committed to the project, involve them as much as possible. Let them "drive". Give them the cable crimper or keyboard and show them how to do the work. Even if you do not have time to explain every detail and even if it will take longer, they need to be involved physically and see not only what has been done, but how much work was done.

The scientific method is taught in many schools and universities.

Many people learn about it by the time they reach high-school science class. Simply put, answers are achieved and problems solved by taking a set of possibilities, then slowly eliminating them through binary tests until you are left with one or only a few possibilities.

With those possibilities in mind, you complete the experiment.

You then test to see if the experiment yields something similar to the expected result. If it did not, you re-calculate your expected result and try again. A young person you might be thinking of employing in your project may have been introduced to the concept, but likely will not have had the opportunity to troubleshoot complex problems.

Even if they are familiar with the scientific method, they might not think to apply it to resolving real problems. This method is very effective, although time consuming. It can be sped up by making logical assumptions. For example, if a long-running access point suddenly stops working after a storm, you might suspect a power supply related problem and thus skip most of the procedure.

People charged with supporting technology should be taught how to troubleshoot using this method, as there will be times when the problem is neither known nor evident. Simple decision trees or flow charts can be made that test variables, and then elimination of the variables to isolate the problem can be suggested.

Of course, these charts should not be followed blindly. It is often easier to teach this method using a non technological problem first.

For example, have your student develop a problem resolution procedure on something simple and familiar, like a battery powered television.

Start by sabotaging the television. Give them a battery that is not charged. Disconnect the aerial. Insert a broken fuse. Test the student, making it clear that each problem will show specific symptoms, and point the way as to how to proceed. Once they have fixed the television, have them apply this procedure to a more complicated problem.

In a network, you can change an IP address, switch or damage cables, use the wrong SSID, or orient the antenna in the wrong direction. It is important that they develop a methodology and procedure to resolve these problems.

Proper troubleshooting techniques

No troubleshooting methodology can completely cover all problems you will encounter when working with wireless networks. But often, problems come down to one of a few common mistakes. Here are a few simple points to keep in mind that can get your troubleshooting effort working in the right direction.

- **Don't panic.** If you are troubleshooting a system, it means that it was working at one time, probably very recently. Before jumping in and making changes, survey the scene and assess exactly what is broken. If you have historical logs or statistics to work from, all the better. If others were using it before it started having problems, ask them to help you by having them tell you what was happening before it stopped working. Be thorough, but don't make it sound like you are accusing them of breaking it.

  They may have important information that will help you fix things and you want them to be on your side. Be sure to collect information first, so you can make an informed decision before making changes.

- **Make a backup**. This applies before you notice problems, as well as after. If you make a complicated software change to a system, having a backup means that you can quickly restore it to the previous settings and start again. When troubleshooting very complex problems, having a configuration that "sort-of" works can be much better than having a mess that doesnt work at all (and that you cant easily restore from memory).

  Even in a broken configuration, make a backup copy of the parts of the system you will be changing before you try to make significant changes. If your changes result in an even worse state than when you first started working on it, you will at least have a known situation to go back to.

- **Is it plugged in?** This step is often overlooked until many other avenues are explored. Plugs can be accidentally (or intentionally) unplugged very easily. Is the lead connected to a good power source? Is the other end connected to your device? Is the power light on? It may sound silly, but you will feel even sillier if you spend a lot of time checking out an antenna feed line only to realise that the AP was unplugged the entire time. Trust me, it happens more often than most of us would care to admit.

- **What was the last thing changed?** If you are the only person with access to the system, what is the last change you made? If others have access to it, what is the last change they made and when? When was the last time the system worked? Often, system changes have unintended consequences that may not be immediately noticed. Roll back that change and see what effect it has on the problem.

- **Look at date/timestamps on files.** Every file on a modern computer system has a date & time associated with it showing when it was created or last changed. On a properly running system, most of the system files will have date/timestamps from months or even years ago. If the system or network was running fine until an hour or so ago, files which have a timestamp within the past few minutes to an hour ago could provide clues about what changed.

- **The known good.** This idea applies to hardware, as well as software. A known good is any component that you can replace in a complex system to verify that its counterpart is in good, working condition. For example, you may carry a tested Ethernet cable in a tool kit. If you suspect problems with a cable in the field, you can easily swap out the suspect cable with the known good and see if things improve. This is much faster and less error-prone than re-crimping a cable, and immediately tells you if the change fixes the problem. Likewise, you may also pack a backup battery, antenna cable, or a CD-ROM with a known good configuration for the system. When fixing complicated problems, saving your work at a given point lets you return to it as a known good, even if the problem is not yet completely solved.

- **Determine what still works.** This will help you "put a fence around the problem". While complex systems like a wireless network can be made up of many different components, it is likely that the problem is only with a very small number of them. If, for example, somebody in a lab complains they can't access the Internet, check to see if others in the same lab are experiencing the same problem. Is there connectivity in another lab or elsewhere in the building? If the problem is just with one user or within one room, you would want to concentrate your efforts on the equipment in just that one space. If the outage were more widespread, perhaps looking at the equipment where your outside connections come in is more appropriate.

- **Do no harm.** If you don't fully understand how a system works, don't be afraid to call in an expert. If you are not sure if a particular change will damage another part of the system, then either find someone with more experience to help you or devise a way to test your change without doing damage. Putting a penny in place of a fuse may solve the immediate problem, but it may also burn down the building.

Your troubleshooting team will need to have good troubleshooting skills, but may not be competent enough to configure a router from scratch or crimp a piece of LMR-400.

It is often much more efficient to have a number of backup components on-hand, and train your team to be able to swap out the entire broken part. This could mean having an access point or router pre-configured and sitting in a locked cabinet, plainly labelled and stored with backup cables and power supplies. Your team can swap out the failed component, and either send the broken part to an expert for repair, or arrange to have another backup sent in.

Assuming that the backups are kept secure and are replaced when used, this can save a lot of time for everyone.

Common network problems

Now we will take a look at some common network problems that you are almost certain to face.

Often, connectivity problems come from failed components, adverse weather, or simple misconfiguration.

Once your network is connected to the Internet or opened up to the general public, considerable threats will come from the network users themselves.

These threats can range from the benign to the outright malevolent, but all will have impact on your network if it is not properly configured. This section looks at some common problems found once your network is in use.

Locally hosted websites

If a university hosts its website locally, visitors to the website from outside the campus and the rest of the world will compete with the university's staff for Internet bandwidth.

This includes automated access from search engines that periodically spider your entire site.

One solution to this problem is to use split DNS and mirroring.

The university mirrors a copy of its websites to a server at, say, a European hosting company, and uses split DNS to direct all users from outside the university network to the mirror site, while users on the university network access the same site locally.

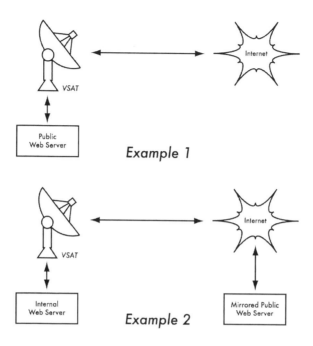

Figure MT 1: In Example 1, all website traffic coming from the Internet must traverse the VSAT. In Example 2, the public web site is hosted on a fast European service, while a copy is kept on an internal server for very fast local access. This improves the VSAT connection and reduces load times for web site users.

Open proxies

A proxy server should be configured to accept only connections from the university network, not from the rest of the Internet.

This is because people elsewhere will connect and use open proxies for a variety of reasons, such as to avoid paying for international bandwidth. The way to configure this depends on the proxy server you are using. For example, you can specify the IP address range of the campus network in your squid.conf file as the only network that can use Squid.

Alternatively, if your proxy server lies behind a border firewall, you can configure the firewall to only allow internal hosts to connect to the proxy port.

Open relay hosts

An incorrectly configured mail server will be found by unscrupulous people on the Internet, and be used as a relay host to send bulk email and spam.

They do this to hide the true source of the spam, and avoid getting caught. To test for an open relay host, the following test should be carried out on your mail server (or on the SMTP server that acts as a relay host on the perimeter of the campus network). Use telnet to open a connection to port 25 of the server in question.

telnet mail.uzz.ac.zz 25

Then, if an interactive command-line conversation can take place (for example, as follows), the server is an open relay host:

MAIL FROM: spammer@waste.com
250 OK - mail from
RCPT TO: innocent@university.ac.zz
 250 OK - rcpt to spammer@waste.com

Instead, the reply after the first MAIL FROM should be something like:

550 Relaying is prohibited.

An online tester is available at sites such as
http://www.mailradar.com/openrelay/ or http://www.checkor.com/

Since bulk emailers have automated methods to find such open relay hosts, an institution that does not protect its mail systems is almost guaranteed to be found and abused.

Configuring the mail server not to be an open relay consists of specifying the networks and hosts that are allowed to relay mail through them in the MTA (eg., Sendmail, Postfix, Exim, or Exchange).
This will likely be the IP address range of the campus network.

Peer-to-peer networking

Bandwidth abuse through peer-to-peer (P2P) file-sharing programs can be prevented in the following way.

Make it impossible to install new programs on campus computers. By not giving regular users administrative access to PC workstations, it is possible to prevent the installation of bandwidth hungry applications. Many institutions also standardise on a desktop build, where they install the required operating system on one PC. They then install all the necessary applications on it, and configure these in an optimal way. The PC is also configured in a way that prevents users from installing new applications. A disk image of this PC is then cloned to all other PCs using software such as Partition Image (see http://www.partimage.org/) or Drive Image Pro (see http://www.powerquest.com/).

From time to time, users may succeed in installing new software or otherwise damaging the software on the computer (causing it to hang often, for example). When this happens, an administrator can simply put the disk image back, causing the operating system and all software on the computer to be exactly as specified.

Programs that install themselves (from the Internet)

There are programs that automatically install themselves and then keep on using bandwidth.

Some programs are spyware, which keep sending information about a user's browsing habits to a company somewhere on the Internet.

These programs are preventable to some extent by user education and locking down PCs to prevent administrative access for normal users. In other cases, there are software solutions to find and remove these problem programs, such as Spychecker (http://www.spychecker.com/).

Windows updates

Microsoft Windows operating systems assume that a computer with a LAN connection has a good link to the Internet, and automatically downloads security patches, bug fixes and feature enhancements from the Microsoft Web site. This can consume massive amounts of bandwidth on an expensive Internet link.

The two possible approaches to this problem are:

- Disable Windows updates on all workstation PCs. The security updates are very important for servers, but whether workstations in a protected private network such as a campus network need them is debatable.

- Install a Software Update Server. This is a free program from Microsoft that enables you to download all the updates from Microsoft overnight on to a local server and distribute the updates to client workstations from there. In this way, Windows updates need not use any bandwidth on the Internet link during the day. Unfortunately, all client PCs need to be configured to use the Software Update Server for this to have an effect.
  If you have a flexible DNS server, you can also configure it to answer requests for windowsupdate.microsoft.com and direct the updater to your update server. This is only a good option for large networks, but can save untold amounts of Internet bandwidth.

Programs that assume a high bandwidth link

In addition to Windows updates, many other programs and services assume that bandwidth is not a problem, and therefore consume bandwidth for reasons the user might not predict. For example, anti-virus packages (such as Norton AntiVirus) periodically update themselves automatically and directly from the Internet. It is better if these updates are distributed from a local server.

Other programs, such as the RealNetworks video player, automatically download updates and advertisements, as well as upload usage patterns back to a site on the Internet. Innocuous looking applets (like Konfabulator and Dashboard widgets) continually poll Internet hosts for updated information. These can be low bandwidth requests (like weather or news updates), or very high bandwidth requests (such as webcams). These applications may need to be throttled or blocked altogether.

The latest versions of Windows and Mac OS X also have a time synchronisation service. This keeps the computer clock accurate by connecting to time servers on the Internet. It is more efficient to install a local time server and distribute accurate time from there, rather than to tie up the Internet link with these requests.

Worms and viruses

Worms and viruses can generate enormous amounts of traffic. It is therefore essential that anti-virus protection is installed on all PCs. Furthermore, user education about executing attachments and responding to unsolicited email is essential. In fact, it should be a policy that no workstation or server should run unused services.

A PC should not have shares unless it is a file server; and a server should not run unnecessary services either.

For example, Windows and Unix servers typically run a web server service by default. This should be disabled if that server has a different function; the fewer services a computer runs, the less there is to exploit.

Email forwarding loops

Occasionally, a single user making a mistake can cause a problem. For example, a user whose university account is configured to forward all mail to her Yahoo account.

The user goes on holiday. All emails sent to her in her absence are still forwarded to her Yahoo account, which can grow to only 2 MB.

When the Yahoo account becomes full, it starts bouncing the emails back to the university account, which immediately forwards it back to the Yahoo account.

An email loop is formed that might send hundreds of thousands of emails back and forth, generating massive traffic and crashing mail servers.

There are features of mail server programs that can recognise loops. These should be turned on by default.

Administrators must also take care that they do not turn this feature off by mistake, or install an SMTP forwarder that modifies mail headers in such a way that the mail server does not recognise the mail loop.

Large downloads

A user may start several simultaneous downloads, or download large files such as 650MB ISO images.

In this way, a single user can use up most of the bandwidth.

The solutions to this kind of problem lie in training, offline downloading, and monitoring.

Offline downloading can be implemented in at least two ways:

- At the University of Moratuwa, a system was implemented using URL redirection. Users accessing ftp:// URLs are served a directory listing in which each file has two links: one for normal downloading, and the other for offline downloading. If the offline link is selected, the specified file is queued for later download and the user notified by email when the download is complete. The system keeps a cache of recently downloaded files, and retrieves such files immediately when requested again. The download queue is sorted by file size. Therefore, small files are downloaded first. As some bandwidth is allocated to this system even during peak hours, users requesting small files may receive them within minutes, sometimes even faster than an online download.

- Another approach would be to create a web interface where users enter the URL of the file they want to download. This is then downloaded overnight using a cron job or scheduled task. This system would only work for users who are not impatient, and are familiar with what file sizes would be problematic for download during the working day.

Sending large files

When users need to transfer large files to collaborators elsewhere on the Internet, they should be shown how to schedule the upload. In Windows, an upload to a remote FTP server can be done using an FTP script file, which is a text file containing FTP commands.

Users sending each other files

Users often need to send each other large files. It is a waste of bandwidth to send these via the Internet if the recipient is local. A file share should be created on the local Windows/Samba/Mac web server, where a user can put the large file for others to access.

Alternatively, a web front-end can be used for a local web server to accept a large file and place it in a download area. After uploading it to the web server, the user receives a URL for the file.

He can then give that URL to his local or international collaborators, and when they access that URL they can download it.

This is what the University of Bristol has done with their FLUFF system. The University offers a facility for the upload of large files (FLUFF) available from *http://www.bris.ac.uk/it-services/applications/fluff/*

There are also tools like SparkleShare (*http://sparkleshare.org/*) and LipSync (*https://github.com/philcryer/lipsync*) which are Open Source packages that you can install and set up yourself to do much the same thing.
There are also new online free services such as Google Drive which can be setup for file sharing and common editing.

Consider using rsync (http://rsync.samba.org/) for those users who need to send each other the same/similar large files on a regular basis.
The Rsync protocol is a synchronisation rather than a straightforward file transfer protocol.
Rather than simply sending a file from start to end, it checks with an rsync server on the destination host to see if the file being sent already exists.
If it does, both sides compare their copies and the sender transmits over only the differences to the destination.

For example, if a 10 MB database of research data only has 23 KB of new data vs. the last version, only the 23 KB of changes will be transmitted.
Rsync can also use the SSH protocol, providing a secure transport layer for sync actions.

Trouble tracking and reporting

Troubleshooting is only one half of the task of problem-solving on a wireless network.
Once a problem has been diagnosed and fixed, it needs to be documented in a permanent way so that others who work on the network either now or in the future, will be able to learn from the incident.

Keeping a record of problems and incidents that occur is also a good way to track and fix long-term problems that may occur, say, once every few months but follow a definite pattern.

You can also reduce the complexity and frustration of troubleshooting a problem if you keep a log of every change made to the network.

The Logbook is where you and your team write down every change made to a system, along with the date and time the change was made. For example:

23 July 10:15AM Changed default route on host alpha from 123.45.67.89 to 123.56.78.1 because upstream ISP moved our gateway.

As your network grows, consider installing a trouble-tracking system like JIRA or Bugzilla to help better keep track of who is working on what problem and what happened during that work. This provides a history of what was worked on and how it was fixed and also provides an orderly method for assigning tasks and helping to prevent things like two people stepping on each other's work while they both try to fix the same problem.

Trouble-ticketing systems are a subject that would fill an entire separate book, so we will only mention them briefly here to make you aware of them. They can also be quite complex to set up, so for simple networks, a logbook will do.

16. NETWORK MONITORING

Introduction

Network monitoring is the use of logging and analysis tools to accurately determine traffic flows, utilisation, and other performance indicators on a network. Good monitoring tools give you both hard numbers and graphical aggregate representations of the state of the network. This helps you to visualise precisely what is happening, so you know where adjustments may be needed. These tools can help you answer critical questions, such as:

- What are the most popular services used on the network?
- Who are the heaviest network users?
- What other wireless channels are in use in my area?
- Are users installing wireless access points on my private wired network?
- At what time of the day is the network most utilised?
- What sites do your users frequent?
- Is the amount of inbound or outbound traffic close to the available network capacity?
- Are there indications of an unusual network situation that is consuming bandwidth or causing other problems?
- Is our Internet Service Provider (ISP) providing the level of service that we are paying for?

This should be answered in terms of available bandwidth, packet loss, latency, and overall availability.
And perhaps the most important question of all:

- Do the observed traffic patterns fit our expectations?

Monitoring and metrics tools are vitally important programs to have on hand to check on the health of your network and diagnose/troubleshoot problems. Throughout previous chapters in this book we've mentioned or given brief examples of using certain tools for specific tasks like configuration and setup, troubleshooting, gathering statistics and metrics data about the health of your network, etc.

This section discusses some of these tools in a more detailed manner. It should be noted that this is by no means an exhaustive list of all the tools available for wired and wireless networks.

It is also important to realise that diagnostic and monitoring tools change just like all other software and hardware does. Staying up to date on the latest versions, bugs in existing versions, new tools in the field, etc. can be an almost full-time job in itself.

For this book, where we've found that a certain tool is no longer being actively maintained between the previous edition and this one, we have left it out of this text. The tools discussed in this section are all being currently developed as of this writing, but it is left as an exercise to the reader to determine if a particular tool is suitable for their situation.

Network monitoring example

Let's look at how a typical system administrator can make good use of network monitoring tools.

An effective network monitoring example

For the purposes of example, let's assume that we are in charge of a network that has been running for three months. It consists of 50 computers and three servers: email, web, and proxy servers.

While initially things are going well, users begin to complain of slow network speeds and an increase in spam emails.

As time goes on, computer performance slows to a crawl (even when not using the network), causing considerable frustration in your users.

With frequent complaints and very low computer usage, the Board is questioning the need for so much network hardware. The Board also wants evidence that the bandwidth they are paying for is actually being used. As the network administrator, you are on the receiving end of these complaints.

How can you diagnose the sudden drop in network and computer performance and also justify the network hardware and bandwidth costs?

Monitoring the LAN (local traffic)

To get an idea of exactly what is causing the slow down, you should begin by looking at traffic on the local LAN. There are several advantages to monitoring local traffic:

1. Troubleshooting is greatly simplified. Viruses can be detected and eliminated.
2. Malicious users can be detected and dealt with.
3. Network hardware and resources can be justified with real statistics.

Assume that all of the switches support the Simple Network Management Protocol (SNMP). SNMP is an application-layer protocol designed to facilitate the exchange of management information between network devices.

By assigning an IP address to each switch, you are able to monitor all the interfaces on that switch, observing the entire network from a single point. This is much easier than enabling SNMP on all computers in a network.

By using a free tool such as MRTG, http://oss.oetiker.ch/mrtg/, you can monitor each port on the switch and present data graphically, as an aggregate average over time. The graphs are accessible from the web, so you are able to view the graphs from any machine at anytime.

With MRTG monitoring in place, it becomes obvious that the internal LAN is swamped with far more traffic than the Internet connection can support, even when the lab is unoccupied. This is a pretty clear indication that some of the computers are infested with a network virus.

After installing good anti-virus and anti-spyware software on all of the machines, the internal LAN traffic settles down to expected levels.

The machines run much more quickly, spam emails are reduced, and the users' morale quickly improves.

Monitoring the WAN (external traffic)

In addition to watching the traffic on the internal LAN, you need to demonstrate that the bandwidth the organisation is paying for is actually what they are getting from their ISP. You can achieve this by monitoring external traffic. External traffic is generally classified as anything sent over a Wide Area Network (WAN). Anything received from (or sent to) a network other than your internal LAN also qualifies as external traffic.

The advantages of monitoring external traffic include:

Internet bandwidth costs are justified by showing actual usage, and whether that usage agrees with your ISP's bandwidth charges.

Future capacity needs are estimated by watching usage trends and predicting likely growth patterns. Intruders from the Internet are detected and filtered before they can cause problems.

Monitoring this traffic is easily done with the use of MRTG on an SNMP enabled device, such as a router. If your router does not support SNMP, then you can add a switch between your router and your ISP connection, and monitor the port traffic just as you would with an internal LAN.

Detecting network outages

With monitoring tools in place, you now have an accurate measurement of how much bandwidth the organisation is using.

This measurement should agree with your ISP's bandwidth charges.

It can also indicate the actual throughput of your connection if you are using close to your available capacity at peak times.

A "flat top" graph is a fairly clear indication that you are operating at full capacity. The following figure NM 1 shows flat tops in peak outbound traffic in the middle of every day except Sunday.

It is clear that your current Internet connection is overutilised at peak times, causing network lag. After presenting this information to the Board, you can make a plan for further optimising your existing connection (by upgrading your proxy server and using other techniques in this book) and estimate how soon you will need to upgrade your connection to keep up with the demand.

This is also an excellent time to review your operational policy with the Board, and discuss ways to bring actual usage in line with that policy.

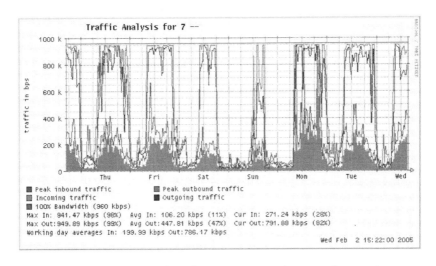

Figure NM 1: A graph with a "flat top" is one indication of overutilisation.

Later in the week, you receive an emergency phone call in the evening. Apparently, no one in the lab can browse the web or send email. You rush to the lab and hastily reboot the proxy server, with no results. Browsing and email are still broken. You then reboot the router, but there is still no success. You continue eliminating the possible fault areas one by one until you realise that the network switch is off - a loose power cable is to blame. After applying power, the network comes to life again.

How can you troubleshoot such an outage without such time consuming trial and error? Is it possible to be notified of outages as they occur, rather than waiting for a user to complain? One way to do this is to use a program such as Nagios (http://www.nagios.org/) that continually polls network devices and notifies you of outages. Nagios will report on the availability of various machines and services, and will alert you to machines that have gone down. In addition to displaying the network status graphically on a web page, it will send notifications via SMS or email, alerting you immediately when problems arise.

With good monitoring tools in place, you will be able to justify the cost of equipment and bandwidth by effectively demonstrating how it is being used by the organisation.
You are notified automatically when problems arise, and you have historical statistics of how the network devices are performing. You can check the current performance against this history to find unusual behaviour, and head off problems before they become critical. When problems do come up, it is simple to determine the source and nature of the problem. Your job is easier, the Board is satisfied, and your users are much happier.

Monitoring your network

Managing a network without monitoring is similar to driving a vehicle without a speedometer or a fuel gauge.
How do you know how fast you are going? Is the car consuming fuel as efficiently as promised by the dealers? If you do an engine overhaul several months later, is the car any faster or more efficient than it was before?
Similarly, how can you pay for an electricity or water bill without seeing your monthly usage from a meter? You must have an account of your network bandwidth utilisation in order to justify the cost of services and hardware purchases, and to account for usage trends.

There are several benefits to implementing a good monitoring system for your network:

- Network budget and resources are justified. Good monitoring tools can demonstrate without a doubt that the network infrastructure (bandwidth, hardware, and software) is suitable and able to handle the requirements of network users.

- Network intruders are detected and filtered. By watching your network traffic, you can detect attackers and prevent access to critical internal servers and services.

- Network viruses are easily detected. You can be alerted to the presence of network viruses, and take appropriate action before they consume Internet bandwidth and destabilise your network.

- Troubleshooting of network problems is greatly simplified. Rather than attempting "trial and error" to debug network problems, you can be instantly notified of specific problems. Some kinds of problems can even be repaired automatically.

- Network performance can be highly optimised. Without effective monitoring, it is impossible to fine tune your devices and protocols to achieve the best possible performance.

- Capacity planning is much easier. With solid historical performance records, you do not have to "guess" how much bandwidth you will need as your network grows.

- Proper network usage can be enforced. When bandwidth is a scarce resource, the only way to be fair to all users is to ensure that the network is being used for its intended purpose.

Fortunately, network monitoring does not need to be an expensive undertaking. There are many freely available open source tools that will show you exactly what is happening on your network in considerable detail. This section will help you identify many invaluable tools and how best to use them.

The dedicated monitoring server

While monitoring services can be added to an existing network server, it is often desirable to dedicate one machine (or more, if necessary) to network monitoring. Some applications (such as ntop http://www.ntop.org/) require considerable resources to run, particularly on a busy network.

But most logging and monitoring programs have modest RAM and storage requirements, typically with little CPU power required. Since open source operating systems (such as Linux or BSD) make very efficient use of hardware resources, this makes it possible to build a very capable monitoring server from recycled PC parts. There is usually no need to purchase a brand new server to relegate to monitoring duties.

The exception to this rule is in very large installations.
If your network includes more than a few hundred nodes, or if you consume more than 50 Mbps of Internet bandwidth, you will likely need to split up monitoring duties between a few dedicated machines.

This depends largely on exactly what you want to monitor.
If you are attempting to account for all services accessed per MAC address, this will consume considerably more resources than simply measuring network flows on a switch port.
But for the majority of installations, a single dedicated monitoring machine is usually enough.While consolidating monitoring services to a single machine will streamline administration and upgrades, it can also ensure better ongoing monitoring.

For example, if you install monitoring services on a web server, and that web server develops problems, then your network may not be monitored until the problem is resolved. To a network administrator, the data collected about network performance is nearly as important as the network itself.

Your monitoring should be robust and protected from service outages as well as possible.
Without network statistics, you are effectively blind to problems with the network.

Where does the server fit in your network?

If you are only interested in collecting network flow statistics from a router, you can do this from just about anywhere on the LAN.
This provides simple feedback about utilisation, but cannot give you comprehensive details about usage patterns.

Figure NM 2 below shows a typical MRTG graph generated from the Internet router. While the inbound and outbound utilisation are clear, there is no detail about which computers, users, or protocols are using bandwidth.

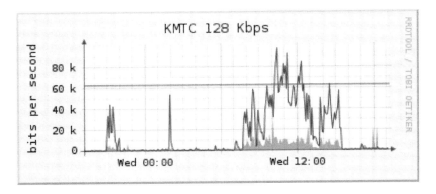

Figure NM 2: Polling the edge router can show you the overall network utilisation, but you cannot break the data down further into machines, services, and users.

For more detail, the dedicated monitoring server must have access to everything that needs to be watched.
Typically, this means it must have access to the entire network.
To monitor a WAN connection, such as the Internet link to your ISP, the monitoring server must be able to see the traffic passing through the edge router. To monitor a LAN, the monitoring server is typically connected to a monitor port on the switch. If multiple switches are used in an installation, the monitoring server may need a connection to all of them.
That connection can either be a physical cable, or if your network switches support it, a VLAN specifically configured for monitoring traffic.

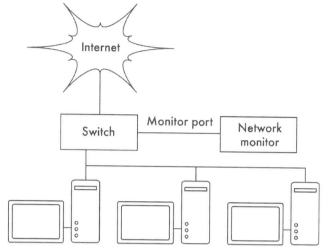

Figure NM 3: Use the monitor port on your switch to observe traffic crossing all of the network ports.

If monitor port functionality is not available on your switch, the monitoring server may be installed between your internal LAN and the Internet. While this will work, it introduces a single point of failure for the network, as the network will fail if the monitoring server develops a problem. It is also a potential performance bottleneck, if the server cannot keep up with the demands of the network.

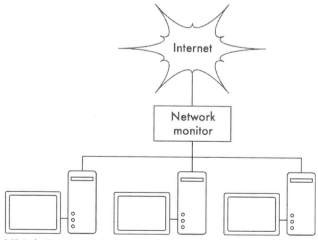

Figure NM 4: By inserting a network monitor between the LAN and your Internet connection, you can observe all network traffic.

A better solution is to use a simple network hub (not a switch) which connects the monitoring machine to the internal LAN, external router, and the monitoring machine.

While this does still introduce an additional point of failure to the network (since the entire network will be unreachable if the hub dies), hubs are generally considered to be much more reliable than routers.

They are also very easily replaced should they fail.

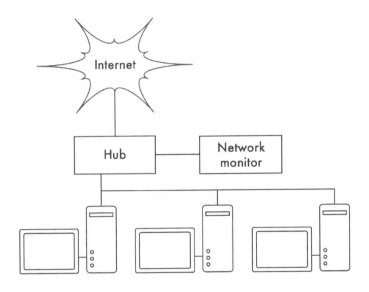

Figure NM 5: If your switch does not provide monitor port functionality, you can insert a network hub between your Internet router and the LAN, and connect the monitoring server to the hub.

Once your monitoring server is in place, you are ready to start collecting data.

What to monitor

It is possible to plot just about any network event and watch its value on a graph over time. Since every network is slightly different, you will have to decide what information is important in order to gauge the performance of your network.

Here are some important indicators that many network administrators will typically track.

Wireless statistics

- Received signal and noise from all backbone nodes
- Number of associated stations
- Detected adjacent networks and channels
- Excessive retransmissions
- Radio data rate, if using automatic rate scaling

Switch statistics

- Bandwidth usage per switch port
- Bandwidth usage broken down by protocol
- Bandwidth usage broken down by MAC address
- Broadcasts as a percentage of total packets
- Packet loss and error rate

Internet statistics

- Internet bandwidth use by host and protocol
- Proxy server cache hits
- Top 100 sites accessed
- DNS requests
- Number of inbound emails / spam emails / email bounces
- Outbound email queue size
- Availability of critical services (web servers, email servers, etc.).
- Ping times and packet loss rates to your ISP
- Status of backups

System health statistics

- Memory usage
- Swap file usage
- Process count / zombie processes
- System load
- Uninterruptible Power Supply (UPS) voltage and load
- Temperature, fan speed, and system voltages
- Disk SMART status
- RAID array status

You should use this list as a suggestion of where to begin. As your network matures, you will likely find new key indicators of network performance, and you should of course track those as well.

There are many freely available tools that will show you as much detail as you like about what is happening on your network.

You should consider monitoring the availability of any resource where unavailability would adversely affect your network users.

Don't forget to monitor the monitoring machine itself, for example its CPU usage and disk space, in order to receive advance warning if it becomes overloaded or faulty. A monitoring machine that is low on resources can affect your ability to monitor the network effectively.

Types of monitoring tools

We will now look at several different classes of monitoring tools.

1. Network detection tools listen for the beacons sent by wireless access points, and display information such as the network name, received signal strength, and channel.

2. Spot check tools are designed for troubleshooting and normally run interactively for short periods of time. A program such as ping may be considered an active spot check tool, since it generates traffic by polling a particular machine.

3. Passive spot check tools include protocol analysers, which inspect every packet on the network and provide complete detail about any network conversation (including source and destination addresses, protocol information, and even application data).

4. Trending tools perform unattended monitoring over long periods, and typically plot the results on a graph.

5. Throughput testing tools tell you the actual bandwidth available between two points on a network.

6. Realtime monitoring tools perform similar monitoring, but notify administrators immediately if they detect a problem. Intrusion detection tools watch for undesirable or unexpected network traffic, and take appropriate action (typically denying access and/or notifying a network administrator).

Network detection

The simplest wireless monitoring tools simply provide a list of available networks, along with basic information (such as signal strength and channel). They let you quickly detect nearby networks and determine if they are in range or are causing interference.

The built-in client.

All modern operating systems provide built-in support for wireless networking. This typically includes the ability to scan for available networks, allowing the user to choose a network from a list. While virtually all wireless devices are guaranteed to have a simple scanning utility, functionality can vary widely between implementations. These tools are typically only useful for configuring a computer in a home or office setting.

They tend to provide little information apart from network names and the available signal to the access point currently in use.

Netstumbler

(http://www.wirelessdefence.org/Contents/NetstumblerMain.htm). This is the most popular tool for detecting wireless networks using Microsoft Windows. It supports a variety of wireless cards, and is very easy to use. It will detect open and encrypted networks, but cannot detect "closed" wireless networks. It also features a signal/noise meter that plots radio receiver data as a graph over time. It also integrates with a variety of GPS devices, for logging precise location and signal strength information.

This makes Netstumbler a handy tool to have for an informal site survey. Macstumbler (http://www.macstumbler.com/). While not directly related to the Netstumbler, Macstumbler provides much of the same functionality but for the Mac OS X platform. It works with all Apple Airport cards.

Spot check tools

What do you do when the network breaks? If you can't access a web page or email server, and clicking the reload button doesn't fix the problem, then you'll need to be able to isolate the exact location of the problem.

These tools will help you to determine just where a connection problem exists.

This section is simply an introduction to commonly used troubleshooting tools.

For more discussion of common network problems and how to diagnose them, see the chapter called **Maintenance and Troubleshooting**.

ping

Just about every operating system (including Windows, Mac OS X, and of course Linux and BSD) includes a version of the ping utility. It uses ICMP packets to attempt to contact a specified host, and tells you how long it takes to get a response.

Knowing what to ping is just as important as knowing how to ping. If you find that you cannot connect to a particular service in your web browser (say, http://yahoo.com/), you could try to ping it:

$ ping yahoo.com

PING yahoo.com (66.94.234.13): 56 data bytes

64 bytes from 66.94.234.13: icmp_seq=0 ttl=57 time=29.375 ms
64 bytes from 66.94.234.13: icmp_seq=1 ttl=56 time=35.467 ms
64 bytes from 66.94.234.13: icmp_seq=2 ttl=56 time=34.158 ms
^C

--- yahoo.com ping statistics ---

3 packets transmitted, 3 packets received, 0% packet loss

round-trip min/avg/max/stddev = 29.375/33.000/35.467/2.618 ms

Hit control-C when you are finished collecting data.

If packets take a long time to come back, there may be network congestion.

If return ping packets have an unusually low Time To Live (TTL), you may have routing problems between your machine and the remote end. But what if the ping doesn't return any data at all?

If you are pinging a name instead of an IP address, you may be running into DNS problems.

Try pinging an IP address on the Internet. If you can't reach it, it's a good idea to see if you can ping your default router:

$ ping 69.90.235.230

PING 69.90.235.230 (69.90.235.230): 56 data bytes

64 bytes from 69.90.235.230: icmp_seq=0 ttl=126 time=12.991 ms
64 bytes from 69.90.235.230: icmp_seq=1 ttl=126 time=14.869 ms
64 bytes from 69.90.235.230: icmp_seq=2 ttl=126 time=13.897 ms
^C

--- 216.231.38.1 ping statistics ---

3 packets transmitted, 3 packets received, 0% packet loss

round-trip min/avg/max/stddev = 12.991/13.919/14.869/0.767 ms

If you can't ping your default router, then chances are you won't be able to get to the Internet either. If you can't even ping other IP addresses on your local LAN, then it's time to check your connection. If you're using Ethernet, is it plugged in? If you're using wireless, are you connected to the proper wireless network, and is it in range?
While it is generally accurate to assume that a machine that does not respond to a ping is likely to be down or cut off from the network this is not always 100% the case.
Particularly on a WAN or the Internet itself, it is also possible that some router/firewall between you and the target host(even the target itself) is blocking pings. If you find a machine is not responding to pings, try another well-known service like ssh or http.
If you can reach the target through either of these services then you know the machine is up and simply blocking pings. It is also worth noting that different systems treat ping differently. The classic UNIX ping utility sends an ICMP ECHO protocol packet to the target host. Some network devices will respond to a ping automatically regardless of whether ICMP is being blocked further up the protocol stack.
This can also be misleading because it can indicate a host is up when actually all that's really going on is that the NIC (Network Interface Card) is powered and the machine itself is not actually up and running.

As we stated above, it's always good to check connectivity with multiple methods. Network debugging with ping is a bit of an art, but it is useful to learn. Since you will likely find ping on just about any machine you will work on, it's a good idea to learn how to use it well.

traceroute and mtr

http://www.bitwizard.nl/mtr/

As with ping, traceroute is found on most operating systems (it's called tracert in some versions of Microsoft Windows). By running traceroute, you can find the location of problems between your computer and any point on the Internet:

$ traceroute -n google.com

traceroute to google.com (72.14.207.99), 64 hops max, 40 byte packets

1 10.15.6.1 4.322 ms 1.763 ms 1.731 ms

2 216.231.38.1 36.187 ms 14.648 ms 13.561 ms

3 69.17.83.233 14.197 ms 13.256 ms 13.267 ms

4 69.17.83.150 32.478 ms 29.545 ms 27.494 ms

5 198.32.176.31 40.788 ms 28.160 ms 28.115 ms

6 66.249.94.14 28.601 ms 29.913 ms 28.811 ms

7 172.16.236.8 2328.809 ms 2528.944 ms 2428.719 ms

*8 * * ***

The -n switch tells traceroute not to bother resolving names in DNS, and makes the trace run more quickly. You can see that at hop seven, the round trip time shoots up to more than two seconds, while packets seem to be discarded at hop eight.

This might indicate a problem at that point in the network.
If this part of the network is in your control, it might be worth starting your troubleshooting effort there.

My TraceRoute (mtr) is a handy program that combines ping and traceroute into a single tool. By running mtr, you can get an ongoing average of latency and packet loss to a single host, instead of the momentary snapshot that ping and traceroute provide.

My traceroute [v0.69⊠

tesla.rob.swn (0.0.0.0) (tos=0x0 psize=64 bitpat Sun Jan 8 20:01:26 2006)

Keys: Help Display mode Restart statistics Order of fields quit

	Packets				Pings		
Host	**Loss%**	**Snt**	**Last**	**Avg**	**Best**	**Wrst**	**StDev**
1. gremlin.rob.swn	0.0%	4	1.9	2.0	1.7	2.6	0.4
2. er1.sea1.speakeasy.net	0.0%	4	15.5	14.0	12.7	15.5	1.3
3. 220.ge-0-1-0.cr2.sea1. Speakeasy.net	0.0%	4	11.0	11.7	10.7	14.0	1.6
4. fe-0-3-0.cr2.sfo1. speakeasy.net	0.0%	4	36.0	34.7	28.7	38.1	4.1
5. bas1-m.pao.yahoo.com	0.0%	4	27.9	29.6	27.9	33.0	2.4
6. so-1-1-0.pat1.dce. yahoo.com	0.0%	4	89.7	91.0	89.7	93.0	1.4
7. ae1.p400.msr1.dcn. yahoo.com	0.0%	4	91.2	93.1	90.8	99.2	4.1
8. ge5-2.bas1-m.dcn. yahoo.com	0.0%	4	89.3	91.0	89.3	93.4	1.9
9.w2.rc.vip.dcn.yahoo.com	0.0%	3	91.2	93.1	90.8	99.2	4.1

The data will be continuously updated and averaged over time.

As with ping, you should hit control-C when you are finished looking at the data. Note that you must have root privileges to run mtr.

While these tools will not reveal precisely what is wrong with the network, they can give you enough information to know where to continue troubleshooting.

Protocol analysers

Network protocol analysers provide a great deal of detail about information flowing through a network, by allowing you to inspect individual packets. For wired networks, you can inspect packets at the data-link layer or above. For wireless networks, you can inspect information all the way down to individual 802.11 frames. Here are several popular (and free) network protocol analysers:

Kismet

http://www.kismetwireless.net/
Kismet is a powerful wireless protocol analyser for many platforms including Linux, Mac OS X, and even the embedded OpenWRT Linux distribution. It works with any wireless card that supports passive monitor mode.
In addition to basic network detection, Kismet will passively log all 802.11 frames to disk or to the network in standard PCAP format, for later analysis with tools like Wireshark. Kismet also features associated client information, AP hardware fingerprinting, Netstumbler detection, and GPS integration.
Since it is a passive network monitor, it can even detect "closed" wireless networks by analysing traffic sent by wireless clients.
You can run Kismet on several machines at once, and have them all report over the network back to a central user interface. This allows for wireless monitoring over a large area, such as a university or corporate campus. Since Kismet uses the radio card's passive monitor mode, it does all of this without transmitting any data.
Kismet is an invaluable tool for diagnosing wireless network problems.

KisMAC

http://kismac-ng.org
Exclusively for the Mac OS X platform, KisMAC does much of what Kismet can do, but with a slick Mac OS X graphical interface. It is a passive scanner that will log data to disk in PCAP format compatible with Wireshark. It supports passive scanning with AirportExtreme cards as well as a variety of USB wireless adapters.

tcpdump

http://www.tcpdump.org/

tcpdump is a command-line tool for monitoring network traffic.

It does not have all the bells and whistles of wireshark but it does use fewer resources. Tcpdump can capture and display all network protocol information down to the link layer. It can show all of the packet headers and data received, or just the packets that match particular criteria.

Packets captured with tcpdump can be loaded into wireshark for visual analysis and further diagnostics. This is very useful if you wish to monitor an interface on a remote system and bring the file back to your local machine for analysis. The tcpdump tool is available as a standard tool in Unix derivatives (Linux, BSD, and Mac OS X). There is also a Windows port called WinDump available at http://www.winpcap.org/windump/

Wireshark

http://www.wireshark.org/

Formerly known as Ethereal, Wireshark is a free network protocol analyser for Unix and Windows.

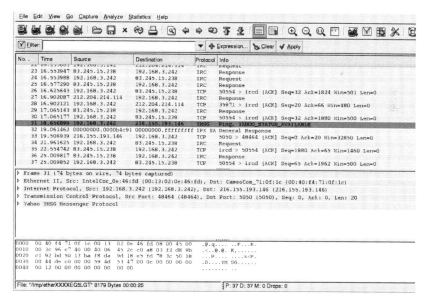

Figure NM 6: Wireshark (formerly Ethereal) is a powerful network protocol analyser that can show you as much detail as you like about any packet.

Wireshark allows you to examine data from a live network or from a capture file on disk, and interactively browse and sort the captured data.

Both summary and detailed information is available for each packet, including the full header and data portions.

Wireshark has several powerful features, including a rich display filter language and the ability to view the reconstructed stream of a TCP session.
It can be daunting to use for first time users or those that are not familiar with the OSI layers.
It is typically used to isolate and analyse specific traffic to or from an IP address, but it can be also used as a general purpose fault finding tool.
For example, a machine infected with a network worm or virus can be identified by looking for the machine that is sending out the same sort of TCP/IP packets to large groups of IP addresses.

Trending tools

Trending tools are used to see how your network is used over a long period of time. They work by periodically monitoring your network activity, and displaying a summary in a human-readable form (such as a graph). Trending tools collect data as well as analyse and report on it.

Below are some examples of trending tools. Some of them need to be used in conjunction with each other, as they are not stand-alone programs.

MRTG

http://oss.oetiker.ch/mrtg/
The Multi Router Traffic Grapher (MRTG) monitors the traffic load on network links using SNMP. MRTG generates graphs that provide a visual representation of inbound and outbound traffic.
These are typically displayed on a web page.

MRTG can be a little confusing to set up, especially if you are not familiar with SNMP.
But once it is installed, MRTG requires virtually no maintenance, unless you change something on the system that is being monitored (such as its IP address).

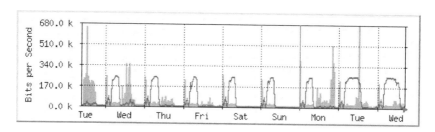

Figure NM 7: MRTG is probably the most widely installed network flow grapher.

RRDtool

RRD is short for Round Robin Database. RRD is a database that stores information in a very compact way that does not expand over time. RRDtool refers to a suite of tools that allow you to create and modify RRD databases, as well as generate useful graphs to present the data. It is used to keep track of time-series data (such as network bandwidth, machine room temperature, or server load average) and can display that data as an average over time.

Note that RRDtool itself does not contact network devices to retrieve data. It is merely a database manipulation tool.
You can use a simple wrapper script (typically in shell or Perl) to do that work for you. RRDtool is also used by many full featured front-ends that present you with a friendly web interface for configuration and display. RRD graphs give you more control over display options and the number of items available on a graph as compared to MRTG.

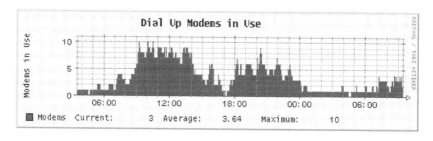

Figure NM 8: RRDtool gives you a lot of flexibility in how your collected network data may be displayed.

RRDtool is included in virtually all modern Linux distributions, and can be downloaded from http://oss.oetiker.ch/rrdtool/

ntop

For historical traffic analysis and usage, you will certainly want to investigate ntop.
This program builds a detailed real-time report of observed network traffic, displayed in your web browser. It integrates with rrdtool, and makes graphs and charts visually depicting how the network is being used. On very busy networks, ntop can use a lot of CPU and disk space, but it gives you extensive insight into how your network is being used. It runs on Linux, BSD, Mac OS X, and Windows.

Some of its more useful features include:

Traffic display can be sorted by various criteria (source, destination, protocol, MAC address, etc.).

Traffic statistics grouped by protocol and port number.

An IP traffic matrix which shows connections between machines Network flows for routers or switches that support the NetFlow protocol Host operating system identification, P2P traffic, identification, numerous graphical charts, Perl, PHP, and Python API.

ntop is available from http://www.ntop.org/ and is available for most operating systems.
It is often included in many of the popular Linux distributions, including RedHat, Debian, and Ubuntu.

While it can be left running to collect historical data, ntop can be fairly CPU intensive, depending on the amount of traffic observed.

If you are going to run it for long periods you should monitor the CPU utilisation of the monitoring machine.

Global TCP/UDP Protocol Distribution

TCP/UDP Protocol	Data	Flows	Accumulated Percentage / Historical Protocol View
FTP	17.8 KB	53	0%
HTTP	40.8 MB	1,410	88.2%
DNS	543.8 KB	4,851	1.1%
NBios-IP	35.4 KB	391	0%
Mail	1.2 MB	166	2.5%
SNMP	0.1 KB	1	0%
NNTP	0.4 KB	4	0%
Gnutella	0.3 KB	2	0%
BitTorrent	16.0 KB	2	0%
Messenger	4.0 KB	3	0%
Other TCP/UDP-based Protocols	3.7 MB	6,129	7.9%

Figure NM 9: ntop displays a wealth of information about how your network is utilised by various clients and servers.

The main disadvantage of ntop is that it does not provide instantaneous information, only long-term totals and averages.
This can make it difficult to use to diagnose a problem that starts suddenly.

Cacti

http://www.cacti.net/
Cacti is a front-end for RRDtool. It stores all of the necessary information to create graphs in a MySQL database. The front-end is written in PHP.
Cacti does the work of maintaining graphs, data sources, and handles the actual data gathering.
There is support for SNMP devices, and custom scripts can easily be written to poll virtually any conceivable network event.

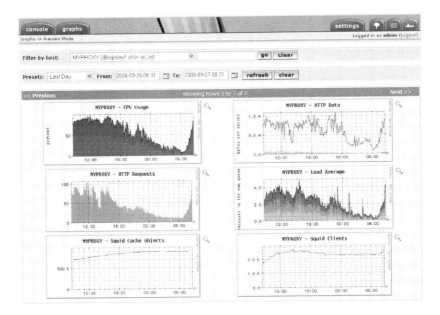

Figure NM 10: Cacti can manage the polling of your network devices, and can build very complex and informative visualisations of network behaviour.

Cacti can be somewhat confusing to configure, but once you work through the documentation and examples, it can yield very impressive graphs.

There are hundreds of templates for various systems available on the cacti website, and the code is under rapid development.

NetFlow

NetFlow is a protocol for collecting IP traffic information invented by Cisco. From the Cisco website:

Cisco IOS NetFlow efficiently provides a key set of services for IP applications, including network traffic accounting, usage-based network billing, network planning, security, Denial of Service monitoring capabilities, and network monitoring.

NetFlow provides valuable information about network users and applications, peak usage times, and traffic routing.

Cisco routers can generate NetFlow information which is available from the router in the form of UDP packets. NetFlow is also less CPU-intensive on Cisco routers than using SNMP. It also provides more granular information than SNMP, letting you get a more detailed picture of port and protocol usage. This information is collected by a NetFlow collector that stores and presents the data as an aggregate over time.

By analysing flow data, one can build a picture of traffic flow and traffic volume in a network or on a connection. There are several commercial and free NetFlow collectors available. Ntop is one free tool that can act as a NetFlow collector and probe. Another is Flowc (see below). It can also be desirable to use Netflow as a spot check tool, by just looking at a quick snapshot of data during a network crisis. Think of NetFlow as an alternative to SNMP for Cisco devices. For more information about NetFlow, see http://en.wikipedia.org/wiki/Netflow .

Flowc

http://netacad.kiev.ua/flowc/. Flowc is an open source NetFlow collector (see NetFlow above). It is lightweight and easy to configure. Flowc uses a MySQL database to store aggregated traffic information. Therefore, it is possible to generate your own reports from the data using SQL, or use the included report generators. The built-in report generators produce reports in HTML, plain text or a graphical format.

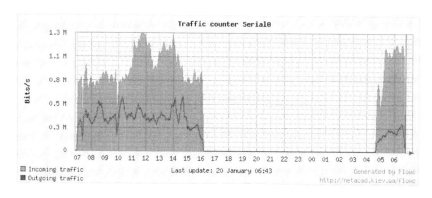

Figure NM 11: A typical flow chart generated by Flowc.

The large gap in data probably indicates a network outage.

Trending tools typically will not notify you of outages, but merely log the occurrence.

To be notified when network problems occur, use a realtime monitoring tool such as Nagios.

SmokePing

http://oss.oetiker.ch/smokeping/. SmokePing is a deluxe latency measurement tool written in Perl.

It can measure, store and display latency, latency distribution and packet loss all on a single graph.

SmokePing uses the RRDtool for data storage, and can draw very informative graphs that present up to the minute information on the state of your network connection.

It is very useful to run SmokePing on a host with good connectivity to your entire network.

Over time, trends are revealed that can point to all sorts of network problems.

Combined with MRTG or Cacti, you can observe the effect that network congestion has on packet loss and latency.

SmokePing can optionally send alerts when certain conditions are met, such as when excessive packet loss is seen on a link for an extended period of time. An example of SmokePing in action is shown in Figure NM 12.

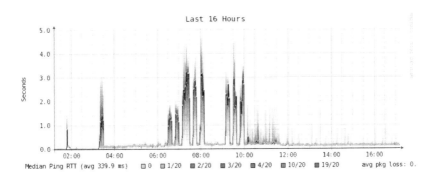

Figure NM 12: SmokePing can simultaneously display packet loss and latency spreads in a single graph.

EtherApe

http://etherape.sourceforge.net/
EtherApe displays a graphical representation of network traffic. Hosts and
links change size depending on the amount of traffic sent and received.
The colours change to represent the protocol most used. As with wireshark
and tcpdump, data can be captured "off the wire" from a live network
connection or read from a tcpdump capture file. EtherApe doesn't show
quite as much detail as ntop, but its resource requirements are much lighter.

iptraf

http://iptraf.seul.org/
IPTraf is a lightweight but powerful LAN monitor. It has an ncurses
interface and runs in a command shell. IPTraf takes a moment to measure
observed traffic, and then displays various network statistics including
TCP and UDP connections, ICMP and OSPF information, traffic flows,
IP checksum errors, and more. It is a simple to use program that uses
minimal system resources. While it does not keep historical data, it is very
useful for displaying an instantaneous usage report.

Figure NM 13: iptraf's statistical breakdown of traffic by port.

Argus

http://qosient.com/argus/.
Argus stands for Audit Record Generation and Utilisation System. Argus is also the name of the mythological Greek god who had hundreds of eyes.

From the Argus website:
Argus generates flow statistics such as connectivity, capacity, demand, loss, delay, and jitter on a per transaction basis. Argus can be used to analyse and report on the contents of packet capture files or it can run as a continuous monitor, examining data from a live interface; generating an audit log of all the network activity seen in the packet stream.
Argus can be deployed to monitor individual end-systems, or an entire enterprises network activity. As a continuous monitor, Argus provides both push and pull data handling models, to allow flexible strategies for collecting network audit data. Argus data clients support a range of operations, such as sorting, aggregation, archival and reporting. Argus consists of two parts: a master collector that reads packets from a network device, and a client that connects to the master and displays the usage statistics. Argus runs on BSD, Linux, and most other UNIX systems.

NeTraMet

http://www.caida.org/tools/measurement/netramet/
NeTraMet is another popular flow analysis tool. Like Argus,
NeTraMet consists of two parts: a collector that gathers statistics via SNMP, and a manager that specifies which flows should be watched.
Flows are specified using a simple programming language that defines the addresses used on either end, and can include Ethernet, IP, protocol information, or other identifiers. NeTraMet runs on DOS and most UNIX systems, including Linux and BSD.

Throughput testing

How fast can the network go?
What is the actual usable capacity of a particular network link?
You can get a very good estimate of your throughput capacity by flooding the link with traffic and measuring how long it takes to transfer the data.

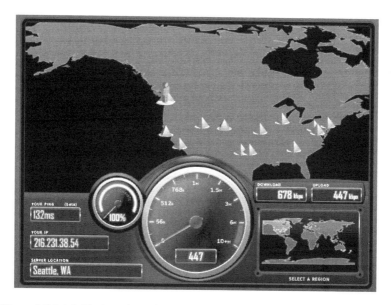

Figure NM 14: Tools such as this one from SpeedTest.net are pretty, but don't always give you an accurate picture of network performance.

While there are web pages available that will perform a "speed test" in your browser (such as http://www.dslreports.com/stest or http://speedtest.net/), these tests are increasingly inaccurate as you get further from the testing source. Even worse, they do not allow you to test the speed of a given link, but only the speed of your link to a particular site on the Internet.
Here are a few tools that will allow you to perform throughput testing on your own networks.

ttcp

Now a standard part of most Unix-like systems, ttcp is a simple network performance testing tool.
One instance is run on either side of the link you want to test.
The first node runs in receive mode, and the other transmits:

node_a$ ttcp -r -s
node_b$ ttcp -t -s node_a
ttcp-t: buflen=8192, nbuf=2048, align=16384/0, port=5001 tcp -> node_a
ttcp-t: socket
ttcp-t: connect

ttcp-t: 16777216 bytes in 249.14 real seconds = 65.76 KB/sec +++
ttcp-t: 2048 I/O calls, msec/call = 124.57, calls/sec = 8.22
ttcp-t: 0.0user 0.2sys 4:09real 0% 0i+0d 0maxrss 0+0pf 7533+0csw

After collecting data in one direction, you should reverse the transmit and receive partners to test the link in the other direction. It can test UDP as well as TCP streams, and can alter various TCP parameters and buffer lengths to give the network a good workout. It can even use a user-supplied data stream instead of sending random data. Remember that the speed readout is in kilobytes, not kilobits. Multiply the result by 8 to find the speed in kilobits per second. The only real disadvantage to ttcp is that it hasn't been developed in years. Fortunately, the code has been released in the public domain and is freely available. Like ping and traceroute, ttcp is found as a standard tool on many systems.

iperf

http://iperf.sourceforge.net/. Much like ttcp, iperf is a command line tool for estimating the throughput of a network connection. It supports many of the same features as ttcp, but uses a "client" and "server" model instead of a "receive" and "transmit" pair.
To run iperf, launch a server on one side and a client on the other:

node_a$ iperf -s
node_b$ iperf -c node_a
--
Client connecting to node_a, TCP port 5001
TCP window size: 16.0 KByte (default)
--
[5] local 10.15.6.1 port 1212 connected with 10.15.6.23 port 5001
[ID] Interval   Transfer   Bandwidth
[5] 0.0-11.3 sec 768 KBytes   558 Kbits/sec

The server side will continue to listen and accept client connections on port 5001 until you hit control-C to kill it. This can make it handy when running multiple test runs from a variety of locations.
The biggest difference between ttcp and iperf is that iperf is under active development, and has many new features (including IPv6 support). This makes it a good choice as a performance tool when building new networks.

bing

http://fgouget.free.fr/bing/indx-en.shtml

Rather than flood a connection with data and see how long the transfer takes to complete, bing attempts to estimate the available throughput of a point-to-point connection by analysing round trip times for various sized ICMP packets. While it is not always as accurate as a flood test, it can provide a good estimate without transmitting a large number of bytes.

Since bing works using standard ICMP echo requests, it can estimate available bandwidth without the need to run a special client on the other end, and can even attempt to estimate the throughput of links outside your network. Since it uses relatively little bandwidth, bing can give you a rough idea of network performance without running up the charges that a flood test would certainly incur.

Realtime tools and intrusion detection

It is desirable to find out when people are trying to break into your network, or when some part of the network has failed.

Because no system administrator can be monitoring a network all the time, there are programs that constantly monitor the status of the network and can send alerts when notable events occur.

The following are some open source tools that can help perform this task.

Snort

Snort (http://www.snort.org/) is a packet sniffer and logger which can be used as a lightweight network intrusion detection system.

It features rule-based logging and can perform protocol analysis, content searching, and packet matching.

It can be used to detect a variety of attacks and probes, such as stealth port scans, CGI attacks, SMB probes, OS fingerprinting attempts, and many other kinds of anomalous traffic patterns.

Snort has a real-time alert capability that can notify administrators about problems as they occur with a variety of methods.

Installing and running Snort is not trivial, and depending on the amount of network traffic, will likely require a dedicated monitoring machine with considerable resources.

Fortunately, Snort is very well documented and has a strong user community.

By implementing a comprehensive Snort rule set, you can identify unexpected behaviour that would otherwise mysteriously eat up your Internet bandwidth.

See http://snort.org/docs/ for an extensive list of installation and configuration resources.

Apache: mod_security

ModSecurity (http://www.modsecurity.org/) is an open source intrusion detection and prevention engine for web applications.

This kind of security tool is also known as a web application firewall.

ModSecurity increases web application security by protecting web applications from known and unknown attacks.

It can be used on its own, or as a module in the Apache web server (http://www.apache.org/).

There are several sources for updated mod_security rules that help protect against the latest security exploits.

One excellent resource is GotRoot, which maintains a huge and frequently updated repository of rules:
http://www.atomicorp.com/wiki/index.php/Atomic_ModSecurity_Rules

Web application security is important in defending against attacks on your web server, which could result in the theft of valuable or personal data, or in the server being used to launch attacks or send spam to other Internet users. As well as being damaging to the Internet as a whole, such intrusions can seriously reduce your available bandwidth.

Nagios

Nagios (http://nagios.org/) is a program that monitors hosts and services on your network, notifying you immediately when problems arise.

It can send notifications via email, SMS, or by running a script, and will send notifications to the relevant person or group depending on the nature of the problem.

Nagios runs on Linux or BSD, and provides a web interface to show up-to-the-minute system status.

Nagios is extensible, and can monitor the status of virtually any network event.

It performs checks by running small scripts at regular intervals, and checks the results against an expected response.

This can yield much more sophisticated checks than a simple network probe.

For example, ping may tell you that a machine is up, and nmap may report that a TCP port responds to requests, but Nagios can actually retrieve a web page or make a database request, and verify that the response is not an error.

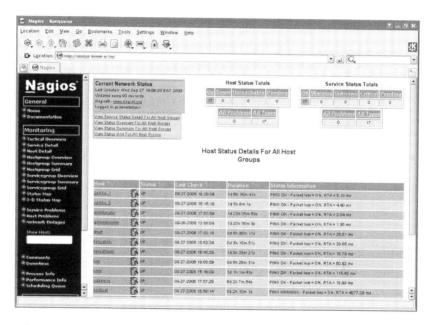

Figure NM 15: Nagios keeps you informed the moment a network fault or service outage occurs.

Nagios can even notify you when bandwidth usage, packet loss, machine room temperature, or other network health indicator crosses a particular threshold.

This can give you advance warning of network problems, often allowing you to respond to the problem before users have a chance to complain.

Zabbix

Zabbix (http://www.zabbix.org/) is an open source realtime monitoring tool that is something of a hybrid between Cacti and Nagios.

It uses a SQL database for data storage, has its own graph rendering package, and performs all of the functions you would expect from a modern realtime monitor (such as SNMP polling and instant notification of error conditions). Zabbix is released under the GNU General Public License.

Other useful tools

There are thousands of free network monitoring tools that fill very specialised needs. Here are a few of our favourites that don't quite fit into the above categories.

ngrep

Ngrep provides most of GNU grep's pattern matching features, but applies them to network traffic. It currently recognises IPv4 and IPv6, TCP, UDP, ICMP, IGMP, PPP, SLIP, FDDI, Token Ring, and much more. As it makes extensive use of regular expression matches, it is a tool suited to advanced users or those that have a good knowledge of regular expressions. You don't necessarily need to be a regex expert to be able to make basic use of ngrep. For example, to view all packets that contain the string GET (presumably HTTP requests), try this:

 # ngrep -q GET

Pattern matches can be constrained further to match particular protocols, ports, or other criteria using BPF filters. This is the filter language used by common packet sniffing tools, such as tcpdump and snoop. To view GET or POST strings sent to destination port 80, use this command line:

 # ngrep -q 'GET|POST' port 80

By using ngrep creatively, you can detect anything from virus activity to spam email. You can download ngrep at http://ngrep.sourceforge.net/.

nmap/Zenmap

nmap is a network diagnostic tool for showing the state and availability of network ports on a network interface. A common use is to scan a network host on a TCP/IP network for what ports are open, thereby allowing one to create a "map" of the network services that the machine provides. The nmap tool does this by sending specially crafted packets to a target network host and noticing the response(s). For example, a web server with an open port 80 but no running web server will respond differently to an nmap probe than one that not only has the port open but is running httpd.

Similarly, you will get a different response to a port that is simply shut off vs. one that is open on a host but blocked by a firewall.

Over time, nmap has evolved from being a simple port-scanner to something that can detect OS versions, network drivers, the type of NIC hardware being used by an interface, driver versions, etc. In addition to scanning individual machines, it can also scan entire networks of hosts. This does mean that nmap is potentially also useful by malicious network users as a way to "scope out" a system before attacking it. Like many diagnostic tools, nmap can be used for good or ill and network administrators would do well to be aware of both sides. The nmap tool is released under the GPL license and the latest version can be found at *http://www.nmap.org.*

Zenmap

Zenmap is a cross-platform GUI for nmap which runs under Linux, Windows, Mac OS X, BSD, etc. and can be downloaded from the nmap.org site as well.

netcat

Somewhat between **nmap** and **tcpdump**, **netcat** is another diagnostic tool for poking and prodding at ports and connections on a network. It takes its name from the UNIX **cat(1)** utility, which simply reads out whatever file you ask it to. Similarly, netcat reads and writes data across any arbitrary TCP or UDP port. The netcat utility is not a packet analyser but works on the data(payload) contained in the packets.

For example, here is how to run a very simple 1-line, 1-time web server with netcat:

```
{ echo -ne "HTTP/1.0 200 OK\r\n\r\n"; cat some.file; } | nc -l 8080
```

The file *some.file* will be sent to the first host that connects to port 8080 on the system running netcat.

The -l command tells netcat to "listen" on port 8080 and wait until it gets a connection.

Once it does, it stops blocking, reads the data off the pipe and sends it to the client connected on port 8080. Some other good examples of netcat usage can be found in the Wikipedia entry for netcat:

https://secure.wikimedia.org/wikipedia/en/wiki/Netcat#Examples

You can download the latest version of netcat from

http://nc110.sourceforge.net/

It is available under a Permissive Free Software License.

What is normal?

If you are looking for a definitive answer as to what your traffic patterns should look like, you are going to be disappointed.

There is no absolute right answer to this question, but given some work you can determine what is normal for your network.

While every environment is different, some of the factors that can influence the appearance of your traffic patterns are:

- The capacity of your Internet connection
- The number of users that have access to the network
- The social policy (byte charging, quotas, honour system, etc.).
- The number, types, and level of services offered
- The health of the network (presence of viruses, excessive broadcasts, routing loops, open email relays, denial of service attacks, etc.).
- The competence of your computer users
- The location and configuration of control structures (firewalls, proxy servers, caches, and so on)

This is not a definitive list, but should give you an idea of how a wide range of factors can affect your bandwidth patterns.

With this in mind, let's look at the topic of baselines.

Establishing a baseline

Since every environment is different, you need to determine for yourself what your traffic patterns look like under normal situations. This is useful because it allows you to identify changes over time, either sudden or gradual. These changes may in turn indicate a problem, or a potential future problem, with your network.

For example, suppose that your network grinds to a halt, and you are not sure of the cause. Fortunately, you have decided to keep a graph of broadcasts as a percentage of the overall network traffic. If this graph shows a sudden increase in the amount of broadcast traffic, it may mean that your network has been infected with a virus.

Without an idea of what is "normal" for your network (a baseline), you would not be able to see that the number of broadcasts had increased, only that it was relatively high, which may not indicate a problem.

Baseline graphs and figures are also useful when analysing the effects of changes made to the network. It is often very useful to experiment with such changes by trying different possible values. Knowing what the baseline looks like will show you whether your changes have improved matters, or made them worse.

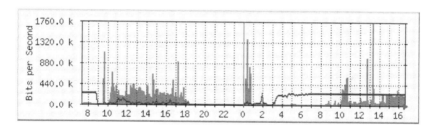

Figure NM 16: By collecting data over a long period of time, you can predict the growth of your network and make changes before problems develop.

In Figure NM 16, we can see the effect of the implementation of delay pools on Internet utilisation around the period of May.

If we did not keep a graph of the line utilisation, we would never know what the effect of the change over the long term was.

When watching a total traffic graph after making changes, don't assume that just because the graph does not change radically that your efforts were wasted. You might have removed frivolous usage from your line only to have it replaced by genuine legitimate traffic.

You could then combine this baseline with others, say the top 100 sites accessed or the average utilisation by your top twenty users, to determine if habits have simply changed. As we will see later, MRTG, RRDtool, and Cacti are excellent tools you can use to keep a baseline.

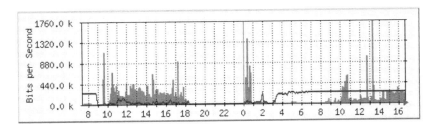

Figure NM 17: The traffic trend at Aidworld logged over a single day.

Figure NM 17 shows traffic on an Aidworld firewall over a period of 24 hours. There is nothing apparently wrong with this graph, but users were complaining about slow Internet access.

Figure NM 18 shows that the upload bandwidth use (blue) was higher during working hours on the last day than on previous days. A period of heavy upload usage started every morning at 03:00, and was normally finished by 09:00, but on the last day it was still running at 16:30. Further investigation revealed a problem with the backup software, which ran at 03:00 every day.

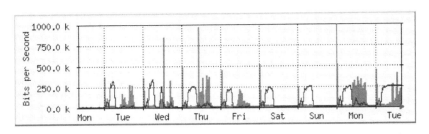

Figure NM 18: The same network logged over an entire week reveals a problem with backups, which caused unexpected congestion for network users.

Figure NM 19 shows measurements of latency on the same connection as measured by the program called SmokePing. The position of the dots shows the average latency, while the grey smoke indicates the distribution of latency (jitter).

The colour of the dots indicates the number of lost packets. This graph over a period of four hours does not help to identify whether there are any problems on the network.

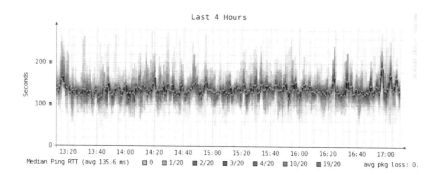

Figure NM 19: Four hours of jitter and packet loss.

The next graph (Figure NM 20) shows the same data over a period of 16 hours.

This indicates that the values in the graph above are close to the normal level (baseline), but that there were significant increases in latency at several times during the early morning, up to 30 times the baseline value.

This indicates that additional monitoring should be performed during these periods to establish the cause of the high latency to avoid problems such as the backup not completing in the future.

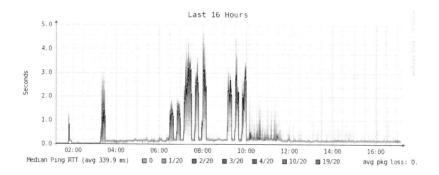

Figure NM 20: A higher spread of jitter is revealed in the 16 hour log.

Figure NM 21 shows that Tuesday was significantly worse than Sunday or Monday for latency, especially during the early morning period.
This might indicate that something has changed on the network.

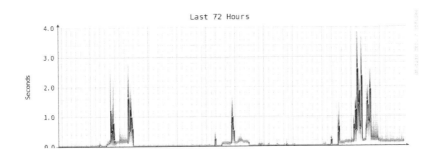

Figure NM 21: Zooming out to the week long view reveals a definite repetition of increased latency and packet loss in the early morning hours.

How do I interpret the traffic graph?

In a basic network flow graph (such as that generated by the network monitor tool MRTG), the green area indicates inbound traffic, while the blue line indicates outbound traffic.

Inbound traffic is traffic that originates from another network (typically the Internet) and is addressed to a computer inside your network. Outbound traffic is traffic that originates from your network, and is addressed to a computer somewhere on the Internet.

Depending on what sort of network environment you have, the graph will help you understand how your network is actually being used. For example, monitoring of servers usually reveals larger amounts of outbound traffic as the servers respond to requests (such as sending mail or serving web pages), while monitoring client machines might reveal higher amounts of inbound traffic to the machines as they receive data from the servers.

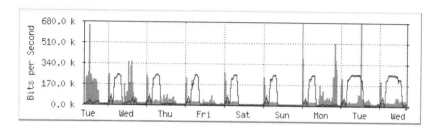

Figure NM 22: The classic network flow graph. The green area represents inbound traffic, while the blue line represents outbound traffic. The repeating arcs of outbound traffic show when the nightly backups have run.

Traffic patterns will vary with what you are monitoring. A router will normally show more incoming traffic than outgoing traffic as users download data from the Internet. An excess of outbound traffic that is not transmitted by your network servers may indicate a peer-to-peer client, unauthorised server, or even a virus on one or more of your clients. There are no set metrics that indicate what outgoing traffic to incoming traffic should look like.It is up to you to establish a baseline to understand what normal network traffic patterns look like on your network.

Detecting network overload

Figure NM 23 shows traffic on an overloaded Internet connection.

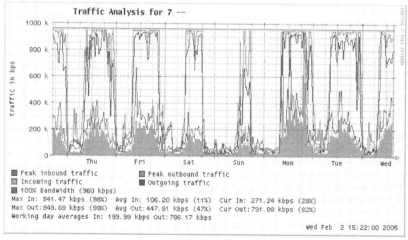

Figure NM 23: Flat-topped graphs indicate overloading of available bandwidth.

The most apparent sign of overloading is the flat tops on outbound traffic during the middle of every day.
Flat tops may indicate overloading, even if they are well below the maximum theoretical capacity of the link.
In this case it may indicate that you are not getting as much bandwidth from your service provider as you expect.

Measuring 95th percentile

The 95th percentile is a widely used mathematical calculation to evaluate regular and sustained utilisation of a network pipe.
Its value shows the highest consumption of traffic for a given period.

Calculating the 95th percentile means that 95% of the time the usage is below a certain amount, and 5% of the time usage is above that amount.
The 95th percentile is a good guideline to show bandwidth that is actually used at least 95% of the time.

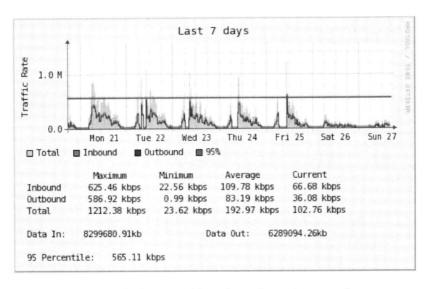

Figure NM 24: The horizontal line shows the 95th percentile amount.

MRTG and Cacti will calculate the 95th percentile for you.
This is a sample graph of a 960 kbps connection.
The 95th percentile came to 945 kbps after discarding the highest 5% of traffic.

Monitoring RAM and CPU usage

By definition, servers provide critical services that should always be available. Servers receive and respond to client machine requests, providing access to services that are the whole point of having a network in the first place. Therefore, servers must have sufficient hardware capabilities to accommodate the work load. This means they must have adequate RAM, storage, and processing power to accommodate the number of client requests. Otherwise, the server will take longer to respond, or in the worst case, may be incapable of responding at all. Since hardware resources are finite, it is important to keep track of how system resources are being used. If a core server (such as a proxy server or email server) is overwhelmed by requests, access times become slow.

This is often perceived by users as a network problem. There are several programs that can be used to monitor resources on a server. The simplest method on a Windows machine is to access the Task Manager using the Ctrl Alt + Del keys, and then click on the Performance tab. On a Linux or BSD box, you can type top in a terminal window. To keep historical logs of such performance, MRTG or RRDtool can also be used.

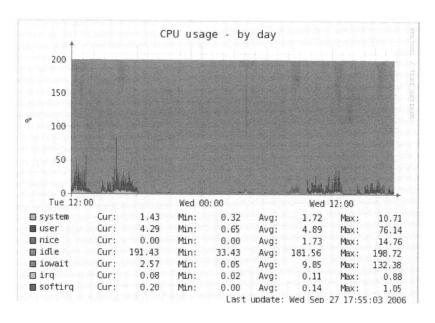

Figure NM 25: RRDtool can show arbitrary data, such as memory and CPU usage, expressed as an average over time.

Mail servers require adequate space, as some people may prefer to leave their email messages on the server for long periods of time.

The messages can accumulate and fill the hard disk, especially if quotas are not in use.

If the disk or partition used for mail storage fills up, the mail server cannot receive mail. If that disk is also used by the system, all kinds of system problems may occur as the operating system runs out of swap space and temporary storage.

File servers need to be monitored, even if they have large disks.

Users will find a way to fill any size disk more quickly than you might think.

Disk usage can be enforced through the use of quotas, or by simply monitoring usage and telling people when they are using too much.

Nagios can notify you when disk usage, CPU utilisation, or other system resources cross a critical threshold.

If a machine becomes unresponsive or slow, and measurements show that a system resource is being heavily used, this may be an indication that an upgrade is required. If processor usage constantly exceeds 60% of the total, it may be time to upgrade the processor. Slow speeds could also be as a result of insufficient RAM. Be sure to check the overall usage of CPU, RAM, and disk space before deciding to upgrade a particular component.

A simple way to check whether a machine has insufficient RAM is to look at the hard disk light.

When the light is on constantly, it usually means that the machine is constantly swapping large amounts of data to and from the disk.

This is known as thrashing, and is extremely bad for performance.

It can usually be fixed by investigating which process is using the most RAM, and killing or reconfiguring that process. Failing that, the system needs more RAM. You should always determine whether it is more cost effective to upgrade an individual component or purchase a whole new machine. Some computers are difficult or impossible to upgrade, and it often costs more to replace individual components than to replace the entire system. Since the availability of parts and systems varies widely around the world, be sure to weigh the cost of parts vs. whole systems, including shipping and taxes, when determining the cost of upgrading.

Summary

In summary in this chapter we've tried to give you an insight into how to monitor your network and computing resources cost effectively and efficiently. We've introduced many of our favourite tools to assist you. Many of them are tried and tested by many network operators.

Hopefully you have understood the importance of monitoring to enable you to both justify upgrades when necessary to those funding the upgrades, plus minimise the impact of problems as they occur.

The end result is keeping your network and computing resources healthy and keeping all of your users happy with the service you are providing them.

17. ECONOMIC SUSTAINABILITY

Introduction

Achieving long-term sustainability is perhaps the most difficult goal when designing and operating wireless networks. The prohibitive cost of Internet connectivity in many countries, particularly those who are heavily regulated by the government, imposes a substantial operating expense that makes these networks sensitive to economic fluctuations and necessitates innovation to attain viability.

Substantial progress in the use of wireless networks for rural communications has been accomplished over the past few years, due in large part to technological breakthroughs.

Long-distance links have been constructed, high bandwidth designs are possible and secure means to access networks are available. In contrast, there have been fewer successes with the development of sustainable business models for wireless networks, particularly for remote areas.

Based on experiences and observations of existing networks, as well as knowledge from entrepreneurial development best practices, this chapter will focus on documenting methods for building sustainable wireless networks.

In the past decade, there has been tremendous growth in Internet access across the world. Most cities now have wireless or DSL networks and fibre optic connections to the Internet, which is a substantial improvement.
Nevertheless, outside urban areas, Internet access is still a formidable challenge. There is little wired infrastructure beyond principal cities in many countries. Therefore, wireless remains one of the few choices for providing affordable Internet access.

There are now proven models for rural access using wireless. This book was written for those wishing to connect their communities.
The models described here are smaller in scale and use affordable designs.

Our aim is to provide examples of how wireless networks can be designed to expand sustainable access where large telecommunications operators have not yet installed their networks into areas that would otherwise not be economically feasible by traditional models.

Two common misconceptions must be dispelled. First, many people assume that there is one preferred business model that will work in every community, and the key to success is to find that one "eureka" solution.

In practice, this is not the case. Each community, town or village is different. There is no prescribed model that meets the needs of all areas. Despite the fact that some places may be similar in economic terms, the characteristics of a sustainable business model vary from community to community.

Although one model may work in one village, another village nearby may not possess the same necessary qualities for this model to be sustainable. In this circumstance, other innovative models must be customised to fit the context of this particular community.

Another misconception is that sustainability has the same definition for all people. Although this term generally means that a system is built to persist indefinitely, this chapter focuses more on the discussion of the economic conditions (financial and managerial) than other aspects of sustainability. Also, instead of the horizon being indeterminate, it centres on a time period of five years – the period in which these ICT infrastructure and wireless technologies are expected to be useful.

Thus, the term sustainability will be used to encapsulate a system designed to persist for approximately five years. As we explained earlier in the book, wireless networks in local communities often stimulate the growth of connectivity and usage, and the installation of fibre begins to become a reality.

So creating a sustainable model for your wireless network may lead to growth of other networks and installation of longer term greater bandwidth fibre links.
Wireless will then likely continue to co-habit alongside fibre in your network as it grows in size and reach.

When determining and implementing the best model for a wireless network, several key factors will help to ensure its success.

This chapter is not meant to be a guide for managing sustainable wireless networks.

Rather, this is a "how-to" guide seeking to present an approach that will enable you to find the model that best fits your situation.

The tools and information contained within this chapter will help people starting wireless networks to ask the right questions and gather the necessary data to define the most appropriate components of their model. Keep in mind that determining the best model is not a sequential process where each step is followed until completion.
In fact, the process is ongoing and iterative.

All of the steps are integrally connected to each other, and often you will revisit steps several times as you progress.

Create a mission statement

What do you want to accomplish by setting up your network?

It seems like a simple question. However, many wireless networks are installed without a clear vision of what they are doing and what they hope to accomplish in the future. The first step involves documenting this vision with the input of your entire team or staff.

- What is the purpose of the wireless network?
- Who does the network seek to serve?
- What does the network do to address the community's needs and to create value?
- What are the principles that guide the network?

A good mission statement expresses the purpose of your network in a concise, meaningful way while articulating your values and services.
Above all, your mission provides a vision of the aspirations for your wireless network.

It is important that every team member working to build the wireless network is included in the process of developing your mission, which helps create further buy-in. It will garner support and commitment not only from your staff, but also from customers, partners and donors, which will further your overall objectives.

In the dynamic world of technology, the needs of customers and the best way to satisfy those needs change rapidly; therefore, the development of your mission is an ongoing process.

After defining the initial mission with your team, you must conduct research to determine whether this first conception is aligned with the realities of your environment. Based on an analysis of the external environment and your internal competencies, you must constantly modify the mission throughout the life-cycle of the wireless network.

Evaluate the demand for potential offerings

The next step in deriving your business model involves assessing the community's demand for the network's products and services. First, identify the individuals, groups and organisations in the community that have a need for information and would benefit from the wireless network's offerings. Potential users could consist of a wide variety of individuals and organisations that include, but are not limited to:

- Farmers' associations and cooperatives
- Women's groups
- Schools and universities
- Businesses and local entrepreneurs
- Health clinics and hospitals
- Religious groups
- International and local non-governmental organizations (NGOs) Local and national government agencies
- Radio stations
- Organisations in the tourist industry

Once you establish a list of all the potential user groups of the network, you must determine their needs for access to information and communication. Often, people confuse services with needs.

A farmer may need to gather information on market prices and climatic conditions to improve his crop yield and sales. Perhaps the way in which he gets this information is through the Internet; however, the farmer could also receive this information through SMS over a mobile phone or through Voice over Internet Protocol (VoIP). It is important to differentiate between needs and services because there may be various ways to satisfy the farmer's needs. Your wireless network should look for the best way to fulfill the farmer's needs, thereby creating value at the lowest cost for the user.

When assessing the needs of the community, it is important to figure out where the network can bring the most value to its users.

For instance, in the small town of Douentza, Mali, a telecentre manager evaluated the potential benefits of establishing a wireless network through discussions with several local organisations. He interviewed one local NGO that discussed its need to send monthly reports to its headquarters office in Bamako. At that time, there was no Internet access in Douentza. In order to email a copy of the report, the NGO sent one of its employees to Mopti once a month, resulting in transportation and lodging costs, as well as the opportunity cost of having the employee out of the office for several days each month.

When the telecentre manager calculated the total monthly costs incurred by the NGO, he was able to demonstrate the value of an Internet connection through cost savings to the organisation.

Assistance from key partners may also be necessary to secure sustainability for your wireless network. During this phase, you should connect with potential partners and explore mutually beneficial collaborations.

You can evaluate the demand in your community by contacting your potential customers and asking questions directly or through surveys, focus groups, interviews or town hall meetings. Conducting research through a review of statistical documentation, industry reports, censuses, magazines, newspapers and other secondary data sources will also help to give you a better picture of your local environment.

The goal of this data collection is to obtain a thorough understanding of the demand for information and communication in your community so that the network being created responds to those needs.

Often, wireless networks that do not succeed forget this key step.

Your entire network should be based on the demand in the community.

If you set up a wireless network in which the community does not find value or cannot afford its services, it will ultimately fail.

Establish appropriate incentives

Often, there is little economic incentive for subsistence-based economic participants to access the Internet.

In addition, the cost of acquiring a computer or mobile smart phone, learning to use it, and getting an Internet connection far outweighs the economic returns that it can provide.

There has recently been some development of applications that address this lack of incentive, such as market information systems, quality standards imposed by importing countries, and commodities exchanges. Internet access becomes an obvious advantage in situations where knowing the day-to-day prices of products can make a significant difference in income. Establishing appropriate economic incentives is paramount to the success of the network. The network must provide economic value to its users in a way that outweighs its costs, or it must be cheap enough that its costs are marginal and affordable to its users. It is crucial to design a network with viable economic uses and with costs that are less than the economic value provided by it.

Additionally, to create a proper incentive structure, you must involve the community in the creation of the network from the beginning of the project, making sure that this initiative is organic and not imposed from the outside.

To begin, you should try to answer the following questions:

1. What economic value can this network generate for the local economy and for whom?
2. How much perceivable economic value can be generated?
3. Can present impediments be overcome to allow the achievement of these economic returns?

By answering these questions, the network will be able to clearly articulate its value proposition for its users. For example, "By using this network you can improve your margins on commodity sales by 2%," or "Internet access will allow you to save $X in phone charges and transportation costs per month." You must figure out how your network can improve efficiencies, reduce costs, or increase revenues for these customers.

For example, if providing market information for the local maize industry, the network should be located near to where farmers bring their crop for sale to merchants.

Your network would then likely need to tie-into market information systems, providing daily price sheets ($1 each), or terminals to sellers and merchants ($2/hr). Your network might also provide the means for farmers to read about new techniques and to buy new products. You might also provide wireless connections to merchants and rent them thin-client terminals for Internet access.

If the market was small, you might be able to reduce costs by limiting access to images and other bandwidth intensive services. Again, knowing how much value your network will create for these merchants will allow you to gauge how much they will be able to afford for your services.

Research the regulatory environment for wireless

The regulatory environment for wireless networks also affects the type of business model that can be implemented.

First, research whether any organisation has the right to use 2.4 GHz frequencies without a license.

In most situations, 2.4 GHz is free to use worldwide; however, some countries restrict who can operate a network or require expensive licenses to do so.

So although wireless networks might be legal in a country, the operator of a network might have to acquire a license to use 2.4 GHz frequencies, which renders this usage prohibitive for anyone other than established Internet Service Providers who have sufficient cash flow to pay the license fees. This restriction makes it difficult for small communities to share a wireless network with other potentially interested parties or organisations.

Other countries are more permissive with no such restrictions on wireless networks, so the possibility to share Internet connectivity in small communities is a viable solution.

The lesson is to do your research at the onset, ensuring your network will comply with the laws of the country and local community. Some project managers have been forced to shut down their wireless networks simply because they were unknowingly breaking the law.

You should also check into the legality of Voice over Internet Protocol (VoIP) services.

In some countries there are complicated rules surrounding VoIP. The rules for VoIP services and VoIP gateways vary a lot so please check in your own country what is legally allowed. You can start by checking wikipedia - http://en.wikipedia.org/wiki/Voice_over_IP

Analyse the competition

The next phase in the evaluation of your community involves an analysis of the wireless network's competition. Competitors include organisations that provide similar products and services (e.g., another wireless Internet Service provider or WISP), organisations viewed as substitutes or alternatives to the products and services your network provides (e.g., a cybercafé), and organisations defined as new entrants to the wireless market.

Once you have identified your competitors, you should research them thoroughly. You can obtain information about your competitors through the Internet, telephone calls, their advertisements and marketing materials, surveys of their customers and visits to their site. Create a file for each competitor. The competitive information you gather can include a list of services (including price and quality information), their target clients, customer service techniques, reputation, marketing, etc.

Be sure to collect anything that will help you determine how to position your network in the community.

It is important to evaluate your competition for many reasons. First, it helps you determine the level of market saturation. Knowing what already exists will allow you to determine how your network can contribute value to the community. In addition, analysing the competition can stimulate innovative ideas for your service offerings. Is there something that you can do better than the competitors to make your services more effectively fit the needs of the community?

Finally, by analysing your competitors from the customers' point of view and understanding their strengths and weaknesses, you can determine your competitive advantages in the community.

Competitive advantages are those which cannot be easily replicated by the competition.

For example, a wireless network that can exclusively offer a faster Internet connection than a competitor is a competitive advantage.

Determine initial and recurring costs and pricing

When you are planning to set up and operate your wireless network, you must determine the resources needed to start your project and the recurring operating costs. Start-up costs include everything you must purchase to start your wireless network.

These expenses can range from the initial investment you make in hardware, access to towers and so on plus equipment for access points, hubs, switches, cables, solar equipment, UPS, etc. to the costs to register your organisation as a legal entity.

Recurring costs are what you must pay to continue to operate your wireless network, including the cost of Internet access, telephones, loans, electricity, salaries, office rental fees, equipment maintenance and repairs, and regular investments to replace malfunctioning or obsolete equipment. Every piece of equipment will eventually break down or become outdated at some point, and you should set aside extra money for this purpose.

An advisable and very common method to deal with this is to take the price of the device and divide it by the period of time you estimate that it will last.

This process is called depreciation. Here is an example. An average computer is supposed to last for two to five years. If the initial cost to purchase the computer was $1,000 USD, and you will be able to use the computer for five years, your annual depreciation will be $200 USD. In other words, you will lose $16.67 USD every month so that you can eventually replace this computer. To make your project sustainable, it is of fundamental importance that you save the money to compensate for the depreciation of equipment each month.

Keep these savings until you finally have to spend them for equipment replacement. Some countries have tax laws that determine the period of depreciation for different types of devices.

In any case, you should try to be very realistic about the life-cycle of all the implemented gear and plan for their depreciation carefully.

It is important to research all your start-up costs in advance, and make realistic estimations of your recurring expenses.

It is always better to over-budget for expenses than to under-budget.

With every wireless project, there are always unforeseen costs, especially during the first year of operations as you learn how to better manage your network. Following, is a non-exhaustive list of categories of costs that you should include, both in start-up phase and for your recurring costs, just to give you an idea of how to get started on calculating your costs:-

Categories of Costs

Labour costs -

- Check ups (analyses) and consultancies
- Development costs for programming, testing, integration etc.
- Installation costs
- Recruiting costs
- Training costs (introduction and on-going)
- Handling costs / salaries for employees or freelancers, including yourself
- Equipment maintenance staff costs
- Software support staff costs
- Security personnel

Non-labour costs -

- Acquisition and production costs (for hardware like PCs, VSAT, radio link equipment and software)
- Ancillary equipment (e.g., switches, cables and cabling, generator, UPS, etc.)
- Data protection and security
- Start-up inventory (chairs, tables, lighting, curtains, tiles and carpeting)
- Premises costs (new building, modification, air conditioning, electrical wiring and boxes, security grills)
- Legal costs, such as business registration
- Initial license costs (VSAT)
- Initial marketing costs (flyers, stickers, posters, opening party)
- Operating costs for hardware and operating systems (Internet access, telephone, etc.)
- Rent or leasing rates (for tower space for example)
- Depreciation of hardware and equipment
- License fees
- Consumables and office supplies (e.g., data media, paper, binds, clips)
- Operational costs to maintain data protection and security

- Insurance premiums
- Costs for energy and to ensure power supply
- Loan payments, capital costs for paying back your setup costs
- Costs for advertising
- Local fees
- Legal and accounting services

To improve your chances of sustainability, it is generally best to maintain the lowest cost structure for your network.

In other words, keep your expenses as low as possible.
Take time to thoroughly research all of your suppliers, particularly the ISPs, and shop around for the best deals on quality service. Once again, be certain that what you purchase from suppliers corresponds with the demand in the community.
Before installing an expensive VSAT, ensure there is a sufficient number of individuals and organisations in your community willing and able to pay for using it.
Depending upon demand for information access and ability to pay, an alternative method of connectivity may be more appropriate. Do not be afraid to think outside the box and be creative when determining the best solution.

Keeping your costs down should not be at the cost of quality. Because low-quality equipment is more likely to malfunction, you could be spending more on maintenance in the long run.
The amount of money you will spend to maintain your ICT infrastructure is hard to guess. The larger and more complicated your infrastructure becomes, the more financial and labour resources you must allocate for its maintenance.
Many times this relation is not linear but exponential. If you have a quality problem with your equipment once it is rolled out, it can cost you an enormous amount of money to fix it. Concurrently, your sales will decrease because the equipment is not up and running.
There is an interesting example of a major wireless Internet Service Provider (WISP) who had more than 3,000 access points in operation for a while. However, the WISP never managed to break-even because it had to spend too much money to maintain all the access points.

In addition, the company underestimated the short life-cycle of such devices. ICT hardware tends to get cheaper and better as time goes on. As soon as the company had invested time and money to install the version of expensive first generation 802.11b access points, the new "g" standard was created.

New competitors designed better and cheaper access points and offered faster Internet access for less money. Finally the first WISP was forced to close down the company, although it was initially the market leader.

Keep in mind the rapid advancement and changes in technology and think about how and when it may be time for you to reinvest in newer and cheaper (or better) devices to keep your infrastructure competitive and up-to-date. As mentioned before, it is highly important that you save enough to be able to do so, when necessary.

Once you have identified and mapped out your costs, you should also determine what and how to charge for your services. This is a complicated and time-consuming process to do correctly. These key tips will assist when making pricing decisions:

- Calculate the prices you charge so that you cover all costs to provide the service, including all recurring expenses
- Examine the prices of your competitors.
- Evaluate what your customers are willing and able to pay for your services, and make sure your prices correspond with these

It is absolutely essential to make a financial plan before you start to find out if your project can be sustainable.

Secure the financing

Once you have determined your initial and recurring costs and created your financial plan, you know how much financing you will need to run a successful wireless network.

The next step is to research and secure the appropriate amount of money to start up and run your wireless network.

The most traditional method of receiving funding for wireless networks is through grants given by donors. A donor typically will contribute funding and other types of donations to an organisation or consortium of organisations to help them manage projects or support causes.

Because this funding is provided in the form of grants or other donations, it is not expected to be repaid by the organisations implementing the wireless projects or by the project's beneficiaries.

Such donors include large international organisations like the United Nations (UN) and various specialised UN agencies like the United Nations Development Program (UNDP) and United Nations Educational, Scientific and Cultural Organization (UNESCO). Government agencies that specialise in international development, such as the United States Agency for International Development (USAID), the United Kingdom's Department for International Development (DFID), and the Canadian International Development Agency (CIDA), are also considered donors. Large foundations like the Gates Foundation and the Soros Foundation and private commercial companies are other types of donors.

Typically, receiving funding involves a competitive or a non-competitive process. The non-competitive process is more infrequent, so we will focus on the competitive process at a very high level. Most donors have complicated procedures surrounding the distribution of funding. During the competitive bid process, the donor creates a request for proposal (RFP) or a request for application (RFA), which solicits various non-governmental organisations, private companies and their partners to submit proposals outlining their plans for projects within the constraints of the donors' objectives and guidelines.

In response to this RFP or RFA, NGOs and other organisations compete through the submittal of their proposals, which are then evaluated by the donors based on specific established criteria. Finally, the donor organisation selects the most appropriate and highest ranking proposal to fund the project.

Sometimes donors also supply funding to support an organisation's operations, but this type of funding is more unusual than the competitive bid process for a specific project.

Another way of accessing the necessary funds to start and maintain a wireless network is through microfinance, or the provision of loans, savings and other basic financial services to the world's poorest people. Pioneered in the 1970's by organisations like ACCION International and Grameen Bank, microcredit, a type of microfinance, enables entrepreneurs to receive loans in small amounts of money to start up small enterprises.

Despite the fact that often individuals lack many of the traditional qualifications needed to obtain loans like verifiable credit, collateral or steady employment, microcredit programs have been highly successful in many countries.

Typically, the process involves an individual or a group completing and submitting a loan application in the hopes of receiving a loan, and the lender, the individual or organisation that provides the loan, giving money on condition that it is returned with interest.

The use of microcredit to fund wireless networks does pose one constraint. Usually, microcredit involves very small sums of money. Unfortunately, because a large amount of capital is needed to purchase the initial equipment for wireless network set up, sometimes a microcredit loan is not sufficient.

However, there have been many successful applications of microcredit that have brought technology to communities. An example includes the story of village phone operators. These entrepreneurs use their microcredit loans to purchase mobile phones and phone credits.

They then rent the use of their mobile phones to community members on a per-call basis and earn enough money to repay their debt and make a profit for themselves and their families.

Another mechanism for getting funding to start a wireless network is angel funding. Angel investors are normally wealthy individuals that provide capital for business start-up in exchange for a high rate of return on their investment.

Because the ventures in which they invest are start ups and, therefore, often high risk, angel investors tend to expect different things in addition to their return.

Many expect a Board position and maybe a role in the organisation.

Some angels want to have a stake in the company, while others prefer shares in the company that can be easily redeemable at face value, thus providing a clear exit for the investor.

To protect their investments, angels frequently ask the businesses not to make certain key decisions without their approval. Because of the high risk involved in developing markets, it is often challenging to find angel investors to help setup a wireless network, but not impossible.

The best way to find potential investors is through your social network and through research online.

Evaluate the strengths and weaknesses of the internal situation

A network is only as good as the people who work and operate it.

The team you put in place can mean the difference between success and failure.

That is why it is important to reflect on your team's qualifications and skills, including those of staff and volunteers, in comparison to the competencies needed for a wireless project.

First, make a list of all the competencies needed to run a wireless project successfully. Capacity areas should include technology, human resources, accounting, marketing, sales, negotiation, legal, and operations, among others. Afterwards, identify local resources to fulfill these skills.

Map your team's skill sets to the competencies needed, and identify key gaps. One tool often used to assist with this self-evaluation is an analysis of strengths, weaknesses, opportunities and threats, called SWOT.

To conduct this analysis, specify your internal strengths and weaknesses, and elaborate upon the external opportunities and threats in your community. It is important to be realistic and honest about what you do well and what you are lacking.

Be sure to distinguish between where your organisation is at the beginning of this endeavour from where it could be in the future. Your strengths and weaknesses allow you to evaluate your capacities internally and better understand what your organisation can do, as well as its limits.

By understanding your strengths and weaknesses and comparing them to those of your competitors, you can determine your competitive advantages in the market.

You can also note the areas where you can improve. Opportunities and threats are external, which enable you to analyse real world conditions and how these conditions influence your network.

The diagram on the next page will help you in creating your own SWOT analysis for your organisation.

Be sure to respond to the questions asked and list your strengths, weaknesses, opportunities and threats in the spaces designated.

Strengths	Weaknesses
What do you do well? What unique resources can you draw on? What do others see as your strengths?	What could you improve? Where do you have fewer resources than others? What are others likely to see as weaknesses?
Opportunities	**Threats**
What good opportunities are open to you? What trends could you take advantage of? How can you turn your strengths into opportunities?	What trends could harm you? What is your competition doing? What threats do your weaknesses expose you to?

Putting it all together

Once you have gathered all of the information, you are ready to put everything together and decide upon the best model for the wireless network in your community. Based on the results of your external and internal analyses, you must refine your mission and service offerings. All of the factors that you researched in the preceding steps come into play when determining your overall strategy. It is essential to employ a model that capitalises on opportunities and works within the constraints of the local environment.

To do this, you must often find innovative solutions to attain sustainability. By exploring several examples and discussing the components of the models implemented in those instances, you will better understand how to arrive at an appropriate model.

In the distant jungles of the Democratic Republic of Congo, there is a rural hospital in a village called Vanga in the province of Bandundu. It is so remote that patients travel for weeks to get there often through a combination of travel by foot and by river.

This village, founded by Baptist missionaries in 1904, has served as a hospital for many years.

Although it is extremely remote, it is renowned for being an excellent facility and has had the support of German and American missionaries who have kept this facility in operation.

In 2004, a project sponsored by USAID established a telecentre in this village to help improve education in this isolated community; this Internet facility was also heavily used by the educated class in the community – the hospital's staff.

The centre had been a great boon to the community, offering access to the world's knowledge and even providing consultation with distant colleagues in Switzerland, France and Canada.

The centre required near total subsidisation to operate and cover its costs, and funding was to end by 2006.

Although the centre added great value to the community, it did have some shortcomings, primarily technical, economic, and political issues that limited its sustainability.

A study was commissioned to consider options for its future.

After reviewing the centre's cost structure, it was determined that it needed to cut its costs and look for new ways to increase its revenues.

The largest expenses were electricity and Internet access; therefore, creative models needed to be constructed to reduce the telecentre's costs and provide access in a way that was sustainable.

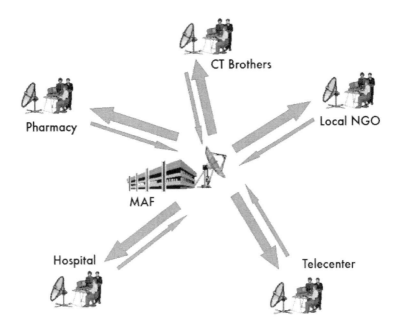

Figure ES 1: Shared Internet over wireless

In this instance, a traditional VSAT had been used for connectivity.

However, there was a unique way of accommodating local community groups' limited ability to pay for Internet services. Various organisations in the community shared this Internet access through a local wireless network; this meant that they also shared the costs associated with the VSAT connection and local wireless network. This model led to greater sustainability for everyone. In Vanga, several organisations, including a hospital, a pharmacy, several missionary groups, a community resource centre, and some non-profit organisations, had a need for Internet access and the means to pay for it. This arrangement enabled the network of organisations to have a higher quality connection at a lower cost. Additionally, one organisation in the village had the capacity and willingness to manage several aspects of the network's operations, including the billing and payment collection, technical maintenance and general business operations of the entire network.

Therefore, this model worked well in Vanga because it had been tailored to meet community demand and leverage local economic resources.

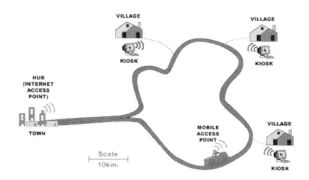

Figure ES 2: DakNet's roaming access point

Another example of a model adapted to fit the local context is that of First Mile Solutions' DakNet. This model has been deployed in villages in India, Cambodia, Rwanda, and Paraguay. By taking into account the limited buying power of villagers, this model addresses their communication needs in an innovative way.

In the DakNet model, there is a franchise that exists in the country, and local entrepreneurs are recruited and trained to operate kiosks equipped with Wi-Fi antennas.

Using pre-paid cards, villagers are able to asynchronously send and receive emails, texts, and voice mails, conduct web searches, and participate in e-commerce. Afterwards, these communications are stored in the local kiosk's server. When a bus or motorcycle with a mobile access point drives past a kiosk, the vehicle automatically receives the kiosk's stored data and delivers any incoming data.

Once the vehicle reaches a hub with Internet connectivity, it processes all requests, relaying emails, messages, and shared files. DakNet integrates both mobile access and franchise models to bring value to people in remote villages. For such a model to be sustainable, several key conditions need to be present.

First, a franchise organisation must exist to provide financial and institutional support, including an initial investment, working capital for certain recurring costs, advice on start-up practices, management training, standardised processes, reporting mechanisms, and marketing tools.

Additionally, this model requires a highly motivated and dynamic individual in the village, with the appropriate skills to manage a business and willingness to accept certain requirements of the franchise organisation. Because these entrepreneurs are often asked to commit their own resources to the start-up costs, they need to have sufficient access to financial resources.

Finally, to ensure this model will sustain itself, there should be sufficient demand for information and communication and few competitors in the community.

Conclusion

No single business model will enable wireless networks to be sustainable in all environments; different models must be used and adapted as the circumstances dictate. Every community has unique characteristics, and sufficient analysis must be conducted at the onset of a project to determine the most appropriate model. This analysis should consider several key factors in the local environment, including community demand, competition, costs, economic resources, etc. Although appropriate planning and execution will maximise the chances of making your network sustainable, there are no guarantees of success. However, by using the methods detailed in this chapter, you will help to ensure that your network brings value to the community in a way that corresponds with the users' needs.

GLOSSARY

Glossary

0 - 9

802.11. While 802.11 is a wireless protocol in its own right, 802.11 is often used to refer to a family of wireless networking protocols used mainly for local area networking. Three popular variants include 802.11b, 802.11g, and 802.11a. See also: ***Wi-Fi.***

A

AC see ***Alternating Current***

access point (AP). A device that creates a wireless network that is usually connected to a wired Ethernet network. See also: ***CPE, master mode***

accumulator. Another name for a ***battery.***

ad-hoc mode. A radio mode used by 802.11 devices that allows the creation of a network without an access point. Mesh networks often use radios in ad-hoc mode. See also: ***managed mode, master mode, monitor mode***

Address Resolution Protocol (ARP). A protocol widely used on Ethernet networks to translate IP addresses into MAC addresses.

address space. A group of IP addresses that all reside within the same logical subnet.

advertised window. The portion of a TCP header that specifies how many additional bytes of data the receiver is prepared to accept.

Alternating Current (AC). An electrical current which varies over time in a cyclic manner. AC current is typically used for lighting and appliances. See also: ***Direct Current***

amortization. An accounting technique used to manage the expected cost of replacement and obsolescence of equipment over time.

amplifier. A device used to increase the transmitted power of a wireless device.

amplitude. The distance from the center of a wave to the extreme of one of its peaks.

anchor clients. Business clients of a subscription system who are reliable and can be considered low-risk.

AND logic. A logical operation that only evaluates as true if all of the items being compared also evaluate as true. See also: **OR logic**.

anonymizing proxy. A network service that hides the source or destination of communications. Anonymizing proxies can be used to protect people's privacy and to reduce an organization's exposure to legal liability for the actions of its users.

anonymity. In computer networks, communications that cannot be linked to a unique individual are said to be anonymous. The trade-off of anonymity versus accountability in communications is an ongoing debate online, and rules about anonymous communications vary widely around the world. See also: **authenticated**

antenna diversity. A technique used to overcome multipath interference by using two or more physically separated receiving antennas.

antenna gain. The amount of power concentrated in the direction of strongest radiation of an antenna, usually expressed in dBi. Antenna gain is reciprocal, which means that the effect of gain is present when transmitting as well as receiving.

antenna pattern. A graph that describes the relative strength of a radiated field in various directions from an antenna. See also: **rectangular plot, polar plot, linear polar coordinates, logarithmic polar coordinates**

AP see **Access Point**

application layer. The topmost layer in the OSI and TCP/IP network models.

Argus see **Audit Record Generation and Utilization System**

ARP see **Address Resolution Protocol**

associated. An 802.11 radio is associated to an access point when it is ready to communicate with the network. This means that it is tuned to the proper channel, in range of the AP, using the correct SSID and other authentication parameters, etc.

at. A Unix facility that allows timed, one-shot execution of programs. See also: **cron**

attenuation. The reduction of available radio power as it is absorbed along a path, such as through trees, walls, buildings, or other objects. See also: **free space loss, scattering**

Audit Record Generation and Utilization System (Argus). An open source network monitoring tool used for tracking flows between hosts. Argus is available from *http://www.qosient.com/argus* .

authenticated. A network user that has proven their identity to a service or device (such as an access point) beyond a shadow of a doubt, usually by some means of cryptography. See also: ***anonymity***

azimuth. The angle that measures deviation with respect to the south in the northern hemisphere, and with respect to the north in the southern hemisphere. See also: ***inclination***

B

bandwidth. A measure of frequency ranges, typically used for digital communications. The word bandwidth is also commonly used interchangeably with ***capacity*** to refer to the theoretical maximum data rate of a digital communications line. See also: ***capacity, channel, throughput***

battery. A device used to store energy in a photovoltaic system. See also: ***solar panel, regulator, load, converter, inverter***

beamwidth. The angular distance between the points on either side of the main lobe of an antenna, where the received power is half that of the main lobe. The beamwidth of an antenna is usually stated for both the horizontal and vertical planes.

benchmarking. Testing the maximum performance of a service or device. Benchmarking a network connection typically involves fboding the link with traffic and measuring the actual observed throughput, both on transmit and receive.

BGAN see ***Broadband Global Access Network***

BNC connector. A coaxial cable connector that uses a "quick-connect" style bayonet lug. BNC connectors are typically found on 10base2 coaxial Ethernet.

bridge. A network device that connects two networks together at the ***data link layer***. Bridges do not route packets at the ***network layer***. They simply repeat packets between two ***link-local*** networks. See also: ***router*** and ***transparent bridging firewall***.

bridge-utils. A Linux software package that is required for creating 802.1d Ethernet bridges. *http://bridge.sourceforge.net/*

Broadband Global Access Network (BGAN). One of several standards used for satellite Internet access. See also: ***Digital Video Broadcast (DVB-S)*** and ***Very Small Aperture Terminal (VSAT).***

broadcast address. On IP networks, the broadcast IP address is used to send data to all hosts in the local subnet. On Ethernet networks, the broadcast MAC address is used to send data to all machines in the same collision domain.

bypass diodes. A feature found on some solar panels that prevents the formation of ***hot-spots*** on shaded cells, but reduces the maximum voltage of the panel.

C

CA see ***Certificate Authority***

Cacti (*http://www.cacti.net/*). A popular web-based monitoring tool written in PHP.

capacity. The theoretical maximum amount of traffic provided by a digital communications line. Often used interchangeably with ***bandwidth***.

captive portal. A mechanism used to transparently redirect web browsers to a new location. Captive portals are often used for authentication or for interrupting a user's online session (for example, to display an Acceptable Use Policy).

cell. Solar panels are made up of several individual cells, which are electrically connected to provide a particular value of current and voltage. Batteries are also made up of individual cells connected in series, each of which contributes about 2 volts to the battery.

Certificate Authority. A trusted entity that issues signed cryptographic keys. See also: ***Public Key Infrastructure, SSL***

channel capacity. The maximum amount of information that can be sent using a given bandwidth. See also: ***bandwidth, throughput, data rate***

channel. A well defined range of frequencies used for communications. 802.11 channels use 22 MHz of bandwidth, but are only separated by 5 MHz. See also: ***Appendix B.***

CIDR see ***Classless Inter-Domain Routing***

CIDR notation. A method used to define a network mask by specifying the number of bits present. For example, the netmask 255.255.255.0 can be specified as /24 in CIDR notation.

circular polarization. An electro-magnetic field where the electric field vector appears to be rotating with circular motion about the direction of propagation, making one full turn for each RF cycle. See also: *horizontal polarization, vertical polarization*

Class A, B, and C networks. For some time, IP address space was allocated in blocks of three different sizes. These were Class A (about 16 million addresses), Class B (about 65 thousand addresses), and Class C (255 addresses). While CIDR has replaced class-based allocation, these classes are often still referred to and used internally in organizations using private address space. See also: *CIDR notation*.

Classless Inter-Domain Routing. CIDR was developed to improve routing efficiency on the Internet backbone by enabling route aggregation and network masks of arbitrary size. CIDR replaces the old class-based addressing scheme. See also: *Class A, B, and C networks*.

client. An 802.11 radio card in *managed mode*. Wireless clients will join a network created by an access point, and automatically change the channel to match it. See also: *access point, mesh*

closed network. An access point that does not broadcast its SSID, often used as a security measure.

coax. A round (coaxial) cable with a center wire surrounded by a dielectric, outer conductor, and tough insulating jacket. Antenna cables are usually made of coax. Coax is short for "of common axis".

collision. On an Ethernet network, a collision occurs when two devices connected to the same physical segment attempt to transmit at the same time. When collisions are detected, devices delay retransmission for a brief, randomly selected period.

conductor. A material that easily allows electric or thermal energy to flow through without much resistance. See also: *dielectric, insulator*

connectionless protocol. A network protocol (such as UDP) that requires no session initiation or maintenance. Connectionless protocols typically require less overhead than session oriented protocols, but do not usually offer data protection or packet reassembly. See also: *session oriented protocol*.

consistent platform. Maintenance costs can be reduced by using a consistent platform, with the same hardware, software, and firmware for many components in a network.

constructive interference. When two identical waves merge and are in phase, the amplitude of the resulting wave is twice that of either of the components. This is called constructive interference. See also: *destructive interference*

controls. In **NEC2**, controls define the RF source in an antenna model. See also: **structure**

converter. A device used to convert DC signals into a different DC or AC voltage. See also: **inverter**

CPE see **Customer Premises Equipment**

cron. A Unix facility that allows timed and repeated execution of programs. See also: **at**

Customer Premises Equipment. Network equipment (such as a **router** or **bridge**) that is installed at a customer's location.

D

data link layer. The second layer in both the OSI and TCP/IP network models. Communications at this layer happen directly between nodes. On Ethernet networks, this is also sometimes called the MAC layer.

data rate. The speed at which 802.11 radios exchange symbols, which is always higher than the available throughput. For example, the nominal data rate of 802.11g is 54 Mbps, while the maximum throughput is about 20 Mbps). See also: **throughput**

dB see **decibel**

DC see **Direct Current**

DC/AC Converter. A device that converts DC power into AC power, suitable for use with many appliances. Also known as an **inverter**.

DC/DC Converter. A device that changes the voltage of a DC power source. See also: **linear conversion, switching conversion**

decibel (dB). A logarithmic unit of measurement that expresses the magnitude of power relative to a reference level. Commonly used units are dBi (decibels relative to an isotropic radiator) and dBm (decibels relative to a milliwatt).

default gateway. When a router receives a packet destined for a network for which it has no explicit route, the packet is forwarded to the default gateway. The default gateway then repeats the process, possibly sending the packet to its own default gateway, until the packet reaches its ultimate destination.

default route. A network route that points to the default gateway.

Denial of Service (DoS). An attack on network resources, usually achieved by flooding a network with traffic or exploiting a bug in an application or network protocol.

depreciation. An accounting method used to save money to cover the eventual break down of equipment.

destructive interference. When two identical waves merge and are exactly out of phase, the amplitude of the resulting wave is zero. This is called destructive interference. See also: **constructive interference**

DHCP see **Dynamic Host Configuration Protocol**

dielectric. A non-conductive material that separates conducting wires inside a cable.

Digital Elevation Map (DEM). Data that represents the height of terrain for a given geographic area. These maps are used by programs such as **Radio Mobile** to model electromagnetic propagation.

Digital Video Broadcast (DVB-S). One of several standards used for satellite Internet access. See also: **Broadband Global Access Network (BGAN)** and **Very Small Aperture Terminal (VSAT)**.

dipole antenna. The simplest form of **omnidirectional antenna**.

Direct Current (DC). An electrical current which remains constant over time. DC current is typically used for network equipment, such as access points and routers. See also: **Alternating Current**

Direct Sequence Spread Spectrum (DSSS). The radio modulation scheme used by 802.11b.

directional antenna. An antenna that radiates very strongly in a particular direction. Examples of directional antennas include the yagi, dish, and waveguide antennas. See also: **omnidirectional antenna, sectorial antenna**

directivity. The ability of an antenna to focus energy in a particular direction when transmitting, or to receive energy from a particular direction when receiving.

diversity see **antenna diversity**

DNS see **Domain Name Service**

DNS caching. By installing a DNS server on your local LAN, DNS requests for an entire network may be cached locally, improving response times. This technique is called DNS caching.

dnsmasq. An open source caching DNS and DHCP server, available from *http://thekelleys.org.uk/*

Domain Name Service (DNS). The widely used network protocol that maps IP addresses to names.

dominant mode. The lowest frequency that can be transmitted by a waveguide of a given size.

DoS see *Denial of Service*

DSSS see *Direct Sequence Spread Spectrum*

DVB-S see *Digital Video Broadcast.*

Dynamic Host Configuration Protocol (DHCP). A protocol used by hosts to automatically determine their IP address.

E

eavesdropper. Someone who intercepts network data such as passwords, email, voice data, or online chat.

edge. The place where one organization's network meets another. Edges are defined by the location of the external *router*, which often acts as a *firewall*.

electromagnetic spectrum. The very wide range of possible frequencies of electromagnetic energy. Parts of the electromagnetic spectrum include radio, microwave, visible light, and X rays.

electromagnetic wave. A wave that propagates through space without the need for a propagating medium. It contains an electric and a magnetic component. See also: *mechanical wave*

elevation see *inclination*

end span injectors. An 802.3af *Power over Ethernet* device that provides power via the Ethernet cable. An Ethernet switch that provides power on each port is an example of an end span injector. See also: *mid span injectors*

end-to-end encryption. An encrypted connection negotiated by both ends of a communications session. End-to-end encryption can provide stronger protection than *link layer encryption* when used on untrusted networks (such as the Internet).

EtherApe. An open source network visualization tool. Available at *http://etherape.sourceforge.net/*

Ethereal see *Wireshark*.

Extended Service Set Identifier (ESSID). The name used to identify an 802.11 network. See also: *closed network*

external traffic. Network traffic that originates from, or is destined for, an IP address outside your internal network, such as Internet traffic.

F

firestarter. A graphical front-end for configuring Linux firewalls available from *http://www.fs-security.com/*.

filter. The default table used in the Linux netfilter firewall system is the filter table. This table is used for determining traffic that should be accepted or denied.

firewall. A router that accepts or denies traffic based on some criteria. Firewalls are one basic tool used to protect entire networks from undesirable traffic.

flush. To remove all entries in a routing table or netfilter chain.

forwarding. When routers receive packets that are destined for a different host or network, they send the packet to the next router closest to its ultimate destination. This process is called forwarding.

forwarding loops. A routing misconfiguration where packets are forwarded cyclically between two or more routers. Catastrophic network failure is prevented by using the TTL value on every packet, but forwarding loops need to be resolved for proper network operations.

free space loss. Power diminished by geometric spreading of the wavefront, as the wave propagates through space. See also: *attenuation, free space loss, Appendix C*

frequency. The number of whole waves that pass a fixed point in a period of time. See also: *wavelength, Hertz*

front-to-back ratio. The ratio of the maximum *directivity* of an antenna to its directivity in the opposite direction.

full duplex. Communications equipment that can send and receive at the same time (such as a telephone). See also: *half duplex*

fwbuilder. A graphical tool that lets you create **iptables** scripts on a machine separate from your server, and then transfer them to the server later. *http://www.fwbuilder.org/*

G

gain. The ability of a radio component (such as an antenna or amplifier) to increase the power of a signal. See also: **decibel**

gain transfer. Comparing an antenna under test against a known standard antenna, which has a calibrated gain.

gasification. The production bubbles of oxygen and hydrogen that occurs when a battery is **overcharged**.

globally routable. An address issued by an ISP or RIR that is reachable from any point on the Internet. In IPv4, there are approximately four billion possible IP addresses, although not all of these are globally routable.

H

half duplex. Communications equipment that can send or receive, but never both at once (such as a handheld radio). See also: **full duplex**.

Heliax. High quality coaxial cable that has a solid or tubular center conductor with a corrugated solid outer conductor which enables it to flex. See also: **coax**

Hertz (Hz). A measure of **frequency**, denoting some number of cycles per second.

HF (**High-Frequency**). Radio waves from 3 to 30 MHz are referred to as HF. Data networks can be built on HF that operate at very long range, but with very low data capacity.

hop. Data that crosses one network connection. A web server may be several hops away from your local computer, as packets are forwarded from router to router, eventually reaching their ultimate destination.

horizontal polarization. An electromagnetic field with the electric component moving in a linear horizontal direction. See also: **circular polarization, vertical polarization**

hot-spot. In wireless networks, a hot-spot is a location that provides Internet access via **Wi-Fi**, typically by use of a **captive portal**. In **photovoltaic systems**, a hot-spot occurs when a single *cell* in a *solar panel* is shaded, causing it to act as a resistive load rather than to generate power.

hub. An Ethernet networking device that repeats received data on all connected ports. See also: *switch*.

Huygens principle. A wave model that proposes an infinite number of potential wavefronts along every point of an advancing wavefront.

Hz see *Hertz*

I

IANA see *Internet Assigned Numbers Authority*

ICMP see *Internet Control Message Protocol*

ICP see *Inter-Cache Protocol*

impedance. The quotient of voltage over current of a transmission line, consisting of a resistance and a reactance. The load impedance must match the source impedance for maximum power transfer (50Ω for most communications equipment).

inbound traffic. Network packets that originate from outside the local network (typically the Internet) and are bound for a destination inside the local network. See also: *outbound traffic*.

inclination. The angle that marks deviation from a horizontal plane. See also: *azimuth*

infrastructure mode see *master mode*

insulator see *dielectric*

Inter-Cache Protocol (ICP). A high performance protocol used to communicate between web caches.

Internet Assigned Numbers Authority (IANA). The organization that administers various critical parts of Internet infrastructure, including IP address allocation, DNS root name servers, and protocol service numbers.

Internet Control Message Protocol (ICMP). A Network Layer protocol used to inform nodes about the state of the network. ICMP is part of the Internet protocol suite. See also: *Internet protocol suite*.

Internet layer see *network layer*

Internet Protocol (IP). The most com-mon network layer protocol in use. IP defines the hosts and networks that make up the global Internet.

Internet protocol suite (TCP/IP). The family of communication protocols that make up the Internet. Some of these protocols include TCP, IP, ICMP, and UDP. Also called the *TCP/IP protocol suite*, or simply *TCP/IP*.

Intrusion Detection System (IDS). A program that watches network traffic, looking for suspicious data or behavior patterns. An IDS may make a log entry, notify a network administrator, or take direct action in response to undesirable traffic.

inverter see *DC/AC Converter*

IP see *Internet Protocol*

iproute2. The advanced routing tools package for Linux, used for traffic shaping and other advanced techniques. Available from *http://linux-net.osdl.org/*

iptables. The primary command used to manipulate netfilter firewall rules.

irradiance. The total amount of solar energy that lights a given area, in W/m2

ISM band. ISM is short for Industrial, Scientific, and Medical. The ISM band is a set of radio frequencies set aside by the ITU for unlicensed use.

isotropic antenna. A hypothetical antenna that evenly distributes power in all directions, approximated by a dipole.

IV characteristic curve. A graph that represents the current that is provided based on the voltage generated for a certain solar radiation.

K

knetfilter. A graphical front-end for configuring Linux firewalls. Available from *http://venom.oltrelinux.com/*

known good. In troubleshooting, a known good is any component that can be substituted to verify that its counterpart is in good, working condition.

L

lag. Common term used to describe a network with high *latency*.

lambda (λ) see *wavelength*

LAN see *Local Area Network*

latency. The amount of time it takes for a packet to cross a network connection. It is often (incorrectly) used interchangeably with Round Trip Time (RTT), since measuring the RTT of a wide-area connection is trivial compared to measuring the actual latency. See also: *Round Trip Time*.

lead-acid batteries. Batteries consisting of two submerged lead electrodes in an electrolytic solution of water and sulfuric acid. See also: *stationary batteries*

lease time. In DHCP, IP addresses are assigned for a limited period of time, known as the lease time. After this time period expires, clients must request a new IP address from the DHCP server.

Line of Sight (LOS). If a person standing at point A has an unobstructed view of point B, then point A is said to have a clear Line of Sight to point B.

linear polar coordinates. A graph system with equally spaced, graduated concentric circles representing an absolute value on a polar projection. Such graphs are typically used to represent antenna radiation patterns. See also: *logarithmic polar coordinates*

linear conversion. A DC voltage conversion method that lowers the voltage by converting excess energy to heat. See also: *switching conversion*

linear polarization. An *electro-magnetic wave* where the electric field vector stays in the same plane all the time. The electric field may leave the antenna in a vertical orientation, a horizontal orientation, or at some angle between the two. See also: *vertical polarization, horizontal polarization*

link budget. The amount of radio energy available to overcome path losses. If the available link budget exceeds the path loss, minimum receive sensitivity of the receiving radio, and any obstacles, then communications should be possible.

link layer encryption. An encrypted connection between *link-local* devices, typically a wireless *client* and an *access point*. See also: *end-to-end encryption*

link-local. Network devices that are connected to the same physical segment communicate with each other directly are said to be link-local. A link-local connection cannot cross a router boundary without using some kind of encapsulation, such as a *tunnel* or a *VPN*.

listen. Programs that accept connections on a TCP port are said to listen on that port.

load. Equipment in a photovoltaic system that consumes energy. See also: *battery, solar panel, regulator, converter, inverter*

Local Area Network (LAN). A network (typically Ethernet) used within an organization. The part of a network that exists just behind an ISP's router is generally considered to be part of the LAN. See also: *WAN.*

logarithmic polar coordinates. A graph system with logarithmically spaced, graduated concentric circles representing an absolute value on a polar projection. Such graphs are typically used to represent antenna radiation patterns. See also: *linear polar coordinates*

long fat pipe network. A network connection (such as VSAT) that has high capacity and high latency. In order to achieve the best possible performance, TCP/IP must be tuned to match the traffic on such links.

LOS see *Line of Sight*

MAC layer see *data link layer*

MAC address. A unique 48 bit number assigned to every networking device when it is manufactured. The MAC address is used for link-local communications.

MAC filtering. An access control method based on the MAC address of communicating devices.

MAC table. A network switch must keep track of the MAC addresses used on each physical port, in order to efficiently distribute packets. This information is kept in a table called the MAC table.

maintenance-free lead-acid batteries see *lead-acid batteries*

Man-In-The-Middle (MITM). A network attack where a malicious user intercepts all communications between a client and a server, allowing information to be copied or manipulated.

managed hardware. Networking hardware that provides an administrative interface, port counters, SNMP, or other interactive features is said to be managed.

managed mode. A radio mode used by 802.11 devices that allows the radio to join a network created by an access point. See also: *master mode, ad-hoc mode, monitor mode*

master browser. On Windows networks, the master browser is the computer that keeps a list of all the computers, shares and printers that are available in **Network Neighborhood** or **My Network Places**.

master mode. A radio mode used by 802.11 devices that allows the radio to create networks just as an access point does. See also: *managed mode, ad-hoc mode, monitor mode*

match condition. In netfilter, a match condition specifies the criteria that determine the ultimate target for a given packet. Packets may be matched on MAC address, source or destination IP address, port number, data contents, or just about any other property.

Maximum Depth of Discharge (DoDmax). The amount of energy extracted from a battery in a single discharge cycle, expressed as a percentage.

Maximum Power Point (Pmax). The point where the power supplied by a solar panel is at maximum.

MC-Card. A very small microwave connector found on Lucent / Orinoco / Avaya equipment.

mechanical wave. A wave caused when some medium or object is swinging in a periodic manner. See also: *electromagnetic wave*

Media Access Control layer see *data link layer*

mesh. A network with no hierarchical organization, where every node on the network carries the traffic of every other as needed. Good mesh network implementations are self-healing, which means that they automatically detect routing problems and fix them as needed.

message types. Rather that port numbers, ICMP traffic uses message types to define the type of information being sent. See also: *ICMP*.

method of the worst month. A method for calculating the dimensions of a standalone photovoltaic system so it will work in the month in which the demand for energy is greatest with respect to the available solar energy. It is the worst month of the year, as this month with have the largest ratio of demanded energy to available energy.

MHF see *U.FL*

microfinance. The provision of small loans, savings and other basic financial services to the world's poorest people.

mid span injectors. A *Power over Ethernet* device inserted between an Ethernet switch and the device to be powered. See also: *end span injectors*

milliwatts (mW). A unit of power representing one thousandth of a Watt.

MITM see *Man-In-The-Middle*

MMCX. A very small microwave connector commonly found on equipment manufactured by Senao and Cisco.

monitor mode. A radio mode used by 802.11 devices not normally used for communications that allows the radio passively monitor radio traffic. See also: *master mode, managed mode, ad-hoc mode*

monitor port. On a managed switch, one or more monitor ports may be defined that receive traffic sent to all of the other ports. This allows you to connect a traffic monitor server to the port to observe and analyze traffic patterns.

Multi Router Traffic Grapher (MRTG). An open source tool used for graphing traffic statistics. Available from *http://oss.oetiker.ch/mrtg/*

multipath. The phenomenon of reflections of a signal reaching their target along different paths, and therefore at different times.

multipoint-to-multipoint see *mesh*

mW see *milliwatt*

My TraceRoute (mtr). A network diagnostic tool used as an alternative to the traditional traceroute program. *http://www.bitwizard.nl/mtr/.* See also: *traceroute / tracert.*

N connector. A sturdy microwave connector commonly found on outdoor networking components, such as antennas and outdoor access points.

Nagios (*http://nagios.org/*) A realtime monitoring tool that logs and notifies a system administrator about service and network outages.

NAT see *Network Address Translation*

nat. The table used in the Linux netfilter firewall system to configure Network Address Translation.

NEC2 see **Numerical Electromagnetics Code**

NetBIOS. A session layer protocol used by Windows networking for file and printer sharing. See also: **SMB**.

netfilter. The packet filtering framework in modern Linux kernels is known as netfilter. It uses the iptables command to manipulate filter rules. *http://netfilter-.org/*

netmask (**network mask**). A netmask is a 32-bit number that divides the 16 million available IP addresses into smaller chunks, called subnets. All IP networks use IP addresses in combination with netmasks to logically group hosts and networks.

NeTraMet. An open source network flow analysis tool available from *freshmeat.net/projects/netramet/*

network address. The lowest IP number in a subnet. The network address is used in routing tables to specify the destination to be used when sending packets to a logical group of IP addresses.

Network Address Translation (NAT). NAT is a networking technology that allows many computers to share a single, globally routable IP address. While NAT can help to solve the problem of limited IP address space, it creates a technical challenge for two-way services, such as Voice over IP.

network detection. Network diagnostic tools that display information about wireless networks, such as the network name, channel, and encryption method used.

network layer. Also called the Internet layer. This is the third layer of the OSI and TCP/IP network models, where IP operates and Internet routing takes place.

network mask see **netmask**

ngrep. An open source network security utility used to find patterns in data flows. Available for free from *http://ngrep.sourceforge.net/*

node. Any device capable of sending and receiving data on a network. Access points, routers, computers, and laptops are all examples of nodes.

Nominal Capacity (CN). The maximum amount of energy that can be extracted from a fully charged battery. It is expressed in Ampere-hours (Ah) or Watt-hours (Wh).

Nominal Voltage (VN). The operating voltage of a photovoltaic system, typically 12 or 24 volts.

ntop. A network monitoring tool that provides extensive detail about connections and protocol use on a local area network. *http://www.ntop.org/*

null. In an antenna radiation pattern, a null is a zone in which the effective radiated power is at a minimum.

nulling. A specific case of **multipath** interference where the signal at the receiving antenna is zeroed by the **destructive interference** of reflected signals.

number of days of autonomy (N). The maximum number of days that a photovoltaic system can operate without significant energy received from the sun.

Numerical Electromagnetics Code (NEC2). A free antenna modeling package that lets you build an antenna model in 3D, and then analyze the antenna's electromagnetic response. *http://www.nec2.org/*

O

OFDM see **Orthogonal Frequency Division Multiplexing**

omnidirectional antenna. An antenna that radiates almost equally in every direction in the horizontal plane. See also: **directional antenna, sectorial antenna**

one-arm repeater. A wireless repeater that only uses a single radio, at significantly reduced throughput. See also: **repeater**

onion routing. A privacy tool (such as **Tor**) that repeatedly bounces your TCP connections across a number of servers spread throughout the Internet, wrapping routing information in a number of encrypted layers.

OR logic. A logical operation that evaluates as true if any of the items being compared also evaluate as true. See also: **AND logic**.

Orthogonal Frequency Division Multiplexing (OFDM)

OSI network model. A popular model of network communications defined by the ISO/IEC 7498-1 standard. The OSI model consists of seven interdependent layers, from the physical through the application. See also: **TCP/IP network model**.

outbound traffic. Network packets that originate from the local network and are bound for a destination outside the local network (typically somewhere on the Internet). See also: **inbound traffic**.

overcharge. The state of a battery when charge is applied beyond the limit of the battery's capacity. If energy is applied to a battery beyond its point of maximum charge, the electrolyte begins to break down. *Regulators* will allow a small amount of overcharge time to a battery to avoid *gasification*, but will remove power before the battery is damaged.

overdischarge. Discharging a battery beyond its *Maximum Depth of Discharge*, which results in deterioration of the battery.

oversubscribe. To allow more users than the maximum available bandwidth can support.

P

packet. On IP networks, messages sent between computers are broken into small pieces called packets. Each packet includes a source, destination, and other routing information that is used to route it to its ultimate destination. Packets are reassembled again at the remote end by TCP (or another protocol) before being passed to the application.

packet filter. A firewall that operates at the Internet layer by inspecting source and destination IP addresses, port numbers, and protocols. Packets are either permitted or discarded depending on the packet filter rules.

partition. A technique used by network hubs to limit the impact of computers that transmit excessively. Hubs will temporarily remove the abusive computer (partition it) from the rest of the network, and reconnect it again after some time. Excessive partitioning indicates the presence of an excessive bandwidth consumer, such as a peer-to-peer client or network virus.

passive POE injector see *Power over Ethernet*

path loss. Loss of radio signal due to the distance between communicating stations.

Peak Sun Hours (PSH). Average value of daily irradiation for a given area.

photovoltaic generator see *solar panel*

photovoltaic solar energy. The use of solar panels to collect solar energy to produce electricity. See also: *thermal solar energy*

photovoltaic system. An energy system that generates electrical energy from solar radiation and stores it for later use. A standalone photovoltaic system does this without any connection to an established power grid. See also: *battery, solar panel, regulator, load, converter, inverter*

physical layer. The lowest layer in both the OSI and TCP/IP network models. The physical layer is the actual medium used for communications, such as copper cable, optic fiber, or radio waves.

pigtail. A short microwave cable that converts a non-standard connector into something more robust and commonly available.

ping. A ubiquitous network diagnostic utility that uses ICMP echo request and reply messages to determine the round trip time to a network host. Ping can be used to determine the location of network problems by "pinging" computers in the path between the local machine and the ultimate destination.

PKI see *Public Key Infrastructure*

plomb. A heavy piece of metal buried in the earth to improve a ground connection.

PoE see *Power over Ethernet*

point-to-multipoint. A wireless network where several nodes connect back to a central location. The classic example of a point-to-multipoint network is an access point at an office with several laptops using it for Internet access. See also: *point-to-point, multipoint-to-multipoint*

point-to-point. A wireless network consisting of only two stations, usually separated by a great distance. See also: *point-to-multipoint, multipoint-to-multipoint*

Point-to-Point Protocol (PPP). A network protocol typically used on serial lines (such as a dial-up connection) to provide IP connectivity.

polar plot. A graph where points are located by projection along a rotating axis (radius) to an intersection with one of several concentric circles. See also: *rectangular plot*

polarization. The direction of the electric component of an electro-magnetic wave as it leaves the transmitting antenna. See also: *horizontal polarization, vertical polarization, circular polarization*

polarization mismatch. A state where a transmitting and receiving antenna do not use the same polarization, resulting in signal loss.

policy. In netfilter, the policy is the default action to be taken when no other filtering rules apply. For example, the default policy for any chain may be set to ACCEPT or DROP.

port counters. Managed switches and routers provide statistics for each network port called port counters. These statistics may include inbound and outbound packet and byte counts, as well as errors and retransmissions.

power. The amount of energy in a certain amount of time.

Power over Ethernet (PoE). A technique used to supply DC power to devices using the Ethernet data cable. See also: *end span injectors, mid span injectors*

PPP see *Point to Point Protocol*

presentation layer. The sixth layer of the OSI networking model. This layer deals with data representation, such as MIME encoding or data compression.

private address space. A set of reserved IP addresses outlined in RFC1918. Private address space is frequently used within an organization, in conjunction with Network Address Translation (NAT). The reserved private address space ranges include 10.0.0.0/8, 172.16.0.0/12, and 192.168.0.0/16. See also: *NAT*.

Privoxy (*http://www.privoxy.org/*). A web proxy that provides anonymity through the use of filters. Privoxy is often used in conjunction with *Tor*.

proactive routing. A *mesh* implementation where every node knows about the existence of every other node in the mesh cloud as well as which nodes may be used to route traffic to them. Each node maintains a routing table covering the whole mesh cloud. See also: *reactive routing*

protocol analyzer. A diagnostic program used to observe and disassemble network packets. Protocol analyzers provide the greatest possible detail about individual packets.

protocol stack. A set of network protocols that provide interdependent layers of functionality. See also: *OSI network model* and *TCP/IP network model*.

PSH see *Peak Sun Hours*

Public key cryptography. A form of encryption used by SSL, SSH, and other popular security programs. Public key cryptography allows encrypted information to be exchanged over an untrusted network without the need to distribute a secret key.

Public Key Infrastructure (PKI). A security mechanism used in conjunction with *public key cryptography* to prevent the possibility of *Man-In-The-Middle* attacks. See also: *certificate authority*

Q

quick blow. A type of fuse that immediately blows if the current flowing through it is higher than their rating. See also: **slow blow**

R

radiation pattern see **antenna pattern.**

radio. The portion of the electromagnetic spectrum in which waves can be generated by applying alternating current to an antenna.

reactive routing. A **mesh** implementation where routes are computed only when it is necessary to send data to a specific node. See also: **proactive routing**

realtime monitoring. A network monitoring tool that performs unattended monitoring over long periods, and notifies administrators immediately when problems arise .

reciprocity. An antenna's ability to maintain the same characteristics regardless if whether it is transmitting or receiving.

recombinant batteries see **lead-acid batteries**

rectangular plot. A graph where points are located on a simple grid. See also: **polar plot**

Regional Internet Registrars (RIR). The 4 billion available IP addresses are administered by the IANA. The space has been divided into large subnets, which are delegated to one of the five regional Internet registries, each with authority over a large geographic area.

regulator. The component of a **photovoltaic system** that assures that the **battery** is working in appropriate conditions. It avoids **overcharging** or **undercharging** the battery, both of which are very detrimental to the life of the battery. See also: **solar panel, battery, load, converter, inverter**

repeater. A node that is configured to rebroadcast traffic that is not destined for the node itself, often used to extend the useful range of a network.

Request for Comments (RFC). RFCs are a numbered series of documents published by the Internet Society that document ideas and concepts related to Internet technologies. Not all RFCs are actual standards, but many are either approved explicitly by the IETF, or eventually become de facto standards. RFCs can be viewed online at *http://rfc.net/*.

return loss. A logarithmic ratio measured in dB that compares the power reflected by the antenna to the power that is fed into the antenna from the transmission line. See also: *impedance*

reverse polarity (RP). Proprietary microwave connectors, based on a standard connector but with the genders reversed. The **RP-TNC** is probably the most common reverse polarity connector, but others (such as RP-SMA and RP-N) are also commonplace.

RF transmission line. The connection (typically *coax*, *Heliax*, or a *waveguide*) between a radio and an antenna.

RIR see *Regional Internet Registrars*

Round Trip Time (RTT). The amount of time it takes for a packet to be acknowledged from the remote end of a connection. Frequently confused with *latency*.

rogue access points. An unauthorized access point incorrectly installed by legitimate users, or by a malicious person who intends to collect data or do harm to the network.

Round Robin Database (RRD). A database that stores information in a very compact way that does not expand over time. This is the data format used by RRDtool and other network monitoring tools.

router. A device that forwards packets between different networks. The process of forwarding packets to the next hop is called *routing.*

routing. The process of forwarding packets between different networks. A device that does this is called a *router*.

routing table. A list of networks and IP addresses kept by a router to determine how packets should be forwarded. If a router receives a packet for a network that is not in the routing table, the router uses its default gateway. Routers operate at the Network Layer. See also: *bridge* and *default gateway*.

RP see *Reverse Polarity*

RP-TNC. A common proprietary version of the TNC microwave connector, with the genders reversed. The RP-TNC is often found on equipment manufactured by Linksys.

RRD see *Round Robin Database*

RRDtool. A suite of tools that allow you to create and modify RRD databases, as well as generate useful graphs to present the data. RRDtool is used to keep track of time-series data (such as network bandwidth, machine room temperature, or server load average) and can display that data as an average over time. RRDtool is available from *http://oss.oetiker.ch/rrdtool/*

rsync (*http://rsync.samba.org/*). An open source incremental file transfer utility used for maintaining mirrors.

RTT see *Round Trip Time*

s

SACK see *Selective Acknowledgment*

scattering. Signal loss due to objects in the path between two nodes. See also: *free space loss, attenuation*

sectorial antenna. An antenna that radiates primarily in a specific area. The beam can be as wide as 180 degrees, or as narrow as 60 degrees. See also: *directional antenna, omnidirectional antenna*

Secure Sockets Layer (SSL). An end-to-end encryption technology built into virtually all web browsers. SSL uses *public key cryptography* and a trusted *public key infrastructure* to secure data communications on the web. Whenever you visit a web URL that starts with https, you are using SSL.

Selective Acknowledgment (SACK). A mechanism used to overcome TCP inefficiencies on high latency networks, such as VSAT.

Server Message Block (SMB). A network protocol used in Windows networks to provide file sharing services. See also: *NetBIOS*.

Service Set ID (SSID) see *Extended Service Set Identifier*

session layer. Layer five of the OSI model, the Session Layer manages logical connections between applications.

session oriented protocol. A network protocol (such as TCP) that requires initialization before data can be exchanged, as well as some clean-up after data exchange has completed. Session oriented protocols typically offer error correction and packet reassembly, while connectionless protocols do not. See also: *connectionless protocol*.

shared medium. A *link-local* network where every node can observe the traffic of every other node.

Shorewall (*http://shorewall.net/*). A configuration tool used for setting up netfilter firewalls without the need to learn iptables syntax.

sidelobes. No antenna is able to radiate all the energy in one preferred direction. Some is inevitably radiated in other directions. These smaller peaks are referred to as sidelobes.

signal generator. A transmitter that emits continuously at a specific frequency.

Simple Network Management Protocol (SNMP). A protocol designed to facilitate the exchange of management information between network devices. SNMP is typically used to poll network switches and routers to gather operating statistics.

site-wide web cache. While all modern web browsers provide a local data cache, large organizations can improve efficiency by installing a site-wide web cache, such as Squid. A site-wide web cache keeps a copy of all requests made from within an organization, and serves the local copy on subsequent requests. See also: *Squid*.

slow blow. A fuse that allows a current higher than its rating to pass for a short time. See also: *quick blow*

SMA. A small threaded microwave connector.

SMB see *Server Message Block*

SmokePing. A latency measurement tool that measures, stores and displays latency, latency distribution and packet loss all on a single graph. SmokePing is available from *http://oss.oetiker.ch/smokeping/*

SNMP see *Simple Network Management Protocol*

Snort (*http://www.snort.org/*). A very popular open source intrusion detection system. See also: *Intrusion Detection System*.

SoC see *State of Charge*

solar module see *solar panel*

solar panel. The component of a *photovoltaic system* used to convert solar radiation into electricity. See also: *battery, regulator, load, converter, inverter*

solar panel array. A set of *solar panels* wired in series and/or parallel in order to provide the necessary energy for a given *load*.

solar power charge regulator see *regulator*

spectrum see *electromagnetic spectrum*

spectrum analyzer. A device that provides a visual representation of the electromagnetic spectrum. See also: *Wi-Spy*

Speed. A generic term used to refer to the responsiveness of a network connection. A "high-speed" network should have low latency and more than enough capacity to carry the traffic of its users. See also: *bandwidth*, *capacity*, and *latency*.

split horizon DNS. A technique used to serve different answers to DNS requests based on the source of the request. Split horizon is used to direct internal users to a different set of servers than Internet users.

spoof. To impersonate a network device, user, or service.

spot check tools. Network monitoring tools that are run only when needed to diagnose a problem. Ping and traceroute are examples of spot check tools.

Squid. A very popular open source web proxy cache. It is flexible, robust, full-featured, and scales to support networks of nearly any size. *http://www.squid-cache.org/*

SSID see *Extended Service Set Identifier*

SSL see *Secure Sockets Layer*

standalone photovoltaic system see *photovoltaic system*

State of Charge (SoC). The amount of charge present in a battery, determined by the current voltage and type of battery.

stateful inspection. Firewall rules that are aware of the the state associated with a given packet. The state is not part of the packet as transmitted over the Internet, but is determined by the firewall itself. New, established, and related connections may all be taken into consideration when filtering packets. Stateful inspection is sometimes called connection tracking.

stationary batteries. Batteries designed to have a fixed location and in scenarios where the power consumption is more or less irregular. Stationary batteries can accommodate deep discharge cycles, but they are not designed to produce high currents in brief periods of time. See also: *lead-acid batteries*

structure. In *NEC2*, a numerical description of where the different parts of the antenna are located, and how the wires are connected up. See also: *controls subnet mask* see *netmask*

subnets. A subset of a range of IP networks, defined by *netmasks*.

switch. A network device that provides a temporary, dedicated connection between communicating devices. See also: *hub*.

switching conversion. A DC voltage conversion method that uses a magnetic component to temporarily store the energy and transform it to another voltage. Switching conversion is much more efficient than *linear conversion*.

T

target. In netfilter, the action to be taken once a packet matches a rule. Some possible netfilter targets include *ACCEPT*, *DROP*, *LOG*, and *REJECT*.

TCP see *Transmission Control Protocol*

TCP acknowledgment spoofing

TCP window size. The TCP parameter that defines how much data that may be sent before an ACK packet is returned from the receiving side. For instance, a window size of 3000 would mean that two packets of 1500 bytes each will be sent, after which the receiving end will either ACK the chunk or request retransmission.

TCP/IP see *Internet protocol suite*

TCP/IP network model. A popular simplification of the OSI network model that is used with Internet networks. The TCP/IP model consists of five interdependent layers, from the physical through the application. See also: *OSI network model*.

tcpdump. A popular open source packet capture and analysis tool available at *http://www.tcpdump.org/*. See also: *WinDump* and *Wireshark*.

Temporal Key Integrity Protocol (TKIP). An encryption protocol used in conjunction with *WPA* to improve the security of a communications session.

thermal solar energy. Energy collected from the sun in the form of heat. See also: *photovoltaic solar energy*

thrashing. The state when a computer has exhausted the available RAM and must use the hard disk for temporary storage, greatly reducing system performance.

throughput. The actual amount of information per second flowing through a network connection, disregarding protocol overhead.

throughput testing tools. Tools that measure the actual bandwidth available between two points on a network.

Time To Live (TTL). A TTL value acts as a deadline or emergency brake to signal a time when the data should be discarded. In TCP/IP networks, the TTL is a counter that starts at some value (such as 64) and is decremented at each router hop. If the TTL reaches zero, the packet is discarded. This mechanism helps reduce damage caused by routing loops. In DNS, the TTL defines the amount of time that a particular zone record should be kept before it must be refreshed. In Squid, the TTL defines how long a cached object may be kept before it must be again retrieved from the original website.

TKIP see *Temporal Key Integrity Protocol*

TNC connector. A common, sturdy threaded microwave connector.

Tor (*http://www.torproject.org/*). An **onion routing** tool that provides good protection against traffic analysis.

traceroute / tracert. A ubiquitous network diagnostic utility often used in conjunction with ping to determine the location of network problems. The Unix version is called traceroute, while the Windows version is tracert. Both use ICMP echo requests with increasing TTL values to determine which routers are used to connect to a remote host, and also display latency statistics. Another variant is tracepath, which uses a similar technique with UDP packets. See also: *mtr*.

traction batteries see *lead-acid batteries*

Transmission Control Protocol (TCP). A session oriented protocol that operates at the Transport Layer, providing packet reassembly, congestion avoidance, and reliable delivery. TCP is an integral protocol used by many Internet applications, including HTTP and SMTP. See also: *UDP*.

transmission power. The amount of power provided by the radio transmitter, before any antenna gain or line losses.

transparent bridging firewall. A firewall technique that introduces a bridge that selectively forwards packets based on firewall rules. One benefit of a transparent bridging firewall is that it does not require an IP address. See also: *bridge*.

transparent cache. A method of implementing a site-wide web cache that requires no configuration on the web clients. Web requests are silently redirected to the cache, which makes the request on behalf of the client. Transparent caches cannot use authentication, which makes it impossible to implement traffic accounting at the user level. See also: *site-wide web cache*, *Squid*.

transparent proxy. A caching proxy installed so that users' web requests are automatically forwarded to the proxy server, without any need to manually configure web browsers to use it.

transport layer. The third layer of the OSI and TCP/IP network models, which provides a method of reaching a particular service on a given network node. Examples of protocols that operate at this layer are *TCP* and *UDP*.

trending. A type of network monitoring tool that performs unattended monitoring over long periods, and plots the results on a graph. Trending tools allow you to predict future behavior of your network, which helps you plan for upgrades and changes.

TTL see *Time To Live*

tunnel. A form of data encapsulation that wraps one protocol stack within another. This is often used in conjunction with encryption to protect communications from potential eavesdroppers, while eliminating the need to support encryption within the application itself. Tunnels are often used conjunction with *VPN*s.

U

U.FL. A very tiny microwave connector commonly used on mini-PCI radio cards.

UDP see *User Datagram Protocol*

unintentional users. Laptop users who accidentally associate to the wrong wireless network.

Unshielded Twisted Pair (UTP). Cable used for 10baseT and 100baseT Ethernet, consisting of four pairs of twisted wires.

Useful Capacity (Cu). The usable capacity of a battery, equal to the product of the *Nominal Capacity* and the *Maximum Depth of Discharge*.

User Datagram Protocol (UDP). A *connectionless protocol* (at the *transport layer*) commonly used for video and audio streaming.

UTP see *Unshielded Twisted Pair*

V

valve regulated lead acid battery (VRLA) see *lead-acid batteries*

vertical polarization. An electro-magnetic field with the electric component moving in a linear vertical direction. Most wireless consumer electronic devices use vertical polarization. See also: *circular polarization, vertical polarization*

Very Small Aperture Terminal (VSAT). One of several standards used for satellite Internet access. VSAT is the most widely deployed satellite technology used in Africa. See also: *Broadband Global Access Network (BGAN)* and *Digital Video Broadcast (DVB-S)*.

video sender. A 2.4 GHz video transmitter that can be used as an inexpensive *signal generator*.

Virtual Private Network (VPN). A tool used to join two networks together over an untrusted network (such as the Internet). VPNs are often used to connect remote users to an organization's network when traveling or working from home. VPNs use a combination of encryption and tunneling to secure all network traffic, regardless of the application being used. See also: *tunnel*.

VoIP (Voice over IP). A technology that provides telephone-like features over an Internet connection. Examples of popular VoIP clients include Skype, Gizmo Project, MSN Messenger, and iChat.

VPN see *Virtual Private Network.*

VRLA see *valve regulated lead acid battery*

VSAT see *Very Small Aperture Terminal*

Very Small Aperture Terminal (VSAT). One of several standards used for satellite Internet access. VSAT is the most widely deployed satellite technology used in Africa. See also: *Broadband Global Access Network (BGN)* and *Digital Video Broadcast (DVB-S)*.

WAN see **Wide Area Network**

War drivers. Wireless enthusiasts who are interested in finding the physical location of wireless networks.

wavelength. The distance measured from a point on one wave to the equivalent part of the next, for example from the top of one peak to the next. Also known as **lambda (λ).**

WEP see **Wired Equivalent Privacy**

wget. An open source command line tool for downloading web pages. *http://www.gnu.org/software/wget/*

Wi-Fi. A marketing brand owned by the Wi-Fi Alliance that is used to refer to various wireless networking technologies (including 802.11a, 802.11b, and 802.11g). Wi-Fi is short for **Wireless Fidelity**.

Wi-Fi Protected Access (WPA). A fairly strong **link layer encryption** protocol supported by most modern **Wi-Fi** equipment.

Wi-Spy. An inexpensive 2.4 GHz spectrum analysis tool available from *http://www.metageek.net/*.

Wide Area Network (WAN). Any long distance networking technology. Leased lines, frame relay, DSL, fixed wireless, and satellite all typically implement wide area networks. See also: **LAN**.

wiki. A web site that allows any user to edit the contents of any page. One of the most popular public wikis is *http://www.wikipedia.org/*

window scale. A TCP enhancement defined by RFC1323 that allows TCP window sizes larger than 64KB.

WinDump. The Windows version of tcpdump. It is available from *http://www.winpcap.org/windump/*

Wired Equivalent Privacy (WEP). A somewhat secure **link layer encryption** protocol supported by virtually all 802.11a/b/g equipment.

Wireless Fidelity see **Wi-Fi**.

wireshark. A free network protocol analyzer for Unix and Windows. *http://www.wireshark.org/*

WPA see **Wi-Fi Protected Access**

Z

Zabbix (*http://www.zabbix.org/*) A realtime monitoring tool that logs and notifies a system administrator about service and network outages.

APPENDICES

APPENDIX A: ANTENNA CONSTRUCTION

Guidelines for building some simple types of antennas

Collinear omni

This antenna is very simple to build, requiring just a piece of wire, an N socket and a square metallic plate. It can be used for indoor or outdoor point-to-multipoint short distance coverage.

The plate has a hole drilled in the middle to accommodate an N type chassis socket that is screwed into place.

The wire is soldered to the centre pin of the N socket and has coils to separate the active phased elements.

Two versions of the antenna are possible: one with two phased elements and two coils and another with four phased elements and four coils. For the short antenna the gain will be around 5 dBi, while the long one with four elements will have 7 to 9 dBi of gain.

We are going to describe how to build the long antenna only.

Parts list and tools required

- One screw-on N-type female connector
- 50 cm of copper or brass wire of 2 mm of diameter
- 10x10 cm or greater square metallic plate
- Ruler
- Pliers
- File
- Soldering iron and solder
- Drill with a set of bits for metal (including a 15 mm diameter bit)
- A piece of pipe or a drill bit with a diameter of 1 cm
- Vice or clamp
- Hammer
- Spanner or monkey wrench

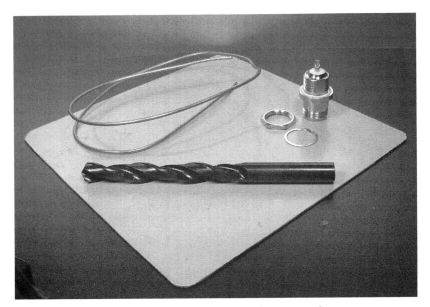

Figure AC 1: 10 cm x 10 cm aluminum plate.

Construction

Straighten the wire using the vice.

Figure AC 2: Make the wire as straight as you can.

With a marker, draw a line at 2.5 cm starting from one end of the wire. On this line, bend the wire at 90 degrees with the help of the vice and of the hammer.

Figure AC 3: Gently tap the wire to make a sharp bend.

Draw another line at a distance of 3.6 cm from the bend.
Using the vice and the hammer, bend once again the wire over this second line at 90 degrees, in the opposite direction to the first bend but in the same plane. The wire should look like a 'Z'.

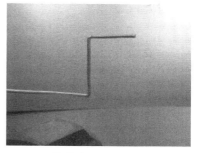

Figure AC 4: Bend the wire into a "Z" shape.

We will now twist the 'Z' portion of the wire to make a coil with a diameter of 1 cm. To do this, we will use the pipe or the drill bit and curve the wire around it, with the help of the vice and of the pliers.

Figure AC 5: Bend the wire around the drill bit to make a coil.

The coil will look like this:

Figure AC 6: The completed coil.

You should make a second coil at a distance of 7.8 cm from the first one. Both coils should have the same turning direction and should be placed on the same side of the wire.

Make a third and a fourth coil following the same procedure, at the same distance of 7.8 cm one from each other.

Trim the last phased element at a distance of 8.0 cm from the fourth coil.

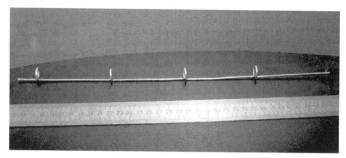

Figure AC 7: Try to keep it as straight possible.

If the coils have been made correctly, it should now be possible to insert a pipe through all the coils as shown.

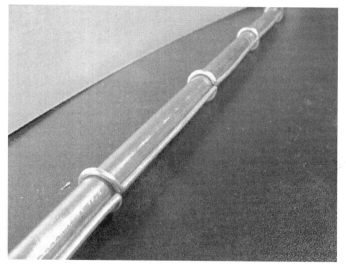

Figure AC 8: Inserting a pipe can help to straighten the wire.

With a marker and a ruler, draw the diagonals on the metallic plate, finding its centre.

With a small diameter drill bit, make a pilot hole at the centre of the plate. Increase the diameter of the hole using bits with an increasing diameter.

Figure AC 9: Drilling the hole in the metal plate.

The hole should fit the N connector exactly. Use a file if needed.

Figure AC 10: The N connector should fit snugly in the hole.

To have an antenna impedance of 50 Ohms, it is important that the visible surface of the internal insulator of the connector (the white area around the central pin) is at the same level as the surface of the plate.

For this reason, cut 0.5 cm of copper pipe with an external diameter of 2 cm, and place it between the connector and the plate.

Figure AC 11: Adding a copper pipe spacer helps to match the impedance of the antenna to 50 Ohms.

Screw the nut to the connector to fix it firmly on the plate using the spanner.

Figure AC 12: Secure the N connector tightly to the plate.

Smooth with the file the side of the wire which is 2.5 cm long, from the first coil.

Tin the wire for around 0.5 cm at the smoothed end helping yourself with the vice.

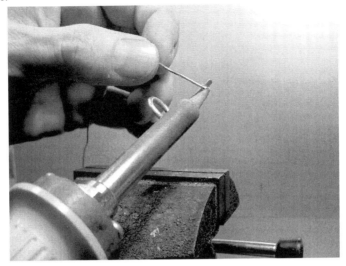

Figure AC 13: Add a little solder to the end of the wire to "tin" it prior to soldering.

With the soldering iron, tin the central pin of the connector. Keeping the wire vertical with the pliers, solder its tinned side in the hole of the central pin. The first coil should be at 3.0 cm from the plate.

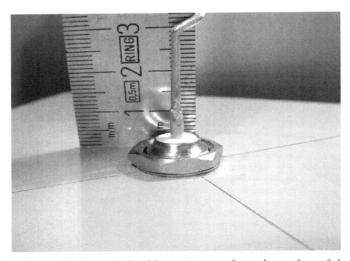

Figure AC 14: The first coil should start 3.0 cm from the surface of the plate.

We are now going to stretch the coils extending the total vertical length of the wire. Using the vice and the pliers, you should pull the cable so that the final length of the coil is of 2.0 cm.

Figure AC 15: Stretching the coils. Be very gentle and try not to scrape the surface of the wire with the pliers.

Repeat the same procedure for the other three coils, stretching their length to 2.0 cm.

Figure AC 16: Repeat the stretching procedure for all of the remaining coils.

At the end the antenna should measure 42.5 cm from the plate to the top.

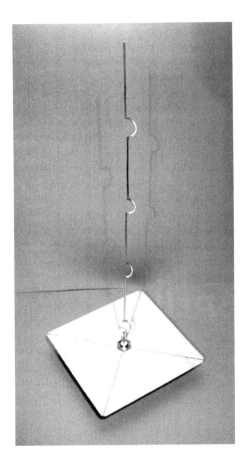

Figure AC 17: The finished antenna should be 42.5 cm from the plate to the end of the wire.

If you have a spectrum analyzer with a tracking generator and a directional coupler, you can check the curve of the reflected power of the antenna.

The following picture shows the display of the spectrum analyzer.

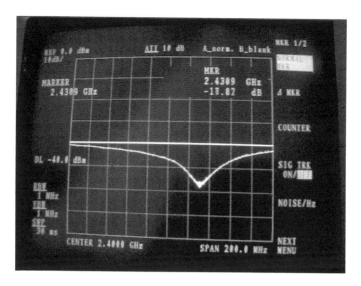

Figure AC 18: A spectrum plot of the reflected power of the collinear omni.

If you intend to use this antenna outside, you will need to weatherproof it. The simplest method is to enclose the whole thing in a large piece of PVC pipe closed with caps. Cut a hole at the bottom for the transmission line, and seal the antenna shut with silicone or PVC glue.

Cantenna

The waveguide antenna, sometimes called a Cantenna, uses a tin can as a waveguide and a short wire soldered on an N connector as a probe for coaxial-cable-to-waveguide transition. It can be easily built at just the price of the connector, recycling a food, juice, or other tin can. It is a directional antenna, useful for short to medium distance point-to-point links. It may be also used as a feeder for a parabolic dish or grid.

Not all cans are good for building an antenna because there are dimensional constraints.

1. The acceptable values for the diameter D of the feed are between 0.60 and 0.75 wavelength in air at the design frequency. At 2.44 GHz the wavelength λ is 12.2 cm, so the can diameter should be in the range of 7.3 - 10 cm.

2. The length L of the can preferably should be at least 0.75 λ_G , where λ_G is the guide wavelength and is given by:

$$\lambda_G = \lambda/(sqrt(1 - (\lambda / 1.706D)^2))$$

For D = 7.3 cm, we need a can of at least 56.4 cm, while for D = 9.2 cm we need a can of at least 14.8 cm. Generally the smaller the diameter, the longer the can should be. For our example, we will use oil cans that have a diameter of 8.3 cm and a height of about 21 cm.

3. The probe for coaxial cable to waveguide transition should be positioned at a distance S from the bottom of the can, given by:

$$S = 0.25\,\lambda_G$$

Its length should be 0.25 λ, which at 2.44 GHz corresponds to 3.05 cm.

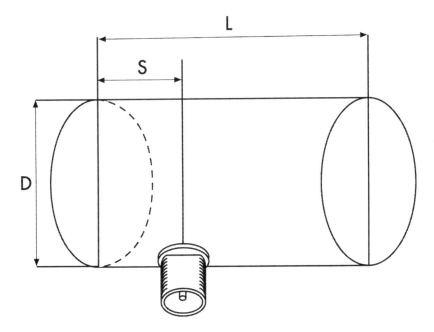

Figure AC 19: Dimensional constraints on the cantenna

The gain for this antenna will be in the order of 10 to 14 dBi, with a beamwidth of around 60 degrees.

Figure AC 20: The finished cantenna.

Parts list

- one screw-on N-type female connector
- 4 cm of copper or brass wire of 2 mm of diameter
- an oil can of 8.3 cm of diameter and 21 cm of height

Figure AC 21: Parts needed for the cantenna.

Tools required

- Can opener
- Ruler
- Pliers
- File
- Soldering iron
- Solder
- Drill with a set of bits for metal (with a 1.5 cm diameter bit)
- Vice or clamp
- Spanner or monkey wrench
- Hammer
- Punch

Construction

With the can opener, carefully remove the upper part of the can.

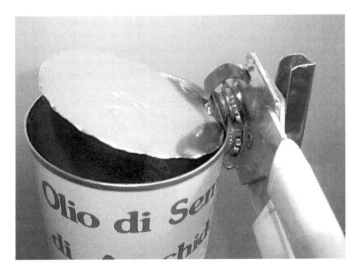

Figure AC 22: Be careful of sharp edges when opening the can.

The circular disk has a very sharp edge. Be careful when handling it!
Empty the can and wash it with soap. If the can contained pineapple, cookies, or some other tasty treat, have a friend serve the food.

With the ruler, measure 6.2 cm from the bottom of the can and draw a point. Be careful to measure from the inner side of the bottom. Use a punch (or a small drill bit or a Phillips screwdriver) and a hammer to mark the point. This makes it easier to precisely drill the hole. Be careful not to change the shape of the can doing this by inserting a small block of wood or other object in the can before tapping it.

Figure AC 23: Mark the hole before drilling.

With a small diameter drill bit, make a hole at the centre of the plate. Increase the diameter of the hole using bits with an increasing diameter. The hole should fit exactly the N connector.
Use the file to smooth the border of the hole and to remove the painting around it in order to ensure a better electrical contact with the connector.

Figure AC 24: Carefully drill a pilot hole, then use a larger bit to finish the job.

Smooth with the file one end of the wire. Tin the wire for around 0.5 cm at the same end helping yourself with the vice.

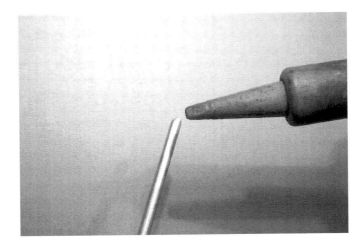

Figure AC 25: Tin the end of the wire before soldering.

With the soldering iron, tin the central pin of the connector.
Keeping the wire vertical with the pliers, solder its tinned side
in the hole of the central pin.

Figure AC 26: Solder the wire to the gold cup on the N connector.

Insert a washer and gently screw the nut onto the connector.
Trim the wire at 3.05 cm measured from the bottom part of the nut.

Figure AC 27: The length of the wire is critical.

Unscrew the nut from the connector, leaving the washer in place.
Insert the connector into the hole of the can.
Screw the nut on the connector from inside the can.

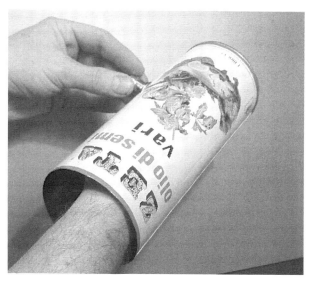

Figure AC 28: Assemble the antenna.

Use the pliers or the monkey wrench to screw firmly the nut on the connector. You are done!

Figure AC 29: Your finished cantenna.

As with the other antenna designs, you should make a weatherproof enclosure for the antenna if you wish to use it outdoors. PVC works well for the can antenna. Insert the entire can in a large PVC tube, and seal the ends with caps and glue.

You will need to drill a hole in the side of the tube to accommodate the N connector on the side of the can.

Cantenna as dish feed

As with the USB dongle parabolic, you can use the cantenna design as a feeder for a parabolic reflector thus attaining a significantly higher gain. Mount the can on the parabolic with the opening of the can pointed at the centre of the dish. Use the technique described in the USB dongle antenna example (watching signal strength changes over time) to find the optimum location of the can for the dish you are using.

By using a well-built cantenna with a properly tuned parabolic, you can achieve an overall antenna gain of 30dBi or more. As the size of the parabolic increases, so does the potential gain and directivity of the antenna. With very large parabolas, you can achieve significantly higher gain.

NEC2

NEC2 stands for Numerical Electromagnetics Code (version 2) and is a free antenna modeling package. NEC2 lets you build an antenna model in 3D, and then analyses the antenna's electromagnetic response. It was developed more than ten years ago and has been compiled to run on many different computer systems. NEC2 is particularly effective for analysing wire-grid models, but also has some surface patch modeling capability.

The antenna design is described in a text file, and then the model is built using this text description. An antenna described in NEC2 is given in two parts: its structure and a sequence of controls.

The structure is simply a numerical description of where the different parts of the antenna are located, and how the wires are connected up. The controls tell NEC where the RF source is connected.

Once these are defined, the transmitting antenna is then modelled. Because of the reciprocity theorem the transmitting gain pattern is the same as the receiving one, so modelling the transmission characteristics is sufficient to understand the antenna's behaviour completely.

A frequency or range of frequencies of the RF signal must be specified. The next important element is the character of the ground. The conductivity of the earth varies from place to place, but in many cases it plays a vital role in determining the antenna gain pattern.

To run NEC2 on Linux, install the NEC2 package from the URL below. To launch it, type nec2 and enter the input and output filenames. It is also worth installing the xnecview package for structure verification and radiation pattern plotting. If all went well you should have a file containing the output. This can be broken up into various sections, but for a quick idea of what it represents a gain pattern can be plotted using xnecview. You should see the expected pattern, horizontally omnidirectional, with a peak at the optimum angle of takeoff. Windows and Mac versions are also available.

The advantage of NEC2 is that we can get an idea of how the antenna works before building it, and how we can modify the design in order to get the maximum gain. It is a complex tool and requires some research to learn how to use it effectively, but it is an invaluable tool for antenna designers.

NEC2 is available from http://www.nec2.org/

APPENDIX B: CHANNEL ALLOCATIONS

The following tables list the channel numbers and centre frequencies used for 802.11a and 802.11b/g.

Note that while all of these frequencies are in the unlicensed ISM and U-NII bands, not all channels are available in all countries. Many regions im☒ pose restrictions on output power and indoor / outdoor use on some chan☒ nels.

These regulations are rapidly changing, so always check your local regula☒ tions before transmitting. Note that these tables show the center frequency for each channel. Channels are 22MHz wide in 802.11b/g, and 20MHz wide in 802.11a.

802.11b / g			
Channel #	Center Frequency (GHz)	Channel #	Center Frequency (GHz)
1	2.412	8	2.447
2	2.417	9	2.452
3	2.422	10	2.457
4	2.427	11	2.462
5	2.432	12	2.467
6	2.437	13	2.472
7	2.442	14	2.484

802.11a	
Channel	Center Frequency (GHz)
34	5.170
36	5.180
38	5.190
40	5.200
42	5.210
44	5.220
46	5.230
48	5.240
52	5.260
56	5.280
60	5.300
64	5.320
149	5.745
153	5.765
157	5.785
161	5805

APPENDIX C: PATH LOSS

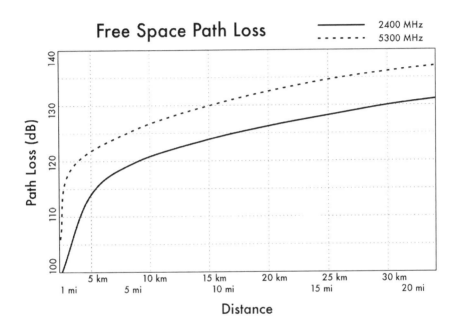

APPENDIX D: CABLE SIZES

Wire gauge, diameter, current capacity, and resistance at 20°C. These values can vary from cable to cable. When in doubt, consult the manufacturer's specifications.

AWG Gauge	Diameter (mm)	Ohms / Meter	Max Amperes
0000	11.68	0.000161	302
000	10.40.00	0.000203	239
00	9.27	0.000256	190
0	8.25	0.000322	150
1	7.35	0.000406	119
2	6.54	0.000513	94
3	5.83	0.000646	75
4	5.19	0.000815	60
5	4.62	0.001028	47
6	4.11	0.001296	37
7	3.67	0.001634	30
8	3.26	0.002060	24
9	2.91	0.002598	19
10	2.59	0.003276	15

APPENDIX E: SOLAR DIMENSIONING

Use these tables to collect the necessary data to estimate the required size of your solar energy system.

General Data

Site Name	
Site Latitude (°)	

Irradiation Data

$G_{dm}(0)$, in kWh / m² per day)

Jan	Feb	Mar	Apr	May	Jun	Jul	Aug	Sep	Oct	Nov	Dec
Worst Irradiation Month											

Reliability and System Operational Voltage

Days of Autonomy (N)	
Nominal Voltage (V_{NEquip})	

Component Characteristics

Solar Panels	
Voltage @ Maximum Power (V_{pmax})	
Current @ Maximum Power (I_{pmax})	
Panel Type/Model and Power (W_p)	
Batteries	
Nominal Capacity @ 100 H (C_{NBat})	
Nominal Voltage (V_{NBat})	
Maximum Depth of Discharge (DoD_{MAX}) or Usable Capacity (C_{UBat})	
Regulator	
Nominal Voltage (V_{NReg})	
Maximum Current (I_{maxReg})	

Solar Panels	
Voltage @ Maximum Power (V_{pmax})	
Current @ Maximum Power (I_{pmax})	
Panel Type/Model and Power (W_p)	
DC/AC Inverter (if needed)	
Nominal Voltage (V_{NConv})	
Instantaneous Power (P_{IConv})	
Performance @ 70% Load	

Loads

Estimated Energy Consumed by the Loads (DC)				
Month of Greatest Consumption				
Description	# of Units	x Nominal Power	x Usage Hours / Day	= Energy (Wh/day)

Estimated Energy Consumed by the Loads (DC)				
Month of Greatest Consumption				
			ETOTAL DC	
Estimated Energy Consumed by the Loads (AC)				
Month of Greatest Consumption				
Description	# of Units	x Nominal Power	x Usage Hours / Day	= Energy (Wh/day)

Estimated Energy Consumed by the Loads (DC)				
Month of Greatest Consumption				
E_{TOTAL} AC (before converter)				
E_{TOTAL} AC (after converter) = E_{TOTAL} AC / 70%				

Finding the Worst Month

Site Name												
Site Latitude (°)												
Nominal Voltage of the Installation V_N												
(Month)	J	F	M	A	M	J	J	A	S	O	N	D
Inclination ß												
G_{dm} (ß) (kWh/m^2 × day)												

Site Name												
Site Latitude (°)												
E_{TOTAL} (DC) (Wh/day)												
E_{TOTAL} (AC) (Wh/day)												
E_{TOTAL} (AC + DC)=												
I_m (A) = E_{TOTAL} (Wh/day) × 1kW/m^2/ (G_{dm}(ß) × V_N)												

Worst Month Summary	
Worst Month	
I_m (A)	
I_mMAX (A) = 1.21 × I_m	
E_{TOTAL} (AC + DC)	

Final Calculations

Panels		
Panels in Series (N_{PS})	$N_{PS} = V_N / V_{Pmax} =$	
Panels in Parallel (N_{PP})	$N_{PP} = I_{mMAX} / I_{Pmax} =$	
Total Number of Panels	$N_{TOT} = N_{PS} \times N_{PP} =$	
Batteries		
Necessary Capacity (C_{NEC})	E_{TOTAL}(WORST MONTH) $/ V_N \times N$	
Nominal Capacity (C_{NOM})	C_{NEC} / DoD_{MAX}	
Number of Batteries in Series (N_{BS})	V_N / V_{NBAT}	
Cables		
	Panels > Batteries	Batteries > Converter	Main Line
Voltage Drop ($V_a - V_b$)			

Panels		
Panels in Series (NPS)	$N_{PS} = V_N / V_{Pmax} =$	
Panels in Parallel (NPP)	$N_{PP} = I_{mMAX} / I_{Pmax} =$	
Thickness (Section) $r \times L \times I_{mMAX} / (V_a - V_b)$			

For cable thickness computation, $r = 0.01286\ \Omega\ mm^2/m$ (for copper) and L is the length in metres.

APPENDIX F: RESOURCES

We recommend these resources for learning more about the various aspects of wireless networking.

Antennas And Antenna Design

Free antenna designs, http://www.freeantennas.com/
Hyperlink Tech, http://hyperlinktech.com/
Pasadena Networks LLC, http://www.wlanparts.com/
SuperPass, http://www.superpass.com/
Unofficial NEC2 code archives, http://www.nec2.org/
USB WiFi dish designs, http://www.usbwifi.orcon.net.nz/

Network Troubleshooting Tools

Bing throughput measurement tool,http://fgouget.free.fr/bing/index-en.shtml
Cacti network monitoring package, http://www.cacti.net/
DSL Reports bandwidth speed tests, http://www.dslreports.com/stest
EtherApe network traffic monitor, http://etherape.sourceforge.net/
Flowc open source NetFlow collector, http://netacad.kiev.ua/flowc/
iptraf network diagnostic tool, http://iptraf.seul.org/
My TraceRoute network diagnostic tool, http://www.bitwizard.nl/mtr/
Nagios network monitoring and event notification tool,
http://www.nagios.org/
NetFlow, the Cisco protocol for collecting IP traffic information,
http://en.wikipedia.org/wiki/Netflow
ngrep network security utility for finding patterns in data flows,
http://ngrep.sourceforge.net/
Network monitoring implementation guides and tutorials,
http://wiki.debian.org/Network_Monitoring
Ntop network monitoring tool, http://www.ntop.org/
SoftPerfect network analysis tools, http://www.softperfect.com/
Squid transparent http proxy HOWTO,
http://tldp.org/HOWTO/TransparentProxy.html
Wireshark network protocol analyzer, http://www.wireshark.org/

MRTG, http://oss.oetiker.ch/mrtg/
rrdtool, http://oss.oetiker.ch/rrdtool/
Smokeping, http://oss.oetiker.ch/smokeping/
Argus, http://qosient.com/argus/
Netramet, http://www.caida.org/tools/measurement/netramet/
Snort, http://www.snort.org/
Mod Security, http://www.modsecurity.org/
Apache, http://www.apache.org/
Zabbix, http://www.zabbix.org/
ngrep, http://ngrep.sourceforge.net/
nmap, http://www.nmap.org
netcat, http://nc110.sourceforge.net/

Security

AntiProxy http proxy circumvention tools and information,
http://www.antiproxy.com/
Anti-spyware tools, http://www.spychecker.com/
Driftnet network monitoring utility,
http://www.exparrot.com/~chris/driftnet/
Introduction to OpenVPN, http://www.linuxjournal.com/article/7949
Linux security and admin software,
http://www.linux.org/apps/all/Networking/Security_/_Admin.html
OpenSSH secure shell and tunneling tool, http://openssh.org/
OpenVPN encrypted tunnel setup guide, http://openvpn.net/howto.html
Privoxy filtering web proxy, http://www.privoxy.org/
PuTTY SSH client for Windows, http://www.putty.nl/
Sawmill log analyzer, http://www.sawmill.net/
Security of the WEP algorithm,
http://www.isaac.cs.berkeley.edu/isaac/wep-faq.html
Stunnel Universal SSL Wrapper, http://www.stunnel.org/
TOR onion router, http://www.torproject.org/
Weaknesses in the Key Scheduling Algorithm of RC4,
http://www.crypto.com/papers/others/rc4_ksaproc.ps
Windows SCP client, http://winscp.net/
Your 802.11 Wireless Network has No Clothes,
http://www.cs.umd.edu/~waa/wireless.pdf
ZoneAlarm personal firewall for Windows, http://www.zonelabs.com/
Logging, http://wigle.net/, http://www.nodedb.com/,
or http://www.stumbler.net/. http://www.isaac.cs.berkeley.edu/isaac/wep-

faq.html http://www.cs.umd.edu/~waa/wireless.pdf
http://dl.aircrack-ng.org/breakingwepandwpa.pdf
http://download.aircrackng.org/wikifiles/doc/enhanced_tkip_michael.pdf
Captive Portals, CoovaChilli, CoovaAP (http://coova.org/CoovaChilli/)
WiFidog (http://www.wifidog.org/)
M0n0wall, pfSense (http://m0n0.ch/wall/)
Putty, http://www.putty.nl/
Win SCP, http://winscp.net/
Cygwin, http://www.cygwin.com/
OpenVPN Journal, http://www.linuxjournal.com/article/7949
Tor, http://www.torproject.org/
Spychecker, http://www.spychecker.com/

Bandwidth Optimisation

Cache hierarchies with Squid,
http://squid-docs.sourceforge.net/latest/html/c2075.html
dnsmasq caching DNS and DHCP server,
http://www.thekelleys.org.uk/dnsmasq/doc.html
Enhancing International World Wide Web Access in Mozambique
Through the Use of Mirroring and Caching Proxies,
http://www.isoc.org/inet97/ans97/cloet.htm
Fluff file distribution utility, http://www.bristol.ac.uk/fluff/
Linux Advanced Routing and Traffic Control HOWTO, http://lartc.org/
Microsoft Internet Security and Acceleration Server,
http://www.microsoft.com/isaserver/
Microsoft ISA Server Firewall and Cache resource site, http://www.isaserver.org/
Optimising Internet Bandwidth in Developing Country Higher
Education, http://www.inasp.info/pubs/bandwidth/index.html
Planet Malaysia blog on bandwidth management,
http://planetmy.com/blog/?p=148
RFC 3135: Performance Enhancing Proxies Intended to Mitigate Link-
Related Degradations, http://www.ietf.org/rfc/rfc3135
Squid web proxy cache, http://squid-cache.org/

Mesh Networking

Freifunk OLSR mesh firmware for the Linksys WRT54G,
http://www.freifunk.net/wiki/FreifunkFirmware
MIT Roofnet Project, http://pdos.csail.mit.edu/roofnet/doku.php

OLSR mesh networking daemon, http://www.olsr.org/
AirJaldi Mesh Router, http://drupal.airjaldi.com/node/9
Open WRT, http://wiki.openwrt.org/toh/start
Village Telco, www.villagetelco.org

Wireless Operating Systems And Drivers

DD-WRT wireless router OS,http://www.dd-wrt.com/
HostAP wireless driver for the Prism 2.5 chipset, http://hostap.epitest.fi/
m0n0wall wireless router OS, http://m0n0.ch/wall/
MadWiFi wireless driver for the Atheros chipset, http://madwifi.org/
Metrix Pyramid wireless router OS,
http://code.google.com/p/pyramidlinux/
OpenWRT wireless router OS for Linksys access points,
http://openwrt.org/
Tomato wireless router OS for Linksys access points,
http://www.polarcloud.com/tomato

Wireless Tools

Chillispot captive portal, http://www.chillispot.info/
Interactive Wireless Network Design Analysis Utilities,
http://www.qsl.net/n9zia/wireless/page09.html
KisMAC wireless monitor for Mac OS X, http://kismac-ng.org/
Kismet wireless network monitoring tool, http://www.kismetwireless.net/
MacStumbler wireless network detection tool for Mac OS X,
http://www.macstumbler.com/
NetStumbler wireless network detection tool for Windows,
http://www.wirelessdefence.org/Contents/NetstumblerMain.htm
Netspot wireless network detection for Mac OS X,
http://www.netspotapp.com/
PHPMyPrePaid prepaid ticketing system,
http://sourceforge.net/projects/phpmyprepaid/
RadioMobile radio performance modeling tool,
http://www.cplus.org/rmw/
Radio Mobile online, http://www.cplus.org/rmw/rmonline.html
Wellenreiter wireless network detection tool for Linux,
http://sourceforge.net/projects/wellenreiter/
WiFiDog captive portal, http://www.wifidog.org/
Proxim, http://www.proxim.com/technology

WiSpy spectrum analysis tool, http://www.metageek.net/
Spectrum Analyser, h t t p : / / w w w. s e e e d s t u d i o. c o m / d e p o t / r f -
e x p l o r e r - m o d e l - w s u b 1 g - p - 9 2 2. h t m l ? c P a t h = 1 7 4
"RF Explorer model 2.4G",
h t t p : / / w w w. s e e e d s t u d i o. c o m / d e p o t / - p - 9 2 4. h t m l c P a t h = 1 7 4
VideoSend. http://www.lightinthebox.com/Popular/Wifi_Video_Transmitter.html

General Wireless Related Information

Homebrew wireless hardware designs, http://www.w1ghz.org/
Linksys wireless access point information, http://linksysinfo.org/
Linksys WRT54G resource guide,
http://seattlewireless.net/index.cgi/LinksysWrt54g
Ronja optical data link hardware, http://ronja.twibright.com/
SeattleWireless community wireless group, http://seattlewireless.net/
SeattleWireless Hardware comparison page,
http://www.seattlewireless.net/HardwareComparison
Stephen Foskett's Power Over Ethernet (PoE) Calculator,
http://www.gweep.net/~sfoskett/tech/poecalc.html
White Spaces project, http://www.wirelesswhitespace.org/projects.aspx

General Computing Tools

File sharing, http://sparkleshare.org, https://github.com/philcryer/lipsync,
http://rsync.samba.org/
Open Relay testing, http://www.mailradar.com/openrelay,
http://www.checkor.com/
Disk imaging, http://www.partimage.org, http://www.powerquest.com/

Networking Services And Training

Wireless Toolkit,
http://wtkit.org/groups/wtkit/wiki/820cb/download_page.html
wire.less.dk consultancy and services, http://wire.less.dk/
Wireless Lab and training at ICTP, http://wireless.ictp.it/
WirelessU, http://wirelessu.org/
Network Startup Resource Center, Oregon, http://www.nsrc.org/
Inveneo, http://www.inveneo.org/
6Deploy (EC FP7 project), http://www.6deploy.org
Association for Progressive Communications wireless connectivity
projects, http://www.apc.org/wireless/

International Network for the Availability of Scientific Publications, http://www.inasp.info/
Makere University, Uganda, http://mak.ac.ug/
Access Kenya ISP, http://www.accesskenya.com/
Broadband Access Ltd. wireless broadband carrier, http://www.blue.co.ke/
Virtual IT outsourcing, http://www.virtualit.biz/
Collection of looking glasses, http://www.traceroute.org/

Regional Internet Registrars

IANA, http://www.iana.org/
AfriNIC, http://www.afrinic.net/
APNIC, http://www.apnic.net/
ARIN, http://www.arin.net/
LACNIC, http://www.lacnic.net/
RIPE NCC, http://www.ripe.net/

IPv6 Transitioning

http://www.petri.co.il/ipv6-transition.htm
http://www.6diss.org/tutorials/transitioning.pdf
http://arstechnica.com/business/2013/01/ipv6-takes-one-step-forward-ipv4-two-steps-back-in-2012/
http://www.6deploy.eu/index.php?page=home
RIPE IPv6 transiton, http://www.ipv6actnow.org/
Test your IPv6, http://tet-ipv6.org
IPv6 Deployment status, http://6lab.cisco.com

Dynamic Routing Protocols

http://www.ciscopress.com/store/routing-tcp-ip-volume-i-ccie-professional-development-9781578700417?w_ptgrevartcl=Dynamic%20Routing%20Protocols_24090
http://www.ciscopress.com/articles/article.asp?p=24090&seqNum=5
http://ptgmedia.pearsoncmg.com/images/9781587132063/samplechapter/1587132060_03.pdf
http://www.inetdaemon.com/tutorials/internet/ip/routing/dyamic_vs_static.shtml https://learningnetwork.cisco.com/docs/DOC-7985
OSPF Design guide: http://www.cisco.com/warp/public/104/1.pdf

Solar Panel Design

Low resolution PSH maps/calculation tools,
http://re.jrc.ec.europa.eu/pvgis/apps4/pvest.php?map=africa&lang=en
Highlands And Islands project,
http://www.wirelesswhitespace.org/projects/wind-fi-renewable-energy-basestation.aspx
PVSYST, http://www.pvsyst.com/
Solar Design, http://www.solardesign.co.uk/

Miscellaneous Links

Cygwin Linux-like environment for Windows, http://www.cygwin.com/
Graphviz network graph visualization tool, http://www.graphviz.org/
ICTP bandwidth simulator, http://wireless.ictp.trieste.it/simulator/
ImageMagick image manipulation tools and libraries,
http://www.imagemagick.org/
NodeDB war driving map database, http://www.nodedb.com/
Partition Image disk utility for Linux, http://www.partimage.org/
RFC 1918: Address Allocation for Private Internets,
http://www.ietf.org/rfc/rfc1918
Rusty Russell's Linux Networking Concepts,
http://www.netfilter.org/documentation/HOWTO/networkig-concepts-HOWTO.html
Ubuntu Linux, http://www.ubuntu.com/
VoIP-4D Primer, http://www.it46.se/voip4d/voip4d.php
wget web utility for Windows, http://users.ugent.be/~bpuype/wget/
ISO Standard, http://standards.iso.org/ittf/PubliclyAvailableStandards

Books

802.11 Networks: The Definitive Guide, 2nd Edition. Matthew Gast, O'Reilly Media. ISBN #0-596-10052-3

802.11 Wireless Network Site Surveying and Installation. Bruce Alexander, Cisco Press. ISBN #1-587-05164-8

The ARRL UHF/Microwave Experimenter's Manual. American Radio Relay League. ISBN #0-87259-312-6

Building Wireless Community Networks, 2nd Edition. Rob Flickenger, O'Reilly Media. ISBN #0-596-00502-4

Deploying License-Free Wireless Wide-Area Networks. Jack Unger, Cisco Press. ISBN #1-587-05069-2

Wireless Hacks, 2nd Edition. Rob Flickenger and Roger Weeks, O'Reilly Media. ISBN #0-596-10144-9

IPv6 Security (Cisco Press Networking Technology). Scott Hogg, Eric Vyncke, Cisco Press. ISBN # 1587055945

LAN Switch Security: What Hackers Know About Your Switches. Eric Vyncke and Christopher Paggen. ISBN #1587052563

Building the Mobile Internet. Mark Grayson, Kevin Shatzkamer, Klaas Wierenga. ISBN # 1587142430

CASE STUDIES

Case Studies - Introduction

No matter how much planning goes into building a link or node location, you will inevitably have to jump in and actually install something. This is the moment of truth that demonstrates just how accurate your estimates and predictions prove to be.

It is a rare day when everything goes precisely as planned. Even after you install your 1st, 10th, or 100th node, you will still find that things do not always work out as you might have intended. This section describes some of our more memorable and recent network projects. Whether you are about to embark on your first wireless project or you are an old hand at this, it is reassuring to remember that there is always more to learn and even the experts who have contributed to this book still learn from each field project they are involved in!

Below are a few last minute tips and thoughts before we embark on telling you about our latest field adventures.

Equipment enclosures

Cheap plastics are often easy to find, but they maybe made of poor materials and are thin, thus mostly unsuitable for enclosing equipment. PVC tubing is far more resilient and is made to be waterproof.

In West Africa, the most common PVC is found in plumbing, sized from 90mm to 220mm. Sometimes Access Points can fit into such tubing, and with end-caps that are torched-on or glued, they can make very robust waterproof enclosures.

They also have the added benefit of being aerodynamic and uninteresting to passers-by. The resulting space left around the equipment assures adequate air circulation.

Also, it is often best to leave an exhaust hole at the bottom of the PVC enclosure, although in one instance ants decided to nest 25 metres above ground inside the PVC holding the access point. A wire mesh cover made from locally available screen material was used to secure the exhaust hole from infestations.

Antenna masts

Recovering used materials for use to build antenna masts is a good plan. Local metal workers may already be familiar with how to make television masts from scrap metal. A few quick adaptations and these same masts can be re-purposed for wireless networks.

The typical mast is the 5 metre pole, comprised of a single 30mm diameter pipe which is then planted into cement. It's best to construct the mast in two parts, with a removable mast that fits into a base which is slightly larger in diameter. Alternatively, the mast may be made with arms that can be securely cemented into a wall.

This type of mast can be augmented by several metres with the use of guy lines.
To steady the pole, plant three lines 120 degrees apart, forming an angle of at least 33 degrees with the tower.

Details of how to earth the mast can be found in the chapter called **Hardware Selection and Configuration.**

Involve the local community

Community involvement is imperative in assuring the success and sustainability of a project. Involving the community in a project can be the greatest challenge, but if the community is not involved the technology will not serve their needs, nor will it be accepted. Moreover, a community might be afraid and could subvert an initiative. Regardless of the complexity of the undertaking, a successful project needs support and buy-in from those it will serve.

Take your time and be selective in finding the right people for your project. No other decision will affect your project more than having effective, trusted local people on your team.

In addition, take note of key players in an institution, or community. Identify those people who are likely to be opponents and proponents of your project. As early as possible, attempt to earn the support of the potential proponents and to diffuse the opponents.

This is a difficult task and one that requires intimate knowledge of the institution or community.

If the project does not have a local ally, the project must take time to acquire this knowledge and trust from the community.

Do not try to introduce a technology to a community without understanding which applications will serve the community.

When gathering information, verify the facts that you are given. Most often, local partners who trust you will be very frank, honest, and helpful.

When looking at payment methods for your new wireless service, pre-paid services are ideal, as they do not require a legal contract. Commitment is assured by the investment of funds before service is given.

Buy-in also requires that those involved invest in the project themselves. A project should ask for reciprocal involvement from the community.

Above all, the "no-go" option should always be evaluated. If a local ally and community buy-in cannot be had, the project should consider choosing a different community or beneficiary.

There must be a negotiation; equipment, money, and training cannot be gifts. The community must be involved and they too must contribute.

Now read on

In the next chapters of this section you will find some of our projects described, which hopefully you will find useful to learn from.

We havent included Case Studies from earlier versions of this book except for the one in Venezuela which was extremely instrumental in establishing the viability of long distance outdoor point to point links for rural connectivity.

One Case Study not included in this section but undertaken by Inveneo who are working closely with the team involved in writing this book can be found at the following URL. Led by Andris Bjornson who is Inveneo's CTO, there is some very useful information in his writeup of the project. http://www.inveneo.org/90km-wireless-link-for-mfangano-island/

We hope you now enjoy reading each of the Case Studies following this intro. You will see who was involved and of course their profile is included in the **Acknowledgements** of this book.

All of the authors have been involved in field deployments and also monitor our Facebook page which you can find here.

https://www.facebook.com/groups/wirelessu

So as you begin to plan your next deployment, please do post any questions you have for the experts to respond to.

Case Study: Long Distance 802.11 in Venezuela

Introduction

Although this Case Study is several years old now, we have left it in this version of the book as it still to this day is the longest ever successful point to point outdoor 802.11 test. Here you will find the pioneering work of those who are still authoring this book. And in fact some of the preparation work involved in the setup of this test is still relevant for those planning long distance outdoor links. Enjoy reading!

Background

Thanks to a favourable topography, Venezuela already had some long range WLAN links, like the 70 km long operated by Fundacite Mérida between Pico Espejo and Canagua.

To test the limits of the WiFi technology, it is necessary to find a path with an unobstructed line of sight and a clearance of at least 60% of the first Fresnel zone.

While looking at the terrain in Venezuela, in search of a stretch with high elevation at the ends and low ground in between, I first focused in the Guayana region. Although plenty of high grounds are to be found, in particular the famous "tepuys" (tall mesas with steep walls), there were always obstacles in the middle ground. My attention shifted to the Andes, whose steep slopes (rising abruptly from the plains) proved adequate for the task. For several years, I have been travelling through sparsely populated areas due to my passion for mountain biking. In the back of my head, I kept a record of the suitability of different spots for long distance communications.

Pico del Aguila is an outstanding place for long distance communication. It has an altitude of 4100 m and is about a two hour drive from my home town of Mérida. For the other end, I finally located the town of El Baúl, in Cojedes State.

Using the free software Radio Mobile
(available at http://www.cplus.org/rmw/english1.html) I found that there
was no obstruction of the first Fresnel zone, over a 280 km path between
Pico del Aguila and El Baúl.

Action Plan

Once satisfied with the existence of a suitable trajectory, we looked at the
equipment needed to achieve the goal.
We have been using Orinoco cards for a number of years.
Sporting an output power of 15 dBm and receive threshold of -84 dBm,
they are robust and trustworthy. The free space loss at 280 km is 149 dB.
So, we would need 30 dBi antennas at both ends and even that would
leave very little margin for other losses.

On the other hand, the Linksys WRT54G wireless router runs Linux.
The Open Source community has written several firmware versions for it
that allow for complete customisation of every transmission parameter.
In particular, OpenWRT firmware allows for the adjustment of the
acknowledgment time of the MAC layer, as well as the output power.
Another firmware, DD-WRT, has a GUI interface and a very convenient
site survey utility.

Furthermore, the Linksys can be located closer to the antenna than a
laptop.
So, we decided to go with a pair of these boxes. One was configured as an
AP (access point) and the other as a client. The WRT54G can operate at
100 mW output power with good linearity, and can even be pushed up to
200 mW.
But at this value, non linearity is very severe and spurious signals are
generated, which should be avoided.
Although this is consumer grade equipment and quite inexpensive, after
years of using it, we felt confident that it could serve our purpose.
Of course, we kept a spare set handy just in case.

By setting the output power to 100 mW (20 dBm), we could obtain a
5dB advantage compared with the Orinoco card.
Therefore, we settled for a pair of WRT54Gs.

Pico del Aguila Site Survey

On January 15, 2006, I went to Pico ⊠guila to check out the site that Radio Mobile had reported as suitable.

The azimuth towards El Baúl is 86°, but since the magnetic declination is 8° 16', our antenna should be pointed to a magnetic bearing of 94°.

Unfortunately, when I looked towards 94°, I found the line of sight obstructed by an obstacle that had not been shown by the software, due to the limited resolution of the digital elevation maps that are freely available.

I rode my mountain bike for several hours examining the surrounding area looking for a clear path towards the East. Several promising spots were identified, and for each of them I took photos and recorded the coordinates with a GPS for later processing with the Radio Mobile software.

This led me to refine my path selection, resulting in the one depicted in Figure CsLD 1 using Google Earth:

Figure CsLD 1: View of the 280 km link. Maracaibo's Lake is to the West, and the Peninsula of Paraguaná is to the North.

The radio profile obtained with Radio Mobile is shown in Figure CsLD 2:

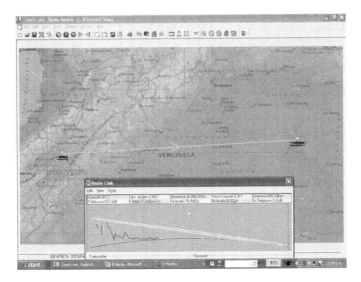

Figure CsLD 2: Map and profile of the proposed path between Pico Aguila, and Morrocoy Hill, near the town of El Baúl.

The details of the wireless link are displayed in Figure CsLD 3:

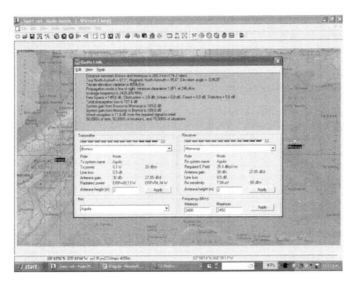

Figure CsLD 3: Propagation details of the 280 km link.

In order to achieve a reasonable margin of some 12 dB for the link, we needed antennas with at least 30 dBi gain at each end.

Antennas

High gain antennas for the 2.4 GHz band were not available in Venezuela.

The import costs are considerable, so we decided instead to recycle parabolic reflectors (formerly used for a satellite service) and replace the feed with one designed for 2.4 GHz. We proved the concept with an 80 cm dish. The gain was way too low, so we tried an offset fed 2.4 m reflector. This offered ample gain, albeit with some difficulties in the aiming of the 3.5° beam. The 22.5° offset meant that the dish appeared to be pointing downwards when it was horizontally aligned.

Several tests were performed using various cantennas and a 12 dBi Yagi as a feed. We pointed the antenna at a base station of the university wireless network that was located 11 km away on a 3500m mountain. The test site sits at 2000m and therefore the elevation angle is 8°. Because of the offset feed, we pointed the dish 14° downward, as can be seen in the following picture:

Figure CsLD 4: 2.4 m offset fed reflector with a 12 dBi antenna at its focus, looking 14° down. The actual elevation is 8° up.

We were able to establish a link with the base station at Aguada, but our efforts to measure the gain of the setup using Netstumbler were not successful. There was too much fluctuation on the received power values of live traffic.

For a meaningful measurement of the gain, we needed a signal generator and a spectrum analyser. These instruments were also required for the field trip in order to align the antennas properly.

While waiting for the required equipment, we looked for an antenna to be used at the other end, and also a pointing system better suited to the narrow radio beam of a high gain antenna.

In February 2006, I travelled to Trieste to participate in the annual wireless training event that I have been attending since 1996. While there, I mentioned the project to my colleague Carlo Fonda, who was immediately thrilled and eager to participate.

The collaboration between the Latin American Networking School (EsLaRed) and the Abdus Salam International Centre for Theoretical Physics (ICTP) goes back to 1992, when the first Networking School was held in Mérida with ICTP support.

Since then, members of both institutions have collaborated on several activities. Some of these include an annual training school on wireless networking (organized by ICTP) and another on computer networks (organized by EsLaRed) that are hosted in several countries throughout Latin America.

Accordingly, it was not difficult to persuade Professor Sandro Radicella, the head of the Aeronomy and Radio Propagation Laboratory at ICTP, to support Carlo Fonda's trip in early April to Venezuela in order to participate in the experiment.

Back at home, I found a 2.75 m parabolic central fed mesh antenna at a neighbours house. Mr. Ismael Santos graciously lent his antenna for the experiment.

Figure CsLD 5 shows the disassembly of the mesh reflector.

Figure CsLD 5: Carlo and Ermanno disassembling the satellite dish supplied by Mr. Ismael Santos.

We exchanged the feed for a 2.4 GHz one, and aimed the antenna at a signal generator that was located on top of a ladder some 30 m away.

With a spectrum analyser, we measured the maximum of the signal and located the focus. We also pinpointed the boresight for both the central fed and the offset antennas. This is shown in Figure CsLD 6:

Figure CsLD 6: Finding the focus of the antennas with the 2.4 GHz feed

We also compared the power of the received signal with the output of a commercial 24 dBi antenna. This showed a difference of 8 dB, which led us to believe that the overall gain of our antenna was about 32 dBi. Of course, there is some uncertainty about this value. We were receiving reflected signals, but the value agreed with the calculation from the antenna dimension.

El Baùl Site Survey

Once we were satisfied with the proper functioning and aim of both antennas, we decided to do a site survey at the other end of the El Baúl link. Carlo Fonda, Gaya Fior and Ermanno Pietrosemoli reached the site on April 8th. The following day, we found a hill (south of the town) with two telecom towers from two cell phone operators and one belonging to the Major of El Baúl. The hill of Morrocoy is some 75 m above the surrounding area, about 155 m above sea level. It provides an unobstructed view towards El Aguila. There is a dirt road to the top, a must for our purpose, given the weight of the antenna.

Performing the experiment

On Wednesday April 12th, Javier Triviño and Ermanno Pietrosemoli travelled towards Baúl with the offset antenna loaded on top of a four- wheel drive truck. Early in the morning of April 13th, we installed the antenna and pointed it at a compass bearing of 276°, given that the declination is 8° and therefore the true Azimuth is 268°. At the same time, the other team (composed by Carlo Fonda and Gaya Fior from ICTP, with assistance of Franco Bellarosa, Lourdes Pietrosemoli and José Triviño) rode to the previously surveyed area at Pico del Aguila in a Bronco truck that carried the 2.7 m mesh antenna.

Figure CsLD 7: Pico del Águila and surrounds map with Bronco truck.

Poor weather is common at altitudes of 4100 m above sea level. The Aguila team was able to install and point the mesh antenna before the fog and sleet began. Figure CsLD 8 shows the antenna and the rope used as an aid in aiming the 3° radio beam.Power for the signal generator was supplied from the truck by means of a 12 V DC to 120 V AC inverter. At 11 A.M in El Baúl, we were able to observe a -82 dBm signal at the agreed upon 2450 MHz frequency using the spectrum analyser. To be sure we had found the proper source, we asked Carlo to switch off the signal. Indeed, the trace on the spectrum analyser showed only noise. This confirmed that we were really seeing the signal that originated some 280 km away. After turning the signal generator on again, we performed a fine tuning in elevation and azimuth at both ends. Once we were satisfied that we had attained the maximum received signal, Carlo removed the signal generator and replaced it with a Linksys WRT54G wireless router configured as an access point. Javier substituted the spectrum analyser on our end for another WRT54G configured as a client.

Figure CsLD 8: Aiming the antenna at el Águila.

At once, we started receiving "beacons" but ping packets did not get through. This was expected, since the propagation time of the radio wave over a 300 km link is 1 ms. It takes at least 2 ms for an acknowledgment to reach the transmitter. Fortunately, the OpenWRT firmware allows for adjusting the ACK timing. After Carlo adjusted for the 3 orders of magnitude increase in delay above what the standard WiFi link expects, we began receiving packets with a delay of about 5 ms.

Figure CsLD 9: El Baúl antenna installation. Actual elevation was 1° upward, since the antenna has an offset of 22.5°.

We proceeded to transfer several PDF files between Carlo's and Javier's laptops. The results are shown in Figure CsLD 10.

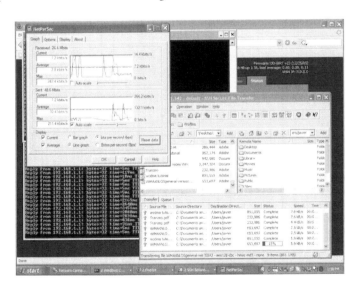

Figure CsLD 10: Screenshot of Javier's laptop showing details of PDF file transfer from Carlo's laptop 280 km away, using two WRT54G wireless routers, no amplifiers.

Note the ping time of a few milliseconds.

Figure CsLD 11: Javier Triviño (right) and Ermanno Pietrosemoli beaming from the El Baúl antenna.

Figure CsLD 12: Carlo Fonda at the Aguila Site

Mérida, Venezuela, 17 April 2006.

One year after performing the experiment just described, we found the time and resources to repeat it.

We used commercial 30 dBi antennas, and also a couple of wireless routers which had been modified by the TIER group led by Dr. Eric Brewer of Berkeley University.

The purpose of the modification of the standard WiFi MAC is to make it suitable for long distance applications by replacing the CSMA Media Access Control with TDMA. The latter is better suited for long distance point-to-point links since it does not require the reception of ACKs.

This eliminates the need to wait for the 2 ms round trip propagation time on a 300 km path.

On April 28th, 2007, a team formed by Javier Triviño, José Torres and Francisco Torres installed one of the antennas at El Aguila site. The other team, formed by Leonardo González V., Leonardo González G., Alejandro González and Ermanno Pietrosemoli, installed the other antenna at El Baúl.

A solid link was quickly established using the Linksys WRT54G routers. This allowed for video transmission at a measured throughput of 65 kbps. With the TDMA routers, the measured throughput was 3 Mbps in each direction.

This produced the total of 6 Mbps as predicted by the simulations done at Berkeley.

Can we do better?

Thrilled by these results, which pave the way for really inexpensive long distance broadband links, the second team moved to another location previously identified at 382 km from El Aguila, in a place called Platillón. Platillón is 1500 m above sea level and there is an unobstructed first Fresnel zone towards El Aguila (located at 4200 m above sea level).

The proposed path is shown in Figure CsLD 13:

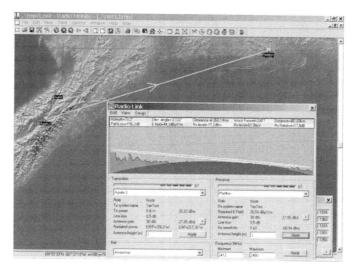

Figure CsLD 13: Map and profile of the 380 km path.

Again, the link was quickly established with the Linksys and the TIER supplied routers.

The Linksys link showed approximately 1% packet loss, with an average round trip time of 12 ms.
The TIER equipment showed no packet loss, with unidirectional propagation times of 1 ms.
This allowed for video transmission, but the link was not stable.

We noticed considerable signal fluctuations that often interrupted the communication.
However, when the received signal was about -78 dBm, the measured throughput was a total of 6 Mbps bidirectional with the TIER routers implementing TDMA.

Figure CsLD 14: The team at el Aguila, José Torres (left), Javier Triviño (centre) and Francisco Torres (right).

Although further tests must be conducted to ascertain the limits for stable throughput, we are confident that Wi-Fi has a great potential for long distance broadband communication. It is particularly well suited for rural areas were the spectrum is still not crowded and interference is not a problem, provided there is good radio line of sight.

Acknowledgments

We wish to express our gratitude to Mr. Ismael Santos for lending the mesh antenna to be installed at El Aguila and to Eng. Andrés Pietrosemoli for supplying the special scaffolding joints used for the installation and transportation of the antennas.

We'd also like to thank the Abdus Salam International Centre of Theoretical Physics for supporting Carlo Fonda's trip from Italy to Venezuela.

Figure CsLD 15: The team at Platillon. From left to right: Leonardo González V., Leonardo González G., Ermanno Pietrosemoli and Alejandro González.

The 2006 experiment was performed by Ermanno Pietrosemoli, Javier Triviño from EsLaRed, Carlo Fonda, and Gaya Fior from ICTP. With the help of Franco Bellarosa, Lourdes Pietrosemoli, and José Triviño.

For the 2007 experiments, Dr. Eric Brewer from Berkeley University provided the wireless routers with the modified MAC for long distance, as well as enthusiastic support through his collaborator, Sonesh Surana. RedULA, CPTM, Dirección de Servicios ULA Universidad de los Andes and Fundacite Mérida contributed to this trial.

The second experiment work was funded by Canada's IDRC. References:

Fundación Escuela Latinoamericana de Redes, Latin American Networking School, http://www.eslared.org.ve/

Abdus Salam International Centre for Theoretical Physics, http://wireless.ictp.it/

OpenWRT Open Source firmware for Linksys, http://openwrt.org/ www.idrc.ca

Fundacite Mérida, http://www.fundacite-merida.gob.ve/

--Ermanno Pietrosemoli

Case Study: Pisces Project

Solar Powered Wifi Links in Micronesia
By Bruce Baikie and Laura Hosman, assisted in the field by Marco
Zennaro and Ermanno Pietrosemoli from Ictp.

Two long-distance, solar-powered wireless point to point connections were set up in the Micronesian Region of the Pacific in early August 2012 as part of the Pacific Island Schools Connectivity, Education, and Solar (PISCES) Project (http://www.piscespacific.org/livesite/),a multi-partnered endeavour focused on training and local capacity building vis-⊠-vis solar-powered information and communications technology (ICT) within the Pacific region.

Figure CSP 1 : Hands On Training

The first half of the PISCES project was a hands-on training workshop on Solar-powered long-distance WiFi technology at the University of Guam.
The workshop focused on WiFi technology, standards, solar power, site surveys, project safety, and link planning tools.
The workshop provided plenty of opportunities for hands-on experiences, with each afternoon's activity consisting of a one-to-three hour lab in which students actualised the practical information from the morning's lecture. On the final day, the students set up a solar powered point to point high bandwidth link in the Centre for Island Sustainability, replacing an older, slower link to the Internet.

Fig. CSP 2: A new faster Internet connection for the Centre for Island Sustainability

For the second half of the PISCES project, the team travelled to Chuuk, one of the Federated States of Micronesia (FSM), and installed both a solar-powered long distance WiFi Internet connection and a Solar-Computer-Lab-in-a-Box at a primary school on a previously un-connected island, Udot, in the Chuuk lagoon.

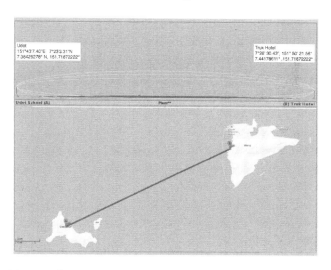

Figure CSP 3 : The Internet connection for the Udot School originated from Chuuk's main Island, Weno, 15 km away.

Figure CSP 4 : Mounting the pole mast in Udot

The Ubiquiti Networks WiFi hardware installed in Weno was mounted on the 3rd story roof of the Truk Stop Hotel, which offered the height necessary for line-of-sight to provide connectivity across the Chuuk Lagoon to Udot Island.

The single-level school on Udot required a 40-foot (12 m) pole mast on which to mount the Ubiquiti Networks WIFI antenna/radio.

Local community members joined in to help when the heft of the pole mast proved too heavy for the PISCES team to raise alone.
With team members on each island, the antennas were lined up and connected to each other.

The network was then routed through a local DSL Internet connection to provide Internet connectivity to the school, and the surrounding local community.

Each WiFi unit is powered with a solar photovoltaic system, which consists of a 30 watt solar panel from SolarLand USA, solar charge controller with Power over Ethernet, and 38 amp hours of battery back up.

Figure CSP 5 : Udot

The unique Solar-Computer-Lab-in-a-Box deployed at Udot school was developed by students at Illinois Institute of Technology.

This turnkey computer lab is designed to be as close to plug-and-play as possible for off-grid environments.

It includes six Intel Classmate laptops, solar panels and mounting gear, a charge controller, wiring, and laptop security equipment, all contained within a uniquely-designed and ready-to-ship box that straightforwardly transforms into the computer lab's table.

Figure CSP 6

The PISCES Project received funding support from **Google**, the **Pacific Telecommunications Council**, and the **Internet Society**.

In addition to the Partners mentioned above, PISCES Project partners include:

- University of Guam
- Illinois Institute of Technology
- Green WiFi, Inveneo
- iSolutions
- International Centre for Theoretical Physics (ICTP)
- University of California, Berkeley's TIER research group.

Case Study: University Of Ghana campus wireless network

Introduction

The University of Ghana is one of six public universities and the premier university in Ghana with a student population of about 41,000. With the growing number of students and faculty it was obvious that we could not continue expanding our computer labs to facilitate learning and research, since we had limited space and limited funds to equip these computer labs.

The solution was to go wireless so that any student with his/her laptop could access the network. This however could not be achieved immediately due the status of the network at the time. It was a big flat, un-managed network, with lots of problems - IP conflicts, rogue DHCP severs, large broadcast domains just to mention a few.

Because the IT unit was not providing a wireless service to the community, users became impatient and started connecting their own wireless routers to the network. This made network management even more difficult. It became obvious that if we did not provide a wireless service to the user community, they would find their own way of doing it.

Our first step in addressing the problem was to redesign our network from a flat network to a more structured one with core, distribution and access layers.

This brought a lot of stability to the network. Because of the managed switches, identifying problems became much easier too. The structured wired network gave us a good foundation to build a wireless network that complements the wired network and meets the growing needs of our users.

Wifi setup and installation

A number of factors were considered in determining the type of Access Point (AP) to deploy. Some of these were:

- Cost
- Support
- Management
- Security

Because of the size of our network we decided to go for an Enterprise Solution which would make management much easier. Due to the high cost of these Enterprise Solutions ($600 per Access Point and above + cost of the controller) we ended up in a long debate over which product to use for the WiFi implementation.

We consulted NSRC (the Network Startup Resource Center at University of Oregon) who were also looking into affordable and scalable wireless solutions and they pointed us to Ubiquiti UniFi which costs about $80 per Access Point and has a free software controller. Together with NSRC personnel we did a survey and ran a successful pilot with 10 Access Points. Because of the cost, functionality, manageability and ease of deployment, University of Ghana decided to scale up from the pilot using Ubiquiti UniFi Access Points and the network quickly grew from 10 Access Points to 90.

For security, UniFi APs support both WPA Personal and WPA Enterprise which allows users to authenticate with a radius server. In addition to all of the deployment advantages of Ubiquiti, we found that there was a large UniFi user community that provided good technical help in the event of problems.

AP Setup

Initial AP setup and configuration involved connecting an AP to a network switch port in the same **vlan** as the controller server.

The AP goes through adoption when connected; this allows the new AP to register itself with the controller for management. After initial setup, the AP is then connected to the designated department's wireless vlan on a switch.

IP Addressing

Private IP addressing was adopted for the wireless network on campus. There are on average 25 /24 wireless subnets in departments, faculties, school and college.

Bandwidth

Bandwidth allocation to the wireless network is 10% of the University's STM1 bandwidth.

Current growing wireless users coupled with growing emerging applications will require more bandwidth to give a good browsing experience.

Security/Authentication

Users and APs authenticate with a radius server using 802.1x. Both student and staff account details are stored in a Mysql database. The campus wireless network exists in separate vlans from the wired for easy identification and management.

Connecting to the UG campus wireless network

The University Of Ghana Campus Wireless Network has three major Networks/SSIDs - STAFF, STUDENT, GUEST.

STAFF
This SSID/Network is available to active staff of the University.
Staff are required to authenticate with their staff ID and PIN as username and password respectively to login.

STUDENT
This SSID/Network is available to students of the University who have registered for a given academic year. Students authenticate with their student ID and PIN as username and password respectively to login.

GUEST
This SSID/Network is available to guests who visit the University within a period of time.
Guests are required to request authentication account details from the IT department by submitting their details.

Pictures of our project and installation

Figure CSG 1: Our campus

Figure CSG 2: Our campus

Figure CSG 3: A classroom

Figure CSG 4: The library

Figure CSG 5: One of our APs

Figure CSG 6: Layout of University of Ghana campus with AP's.

Figure CSG 7: With UniFi we are able to simulate the coverage of the wireless signal as shown above.

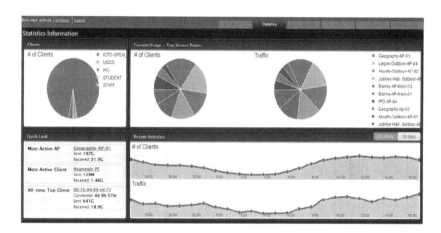

Figure CSG 8: Above is the UniFi controller showing some statistics of our usage.

Figure CSG 9: Above are statistics per Access Point.

Figure CSG 10: We are also able to get statistics per user from the controller.

Challenges we faced

One of our main challenges was getting a good Cat5 cable for the installation. In addition it was somewhat challenging to get the cable to the right location since the buildings were not designed with this in mind. Bandwidth is also a challenge but we are trying to limit peer-to-peer activity on the network using cyberoam.

Next steps

We (the IT Department of the University of Ghana) operate the wireless network ourselves. We have immediate plans to expand the wireless network until we get as much coverage of our campus buildings as we can. This will then reduce the need for our students to setup their own APs making our job managing the network much easier!

Author: Emmanuel Togo, Head of the Networking Unit of the University of Ghana's Computing Systems (UGCS).

Case Study: Airjaldi's Garhwal Network, India

Introduction

In version 2 of this book we included a case study about the Dharamsala Community Wireless Mesh Network. Following on from this original deployment in the subsequent years, a new set of networks and a commercial wireless ISP has emerged in the same region led by the same people. Here we describe one of their major projects.

Airjaldi's Garhwal Network:

Working Towards Technical and Economic Viability in the Himalayan Range

About Rbb/Airjaldi

Rural Broadband Pvt. Ltd., is a leading innovator and implementer of technically and economically viable connectivity solutions for rural areas. We design, build and operate broadband networks in rural areas in India. Incorporated in India in 2009, RBB presently owns and operates networks in the Indian states of Himachal Pradesh, Uttarakhand, Jharkhand, and Karnataka. RBB uses AirJaldi as a brand name for its network and other connectivity-related initiatives.

The company's activities are carried out from our management office in Delhi, our operations office in Dharamsala, Himachal Pradesh and offices in each network location.

Our diverse team includes local Indian villagers, Tibetan refugees, skilled professionals from metropolitan areas in India and people from outside India.

We believe that rural networks need to be technically viable – they need to provide a quality and consistency of services that is at the very least similar to that offered anywhere else.

They also need to be economically sustainable – they need to be able to stand on their own within a relatively short period of time (around 18 months) while offering clients services at reasonable prices.

Further, we are strong advocates of an all-inclusive "retail ecosystem" approach – once we reach an area, we seek to connect all clients who are in need of connectivity, regardless of the size of their operation or package demand. We aspire to make all our clients pay for their connectivity, although subsidies are offered to selected clients, mainly those dedicated to social and developmental causes and showing financial need.

RBB works closely with AirJaldi Research and Innovation, a non-for-profit (section 25) enterprise registered in India. Created in 2007, AirJaldi Research & Innovation identifies suitable and affordable networking solutions for rural areas, tests them in real-life environments and shares its learning with like-minded organizations and individuals. AirJaldi also operates a training and capacity-building centre in Dharamsala, where network operators and activists can acquire the skills to build and manage rural wireless networks.

Most of our deployment team members have been trained by AirJaldi Research and Innovation.
Team members normally take the basic one-month "Wireless 108" and the more advanced "Wireless 216" courses offered at the AirJaldi Network Academy. After an additional 3-month internship where they work under the close supervision of senior team members in one of our networks, they are enrolled as permanent team members.

The Airjaldi Garhwal Network – Vital Statistics

Date of initiation/inception | January 2010
Size/spread | about 100 km², ranging from the Dehradun valley to the heights of Tehri Garwal Mountains (height of about 2,000 metres).
Primary clients | Micro banking enterprise, Schools, Community based Organizations, Business and Private users
Longest link | 55 Km.
Population density | 169/km² (in comparison: Delhi: 9,294/km²; Overall India: 363/km²; USA: 33.7/km²)

Realities, needs

The Tehri and Pauri Garhwal districts of Uttarakhand, which stretches from the Himalayan peaks of Thalaiya Sagar, Jonli and the Gangotri group to the Dheradun Valley and Rishikesh, are one of India's most mountainous areas. Renowned for its many religious temples, located on the banks of the River Ganges and on hilltops leading to the Himalayan range, the region is known for its rugged beauty. This ruggedness is however also a cause for the relative poverty of the region: its inhabitants live mostly in small villages separated from each other by tall mountains and deep valleys. The main local sources of income are subsistence agriculture and cottage industries. Many work outside the region, in large cities, in the plains and in military and government services across India.

In 2009, KGFS Rural Services[1], a micro-banking institution affiliated with IFMR Trust[2], decided to set up operations in this hilly area. Their aim was to reach potential clients in the villages of Tehri and Pauri who until then had little or no access to regular banking services and were also considered "barely bankable" by most banks.

Using population density mapping, KGFS sought locations for its bank branches at the centre of "catchment areas" reaching about 10,000 people. It soon became clear to KGFS that, having solved the density and accessibility pre-requisites, they were faced with severe connectivity limitations, as most of their proposed locations had no Internet infrastructure. initial deployment of VSATs and use of local ADSL services proved expensive, slow and prone to breakdowns.

That was when we received a call from IFMR's IT team, asking if we'd be interested in proposing a connectivity solution for their first 15 branches.

The initial deployment: approach, design, deployment

When responding to the call, as we do with similar requests, AirJaldi forms its response based on information on the following network essentials:

how far is the closest available backhaul location (the place from where we can avail access to the Internet cloud) from the proposed deployment area? Our surveys suggested that there were relatively few backhaul locations in the area.

---

1   http://ruralchannels.ifmr.co.in/kgfs-model/what-is-a-kgfs/
2   http://www.ifmr.co.in/

Most were located in Dehradun and neighbouring towns (see Figure CSD 1). Reaching from there to the proposed deployment area proved problematic. After much effort we located a backhaul "drop" at a BTS in the town of Narrandar Nagar (see Figure CSD 1). Although very close to some of the branches, there was no line of site from the drop to any of the branches. This led us to come up with a somewhat counter-intuitive solution: use the Narrandar Nagar location as backhaul to connect a Network Operation Center (NOC) that will be placed in the valley below, where better LoS to favourable mountain tops for potential relays was available.

Figure CSD 1: Garhwal Network Backhaul, NOC and some IFMR branches, 2009

Can we propose a deployment plan that is technically solid but at the same time affordable to the "anchor client" [3] and additional future clients? Our initial survey of the area consisted of collecting latitude and longitude data of proposed branch locations, identification of possible relay locations, assessment of overall client potential in the area and assessment of the public infrastructure of the area.

---

3 Each of our networks has one or more clients which are referred to as "anchor clients". The term denotes their role as the first clients in a network, and the ones that contribute mostly to its initial economic feasibility.

We were not surprised to find that the branches were mostly located in valleys (easier access for clients, close to roads) and separated from each other by high mountain ranges preventing direct LoS between them.

Power supply to relays promised to be a challenge – most potential relay locations were either away from the grid or located at weak points where power was at times down for days and wild power fluctuations caused routers to "hang" or burn when power spikes hit them.

Having decided that we will pursue this deployment, our deployment team had to make sense of it all.

The solution we reached was based on placing solar-powered relays on as few strategic locations as possible.

This led to additional field surveys, involving hours of preliminary office research using topographic maps, Google Earth and other tools, followed by days of trekking to identified locations in order to see if a location could be secured through a rent deal, and the safety and integrity of the relay guaranteed by the land owners.

What is the overall client potential in the area and will it be sufficient to, at a minimal level, ensure that the network is economically self-sustainable within around 18 months?

Client potential did not seem bright as well.

Besides the proposed bank branches we found few additional clients – mainly schools and other organizations.

We were sure that the client base will grow in time, but had to find a way to ensure viability in a relatively short time.

We decided to extend the network to more densely-populated areas at the foothills of the Garhwal range.

The actual deployment work began in October 2009.

After around two months of deployment work, the core network was ready.

Its size was around 50x70km.

It provided connectivity to 15 bank branches and around five schools and institutions in the valley.

Most of its stand-alone relays were solar-powered and its longest single-hop link was 54 km.

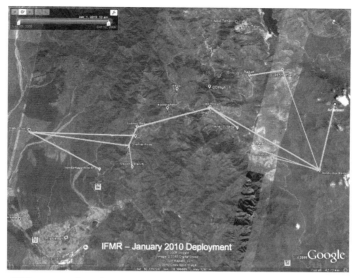

Figure CSD 2: Garhwal network, partial deployment topology, Early 2010

Figure CSD 3: Field survey in the Garhwal region

Figure CSD 4: Field survey in the Kumaon region

Figure CSD 5: Preparing for relay installation, Kumaon Network

Figure CSD 6: Relay troubleshooting, Kumaon Network

Figure CSD 7: Last touches to solar relay, Kumaon Network

Figure CSD 8: Backhaul Relay, Garhwal Network

Figure CSD 9: Client Relay, Kumaon Network

Figure CSD 10: Client Relay, Kangra Valley Network

Three years on – operation, economic viability, challenges and responses.
Three years on, the present network still caters to the initial 15 branches and original clients. It has also grown significantly. Its present size is around 120x100 km.

Travel time between our NOC/Office to its furthest reaches takes around seven hours (!), and the number of clients on the mountain ranges and valley has grown significantly. In its ongoing efforts to maintain this challenging network, our team has had to deal with land slides, snow, rain and thunder storms, wild elephants, leopards (yes!) and of course clients for whom much of this mattered little when their line is down.

We are very proud of the fact that our average uptime in our years of operation in the area exceeds 95% and that the network has attracted much attention and praise from its users, other Internet players and the media.

This said, challenges still abound and the struggle to keep it all running is still very much an ongoing effort. The main challenges we faced since the inception of the network are:

Energy – power is still a major challenge. The grid-provision is erratic and problematic. Using the grid as the power source for our relays (even with battery backup) has led to burnt routers and hours of travel to troubleshoot links drained of their power backup after the line went down.

Solar power, on the other hand, although almost trouble-free[4] is however still costly.
Bearing the capital costs for these relays ourselves stretches a challenging economic proposition even further, while having clients pay the full costs for a relay limits the type of clients who can avail our connections.

Relays – a good relay is one that is placed in a location covering as much space as possible. In the Garhwal area, this means mountain tops. These places are normally and naturally hard to get to and quite isolated. Building, maintenance and security in these sites is an ongoing challenge. At the time of writing we are busy rebuilding a relay whose solar panels, battery, charger and other equipment were stolen.

---

4 The monsoon months, with their very heavy rains and perpetually overcast skies create limiting charging conditions, and have forced us to equip most of our relay with extra charging capacity, which further increases the costs of relays

Some of our new connections have proven to be so challenging as to require a relay for each new locations, clearly not an easy economic proposition.

<u>Size</u> – as proud as we are in the size of the network, a travel distance of over 10 hours between its to extreme points is stretching out team, who move mostly on bikes, to the limit, with some trouble-shooting trips lasting over two days on account of travel.

<u>Economic viability</u> – the dual mountain-valley/sparse population-high density model, which is an implicit a cross-subsidy has proven successful, but has not fully solved the challenge of economic viability in marginal mountainous areas. Our responses to these challenges evolve constantly. Present observations and steps implemented include:

<u>Energy</u> – we have learnt that solar-powered relays are the only real option for the Garhwal network (as for many other locations).

In trying to rationalize the costs for clients we have limited our deployment to private/small business clients to clusters where a minimum of 10 clients with a demand of at least 1Mbps per client can be identified within a 5-6 month period. A lower density means higher prices and is usually requested by clients for whose operations connectivity is essential.

We have also begun to offer a "payment spread for time commitment" deals: clients pay for the cost of equipment and relays (or their share of it) over a period of time, if they agree to a contractual lock-in period of over two years.

<u>Relays</u> – the best location is not necessarily the one with the best coverage ONLY, but one that present an optimal combination of visibility, security, accessibility and price. In certain cases this means a higher number of relays, but we believe that on the whole this is a cheaper and more rational option.

<u>Size</u> – a simple solution to the size problem would be to break the network to smaller sub or autonomous network. We are using this approach in other networks, where the size limit is considered by a role of thumb of a max. of three hour travel to a site.

This however makes little sense in a sparse network such as the upper Garhwal reaches.

The compromise solution we came across was placement of "staging posts" in strategic network junctions.

Essentially, these "posts" are storage spaces where equipment is kept in sufficient quantities to serve a defined catchment area.

While not saving the travel time, it allows our team to move quickly without having to carry along heavy and bulky equipment. The evolution of a staging post to an operational base will come when keeping a team of two people at such a location could be justified by its client density and revenues.

Economic viability – AirJaldi's over-arching goal is to reach un-served and under-served areas with high quality internet connectivity. As such, the cross-subsidy model is justified as a partial solution to the challenges faced by many aspiring rural internet providers lacking the skills or desire to build networks of the scale of the Garhwal network.

That is however cannot be the only economic avenue for deployment in such challenging environments.

Our attempts to increase the viability of "mountain provision" include aggressive pricing of higher bandwidth packages – the marginal cost for higher bandwidth is relatively low, allowing us higher marginal revenues while still being able to offer very reasonable deals for such packages.

This stands in major contrast to the non-scalable dongle and VSAT solutions that are often the alternatives to our services in these areas. At the other end of the spectrum, we have begun offering limited time/bandwidth packages.

Although more expensive on a per-unit basis, these packages are appealing to clients who wish to limit their consumption to their economic willingness to pay.

Future plans include introduction of "4-C": local centers where users can utilize bandwidth for Classroom setup (online individual learning, live teaching to a classroom etc.) Cinema (watching films or other online content at a local center), Café (internet café where people can use individual computers) and Connectivity (through local hotspot in the area of the center or through connectivity deals to houses and offices sold at the Café).

Summary

The Garhwal network was created in response to a request from an "anchor client", a micro-banking enterprise for whom connectivity was a necessary condition for a successful implementation of its vision of rural banking. AirJaldi responded to the deployment challenge with the hope of ensuring high-quality and high-availability broadband internet connection and long-term economic sustainability. These objectives were reached through a combination of carful deployment planning, utilization of natural topographic assets, expansion of the network from the sparsely-populated mountain terrains to the more densely inhabited Dherdun valley, and aggressive pricing on higher bandwidth packages. Future plans include further expansion of the network, "thickening" of the client density in existing areas and enriching of packages and services offered by AirJaldi.

Case Study: Open Technology Institute

Red Hook initiative Wifi & Tidepools

Red Hook Initiative WiFi is a collaboratively designed mesh network. It provides Internet access to the Red Hook section of Brooklyn, NY, and serves as a platform for developing local applications and services. Red Hook Initiative has built the network in partnership with the Open Technology Institute, putting human-centered design and community engagement at the core of the project.

The community expanded the network significantly following a natural disaster in the Fall of 2012.

Key aspects

1. Main network anchors are trusted community organizations.
2. Solid relationship with technical support provider from outside the community.
3. Community-led design process emphasizes local needs and enhances engagement.
4. Rapid prototyping of applications designed for the local area network.

History of the network

Beginning in Fall 2011, the Red Hook Initiative (RHI), a Brooklyn non-profit focused on creating social change through youth engagement, approached the Open Technology Institute about collaborating on a community wireless network. RHI wanted a way to communicate with the residents immediately around its community center. OTI was initially unable to support the effort directly, but introduced Anthony Schloss, RHI Media Programs Coordinator, to Jonathan Baldwin, a Parsons School of Design graduate student who had been experimenting with wireless mesh as a local digital platform.

Figure CsOTI 1: Red Hook West Houses - public housing buildings

Red Hook is the northwestern corner of Brooklyn, jutting out into the Hudson Bay. It is cut off from the rest of the borough by the Gowanus Expressway, which carries traffic from points south into lower Manhattan.

The neighborhood is home to approximately 5000 residents of Red Hook Houses public housing and other low income areas near an expressway, as well as a gentrifying section with many small businesses closer to the water. Many industrial sites, an Ikea and a number of public parks fill out the area.

Figure CsOTI 2: Walking from RHI Offices to Gowanus Expressway

Figure CsOTI 3: Gowanus Expressway which divides the Red Hook
neighborhood from the rest of Brooklyn.

RHI WiFi's initial plan was to provide wireless access to the Internet in and around RHI's building, which is near the expressway and the Red Hook Houses.

Schloss and Baldwin installed a Ubiquiti Nanostation on the roof and a Linksys router inside the building, connected via ethernet.
They connected the Linksys router to the center's modem. This installation provided an opportunity to prototype early versions of RHI WiFi local applications. When residents or visitors to RHI connected to the wireless access point named "Red Hook Initiative WiFi," they were directed to a website on a local server.

On this website was a "Shout Box," a local digital message board allowing everyone to leave a comment or a note behind and participate in the project.

Figure CsOTI 4: First RHI WiFi node (Ubiquiti Nanostation) installed on the rooftop of the building that houses the RHI offices.

In March of 2012, Baldwin and Schloss installed an additional Ubiquiti Nanostation on the roof of an apartment building overlooking Coffey Park and much of the rest of the neighborhood.

A resident of the building with social ties to RHI donated the electricity and rooftop access.

With this vantage point on the neighborhood, the possibility of a wireless network connecting public spaces began to take shape.

Initially, the Coffey Park wireless access point was not connected to the Internet, but was connected to a <u>GuruPlug Server.</u>

The basic, low power server hosted a local web page on the network and a "Shout Box" similar to the one running in RHI.

Figure CsOTI 5: Running cable to install a node on the roof of an apartment building north of Coffey Park.

Figure CsOTI 6: Node installed on the roof of an apartment building.

RHI WiFi uses OTI's <u>Commotion Wireless</u> firmware running on Ubiquiti routers.

Commotion is a free and open-source communication tool that uses mobile phones, computers, and other wireless devices to create decentralized mesh networks.

Most importantly, Commotion allows network development to occur dynamically and organically - so the community can decide where and how the network should grow.

Commotion networks are sustainable without a connection to the Internet, which makes them resilient to outages; they can distribute access to applications hosted on local servers or on the routers themselves.

Social software and growth of the network

Based on research into community wireless networks around the world, Baldwin had identified a need for social software that would add value and a distinct identity to the community wireless network, specifically to:

- Spark civic and community engagement by addressing local needs, interests and culture.
- Foster trust, interdependence, and reciprocity throughout the community.
- Merge digital and physical community spaces.
- Ensure that people know about mesh / have software installed before a communication outage occurs.

Schloss and Baldwin began working with participants in RHI's established media programs on a collaborative, human-centered design process that called upon the knowledge and interests of local residents.

Throughout the first year, Baldwin and Schloss held workshops with community members to determine local needs and gather design ideas for <u>Tidepools</u>, developed by Baldwin for piloting on the RHI network. <u>Tidepools</u> is an open-source customizable local mapping platform built using Javascript, <u>LeafletJS</u>, PHP and <u>MongoDB.</u>
Baldwin designed it for local communication, placemaking, and organizing around events, issues, and community assets.

Figure CsOTI 7: Workshop to identify local needs for RHI WiFi network.
(Photo by Becky Kazansky).

Figure CsOTI 8: Tidepools map of RHI WiFi network.

The community workshops produced ideas for local applications that would address specific needs identified by the community.
The needs identified in the community workshops were:

- Access to the Internet (at home, via mobile, and at neighborhood kiosks).
- Accountable community participation (FAQs, electronic bulletin boards, SMS enabled features).
- Access to resources (employment and skills sharing).
- Local Information System (historical archive, landmarks).
- Multilingual (Spanish, Arabic and Tagalog).
- Playful interface to promote exploration.

In the summer of 2012, Baldwin joined OTI's staff, and OTI brought additional tech expertise to the collaboration along with experience closing digital divides and developing community-controlled infrastructure.

The organization's experience in <u>Detroit</u> and <u>Philadelphia</u> provided guidance on how to collaborate with communities that have been socially, geographically and technologically isolated within cities.

During the months following the initial tests of the local network, OTI and RHI focused on realizing three initial applications that would use the Tidepools platform and run over the local wireless network:

- <u>Where's the B61 Bus?</u> - An application for accessing real-time bus locations and arrival times using data from the Metropolitan Transit Authority's BusTime API (*launched October 9, 2012*).
- <u>Stop & Frisk Survey </u> - A survey application that residents can use to document police interactions in Red Hook and improve public safety (*launched October 17, 2012*).
- <u>RHI Radio </u> - An online radio station, streaming content produced by the Youth Radio Group at RHI (*under development*).

Figure CsOTI 9: Magnet advertising "Where's the B61 Bus". Tidepools application.

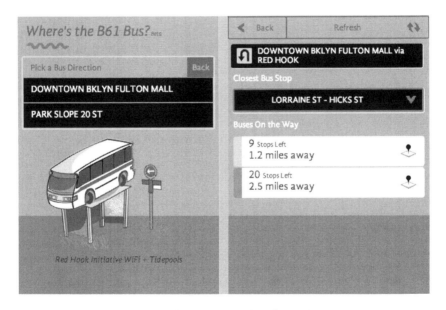

Figure CsOTI 10-11: Mobile user interface for "Where's the B61 Bus?"
application.

Expansion after superstorm Sandy

On October 29, 2012, Superstorm Sandy devastated low-lying Red Hook along with much of the surrounding region.
Amid power outages and flooding, the need for access to communications systems for information about what was happening and where help was needed became crucial.

The RHI building was one of the few locations that had managed to keep power and, as a result, RHI WiFi had stayed up through the storm. In the days immediately following the storm, up to 300 people per day were accessing the network to communicate with loved ones, learn what was happening in the rest of the city, and seek recovery assistance.

"We immediately saw communications as one of the critical needs in the community," says Tony Schloss. *"We wanted it to be as easy as possible for people to contact their networks to find housing, gain access to information, and report their safety status."*

Text messaging was the most widely – and in some cases the only – means of communication for neighborhood residents after the storm, so in a matter of days, OTI developed RHI Status - an SMS to Map plugin for Tidepools using the Tropo Application Programming Interface (API) for handling SMS messages and the Google Geocoding API for handling natural language addresses.

RHI Status provided a means for residents to text their location and needs to a contact number, which automatically maps the information in Tidepools with threaded discussion so others in the community can respond.

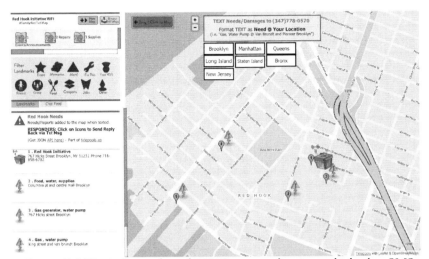

Figure CsOTI 12: Screenshot of RHI/Status application, which plots SMS messages on a Tidepools map.

As recovery progressed, Frank Sanborn, a Federal Emergency Management Administration (FEMA) Innovation Fellow, reached out to RHI about expanding the network to further support recovery efforts in Red Hook. Sanborn recruited volunteers from NYC Mesh and HacDC, a Washington, DC based hackerspace, and coordinated with the International Technology Disaster Resource Center (ITDRC).

OTI already had a store of routers at RHI from before the storm.

With technical direction from OTI and operating according to the goals established by RHI, the team set up a FEMA satellite link on the roof of RHI and installed a Commotion router on the roof of an auto body shop down the block from RHI.

Previously, the owner of the shop had been reluctant to host a router, as he did not see a benefit in doing so. However, as the community rallied in response to the crisis, the auto body shop became a key link between the Internet gateway at RHI and the router overlooking Coffey Park, which had by then become an important aid distribution point for Red Hook.

Although the satellite uplink was offered for only 30 days and provided modest bandwidth, the mesh network could distribute the Internet connection to key locations where residents, first responders, and recovery volunteers needed it most.

As the community came together to respond to the storm, the need to grow this resilient communications infrastructure became clear.

With power and water still off in much of Red Hook in the following month, many local organizations and residents reached out to help. Brooklyn Fiber, a local Internet service provider (ISP), volunteered an additional gateway to RHI WiFi.

To add the gateway into the mesh, OTI, RHI and Brooklyn Fiber installed a 5 GHz Ubiquiti Nanostation Loco router running AirOS (to receive the fiber signal), and a Ubiquiti Nanostation running Commotion (as a wireless access point), on the 3rd floor of the Visitation Church Rectory on the west side of Coffey Park.
The church was also without power at the time, but the team installed an uninterruptible power supply that could run the routers for 12 hours at a time.

Figure CsOTI 13: Rooftop node. In the aftermath of Superstorm Sandy, additional community members offered to host RHI WiFi nodes and a local ISP donated Internet connectivity.

Figure CsOTI 14: Rooftop node installed after Superstorm Sandy.

Since the storm, RHI WiFi has supported approximately 100 users per week even without promotion of the resource.

Data collected by Commotion on current DHCP leases, as well as the Google Analytics on the landing webpage, show that residents appear to be primarily connecting using Android devices and Apple iPod Touches.

In addition, many residents use the computer workstations in the RHI media lab as well as the wireless available in RHI. RHI is serving as both a physical and social anchor for the wireless network, driving digital adoption, educating the neighborhood, and coordinating relief efforts.

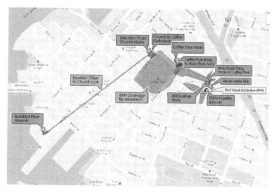

Figure CsOTI 15: RHI WiFi network map. Base map (c) Google Maps 2013.

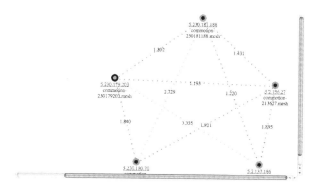

Figure CsOTI 16: Mesh network topology viewed in OLSRViz.

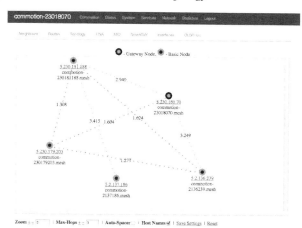

Figure CsOTI 17: Mesh network topology viewed in OLSRViz.

Sustainability & future goals

RHI will continue to develop the project with the goal of supporting community-wide recovery from Superstorm Sandy. With support from New York City Workforce Development funding, RHI and OTI are launching a training program in January 2013 to engage neighborhood residents in maintaining and growing the wireless network. Modeled on the "Digital Stewards" curriculum developed by OTI and Allied Media Projects in Detroit, Michigan, the curriculum will train young adults to install new routers, maintain existing ones, and promote adoption of the RHI WiFi network throughout Red Hook. The RHI Digital Stewards will prioritize additional public spaces for network expansion and work with other residents to design new, local applications. OTI will continue to assist in the development of the applications and support the engineering of the network in close collaboration with the community.

Cost of the network

Donated labor from local residents and technologist. Institutional support from RHI and OTI. Hardware (~$50 to ~$85 each router).
Installation (3-5 work hours for two people per site). Bandwidth (donated by RHI, Brooklyn Fiber, and FEMA). Training program for local residents to maintain and expand network as part of a municipal employment program.

Lessons learned

Having relationships and anchor wireless nodes in place prior to a disaster facilitates rapid network deployment through:

- Already-established relationships with key community stakeholders.
- A heightened level of technological literacy in the community.
- Pre-positioned wireless network equipment in the neighborhood.

The most challenging investment is in the initial organizing and design phase before any value is realized. Community-designed applications add value to a local network, even at a small scale.

Related articles & websites

RELEASE: New Community-Tech Tool to Help in Sandy's Aftermath
http://oti.newamerica.net/pressroom/2012/release_new_community_tech_t
ool_to_help_in_sandys_aftermath

What Sandy Has Taught Us About Technology, Relief and Resilience
http://www.forbes.com/sites/deannazandt/2012/11/10/what-sandy-has-taught-us-about-technology-relief-and-resilience/

A Community Wireless Mesh Prototype in Detroit, MI
http://www.newamerica.net/node/34925

Tidepools
http://tidepools.co
http://www.animalnewyork.com/2012/tidepools-a-social-networktool-in-the-service-of-the-community/
http://wlan-si.net/en/blog/2012/05/26/introducing-tidepools-social-wifi
http://www.core77.com/blog/social_design/a_community-owned_map_accessed_via_mesh_networks_23319.asp

http://www.jrbaldwin.com/tidepoolswifi/

Stop & Frisk App
http://animalnewyork.com/2012/stop-and-frisk-app-launched-by-red-hook-initiative
http://www.dnainfo.com/new-york/20121017/red-hook/stop-and-frisk-app-launched-by-red-hook-initiative

Red Hook
http://www.nycgovparks.org/parks/redhookpark/history

Made in the USA
San Bernardino, CA
20 July 2013